The Mission

or: *How a Disciple of Carl Sagan, an Ex-Motocross Racer, a Texas Tea Party Congressman, the World's Worst Typewriter Saleswoman, California Mountain People, and an Anonymous NASA Functionary Went to War with Mars, Survived an Insurgency at Saturn, Traded Blows with Washington, and Stole a Ride on an Alabama Moon Rocket to Send a Space Robot to Jupiter in Search of the Second Garden of Eden at the Bottom of an Alien Ocean Inside of an Ice World Called Europa* **(A True Story)**

The Mission

David W. Brown

ch.
CUSTOM
HOUSE

HarperCollins books may be purchased for educational, business, or sales promotional use. For information please email the Special Markets Department at SPsales@harpercollins.com.

FIRST EDITION

DESIGNED BY *Lucy Albanese*
PHOTOGRAPH © *NASA / JPL-Caltech / SETI Institute (title page)*

Library of Congress Cataloging-in-Publication Data has been applied for.

ISBN 978-0-06-265442-7

21 22 23 24 25   LSC   10 9 8 7 6 5 4 3 2 1

TO MY MOM,
the first reader I ever knew

Beware, Diomedês! Forbear, Diomedês!
Do not try to put yourself on a level with the gods;
that is too high for a man's ambition.

—ILIAD

Contents

Dramatis Personœ

Robert Pappalardo, a planetary scientist

Louise Prockter, a geomorphologist

Galileo, a spacecraft orbiting Jupiter from 1995 to 2003

Carl Sagan, an astrophysicist

Viking 1
Viking 2 } Two Mars probes, 1976 to 1982

Edward Weiler, the head of NASA science missions

Curt Niebur, a program scientist at NASA headquarters

Karla Clark, an engineer at Jet Propulsion Laboratory

Wernher von Braun, a rocket scientist

Daniel Goldin, the NASA administrator from 1992 to 2001

Pathfinder, a Mars rover mission in 1997

James Green, the head of planetary science at NASA

Fran Bagenal, a planetary scientist

Ralph Lorenz, a planetary scientist

Jonathan Lunine, a planetary scientist

Huygens, a probe that landed on Titan in 2004

Cassini, a spacecraft orbiting Saturn from 2004 to 2017

Susan Niebur, an astrophysicist

John Culberson, a U.S. congressman from Texas

Todd May, a materials engineer

Alan Stern, a planetary scientist and NASA science missions lead

MESSENGER, a spacecraft orbiting Mercury from 2011 to 2015

Mike Griffin, the NASA administrator from 2005 to 2009

Don Blankenship, a geophysicist

Ron Greeley, a geologist

Cynthia Greeley, a historian

Europa Orbiter, a spacecraft that never was, in 1999

JIMO, a Europa nuclear spacecraft that never was, in 2003

Europa Explorer, a spacecraft that never was, in 2007

Jupiter Europa Orbiter, a spacecraft that almost was, in 2010

Europa Clipper, a spacecraft that will orbit Jupiter

Mars Science Laboratory, a rover mission in 2012

Tom Gavin, an engineer

Dave Senske, a planetary scientist

Spirit
Opportunity } Two Mars rovers, landed in 2004

Lori Garver, the deputy administrator of NASA from 2009 to 2013

Joan Salute, a program executive at NASA headquarters

Barry Goldstein, an engineer

Brian Cooke, an engineer

CHAPTER 1

The Course of Icy Moons

IT WASN'T ENOUGH TO MAKE THE STYROFOAM SOLAR system over his bed. The boy needed toothpicks, too, for the moons, and he pressed them into each planet, none for Mercury or Venus, one for Earth, two for Mars. He puzzled over Jupiter, the dozen discovered being too many toothpicks, so he accounted only for the four found by Galileo in 1610. The moons were made of crumpled masking tape, which he skewered and placed into orbit. And while little Europa wasn't labeled or anything, it was with young Robert Pappalardo even then, on a two-inch wooden spike in a foam world of orange and red.

In 1970 Robert's dad packed the clan Pappalardo into the family car and steered south to Virginia, one long, continuous drive from Jericho, New York, on the north shore of Long Island, west through the city and then down along the coast. The trip ended on the Chesapeake Bay Bridge-Tunnel, which intersected the "path of totality" of a solar eclipse, the reason for the excursion. A total eclipse. A TOTAL SOLAR ECLIPSE. The sun blocked by the moon but for phantasmal arms of a shimmering corona, a black hole ripped in the daytime sky like some portal to another dimension, day turned to night for minutes and animals going crazy because what the hell is going on up there?—but across only a spaghetti strand of the United States, everywhere not in the path experiencing a miserable and pathetic partial eclipse, the moon taking a bite from the sun but nothing else, the animals not even noticing.

Anyway and either way, he didn't get that lesson, young Robert, five at the time and not yet of the toothpick moons, the sedan seating only so many, and at some point, Bob—we're sorry, really, and we'll make it up to you—we've got to make a cut and, Bob, look, you would miss an awful lot of school for this trip and you love staying at your grandparents' so what do you say? We'll catch the next one? OK? OK.

Thirty-four years later, he thought of that, the eclipse, the craft store solar system, the masking tape Europa, when they offered him the moon from the ninth floor of NASA headquarters. Europa. It was discreet, but it was an offer. It happened between meetings on October 22, 2004. He was there that day on loan from the University of Colorado, where he was an assistant professor, to advise the agency on JIMO—the Jupiter Icy Moons Orbiter—and what could be achieved from further expeditions to the Jovian worlds. Pappalardo was part of the mission's science definition team, an ad hoc group of scientists who kneaded the knowns and knowledge gaps of a celestial object or system and determined the scientific objectives that should thus drive a prospective mission there. The simple ability to go somewhere started conversations, but before the National Aeronautics and Space Administration would pay engineers to spark blowtorches and build spaceships, there had to be a *reason*, some overriding and scientifically compelling purpose beyond Because It's There.

The Jupiter system was the destination, but Europa was the goal, the target, the quarry. Beneath the frozen moon's granite-hard ice shell was a liquid saltwater ocean, long hypothesized[1] but at last discovered by physicists at the University of California, Los Angeles.[2] If life existed anywhere else in the solar system, it was in that water—and not fossilized microbes like they scratched and sniffed for fruitlessly (and relentlessly) on Mars, but conceivably complex life—fish!—and few causes could be more compelling.

Still, no way it would ever fly, JIMO.[3] Bob knew that. It was too big, too expensive. Bob had been part of previous Europa mission studies far more feasible, financially and technologically, that died ultimately and unceremoniously in agency filing cabinets. JIMO would have been a thrilling, monumental mission, certainly— would have transformed not only Europan science, but the very nature of planetary exploration itself—but something so big needed the sort of sustained support unlikely to last long, and the probability of a pulled plug was why a manager from Jet Propulsion Laboratory, NASA's research and development center in Pasadena, California, pulled him aside.[4]

Bob, began the furtive conversation, *I hear that maybe you're not satisfied teaching.*[5]

WHEN PAPPALARDO WAS a child, astronauts worked and played on the moon. It was just normal, a thing people did, like work in a factory, bank, or bakery. They set up seismometers. They collected rocks. They explored this strange new world and encountered mystifying phenomena: spots of orange dirt on its plaster landscape, jarring flashes of light when they closed their eyes. Scientists on Earth solved these puzzles: the former formed by fire fountains in Luna's primordial past;[6] the latter, cosmic rays colliding with retinas.[7] Astronauts drove cars up there. They played golf. They brought mice with them. Pets! In space!

Bob was never going to be an astronaut—that just wasn't in

the cards—but he was a child of *Star Trek,* and across five-year missions, Spock pointed the way in first run, and then in syndication, and then in animation (the Vulcan really drove the point home): science. During high school, Bob worked at Vanderbilt Planetarium in Centerport, Long Island, not far from his house. A few years later at Cornell University in Ithaca, he found the field of planetary science the way everyone else in the wider discipline did: through a Carl Sagan Moment. Upon enrolling, Bob had intended to take up the ancient art of astronomy, but problem the first: there was no astronomy major proper at Cornell; the field was taught under physics. Problem the second: physics. Bob therefore soon switched studies, his base camp now the university's new geology building. Myriad mineralogy labs lined its hallways, and to study the stuff of the Earth was to spend long days and late nights in them, over big, binocular-type microscopes adorning tall black lab tables. Pappalardo was seated in one such lab on one such night behind one such microscope when another student, first name John, last name Berner, and so of course they called him Bunsen, walked in.

Hey, Bob, asked Bunsen. You know about this course Sagan is offering? This course on icy moons?[8]

Bob did not know about this course Sagan was offering, this course on icy moons, but he went home that night, pulled out the course catalog, and looked it up. ICES AND OCEANS IN THE OUTER SOLAR SYSTEM. Instructor: SAGAN, CARL.

That Carl Sagan, yes. He of Cornell, yes, of course. The prominent professor was for Bob part of the university's initial allure, but Sagan didn't always teach, had a lot going on. Carl Sagan, recently of *Cosmos* (book and television series) and *The Tonight Show* with Johnny Carson (television series). Carl Sagan, of the NASA spacecraft Voyager 1 and 2, the robotic explorers that in 1979 transformed the four telescopic dots of Jupiter's moons into jaw-dropping globes of fire and ice, and later, around Saturn, revealed worlds diverse, living, geologically active, almost fully-fledged and practically *planets* (further study needed). Carl Sagan of the spacecraft Viking 1 and 2, which in 1976 revealed the Martian surface to be barren, though certainly in possession of everything necessary for life to exist (further study needed). Carl Sagan, a superstar in a profession bereft even of minor celebrities, who brought astrophysics into blue-collar living rooms, who in turtlenecks and with windswept hair spoke like a philosopher, practically in verse, and presented the pursuit of knowledge as something intrinsic and vital to the soul of every human being. Carl Sagan, who used his platform and lilting baritone to advance social causes, who planned protests against nuclear weapons test sites, arms linked with fellow activists, who crossed police lines and was placed gladly under arrest—Armageddon's Thoreau—in order to advance the cause. Page one the next day: a rebel astronomer in handcuffs! Carl Sagan, with a Ph.D. behind his name and a childlike imagination inside his head, who had, for example, lobbied for a light to be added to the Viking lander so that at night Martian animals might be attracted to it and scurry up to investigate.[9] Well, why not? Nobody knew what waited on the surface of the Red Planet.

That Carl Sagan.

It was a graduate-level course. Bob the undergrad had to request permission, and Carl the professor had to grant it. Bob did. Carl did. The peculiar part of taking a celebrity scientist's class, Bob soon learned, was that you began to see him as a standard-issue human being. There is Carl Sagan with a coffee stain on his shirt. There is Carl Sagan making a caustic and callous remark about the student operating the slide projector. There is Carl Sagan being boring, writing another excruciating equation on the blackboard. There is Carl Sagan, exemplary but ordinary teacher, whose class, unbeknownst to Bob, would change the trajectory of his (i.e., Bob's) life. It was in that classroom—conventional inventory: desks, black-board, snapped chalk, and tile flooring—that Bob first learned about the most mesmerizing moon revealed by the two spacecraft Voyager: the world Europa, this icy-blue eyeball circling Jupiter, etched mysteriously with crazy brown scratches. Some speculated that an ocean might exist beneath all that ice—the physics suggested it—but, again, more study was needed.

Bob completed his bachelor of arts at Cornell and pursued a graduate degree at Arizona State University. He hit a wall there, however, and left before completing his master's in geology. The 105-degree summers didn't help, the Tempe sun so severe, so low in the sky that you could almost reach your arm into it—how he *hated* Arizona, and the conservatism of the place, and the corrupt governor Evan Mecham, who was just thoroughly reprehensible: canceling the state's paid Martin Luther King Jr. Day holiday, defending the use of racial slurs. (Bob had even worked on the Mecham recall, though the governor was impeached for, and convicted of, obstruction of justice and misusing government funds, and removed from office before voters could do it themselves.) Bob had friends in the simmering state, of course, followed local bands and played guitar at open mic nights and sang Dylan and the Dead, but graduate school was just dispiriting, basically hopeless, and with little promise of hope ahead. He realized a little too late in his program that his master's thesis was overly aggressive in scope and that he would never be able to finish it. His graduate advisor, recognizing a lost cause, seemed to have moved on. So when Bob learned from a friend about a one-year internship at Vanderbilt Planetarium, where he had worked in high school, he applied and was hired. A one-year sabbatical from school—that was the plan—it would be a kind of collegiate convalescence—but maybe he was just finished with the enterprise entirely and would never return.

He liked life at the planetarium, stepped right back into it, and he was teaching the stars to students who were captivated, eager to learn, and Bob talked daily in the dark to the backs of two hundred heads, and it was rewarding, worthwhile work. That master's had just stretched on and on and on, and he looked at what he was doing now, and maybe he didn't need graduate school after all.

The internship paid ten thousand dollars, which stretched just enough to cover rent for a moldy basement apartment on Jackson Avenue, a little dead-end street in Huntington, and not far from his new job. He shared said quarters with a Vietnam vet who would sometimes have his homeless friends over to sleep on the floor. The substandard accommodations were worth it because Bob enjoyed what he was doing, and the joy increased with each passing week and month. Only twelve of those months were funded, however, and after enough pages from the calendar fell to the floor, a grim reality set in: Bob was running out of time.

Then one day, while he was practicing at the planetarium dome console for the next show, the phone rang. He answered.

Bob, said the voice, how would you like to come back?[10]

It was Ron Greeley, his graduate advisor and mentor at ASU.

Before that call, Bob had felt that maybe Ron had given up on him. It's part of what made the planetarium job such a relief: that lack of disappointment looming gloomily from above. Ron was a professional, a professor's professor, a founding father of the field of planetary science, and was above all a soft-spoken gentleman. But you felt his disapproval. The call was evidence that maybe Ron still believed, still saw potential in wayward Bob. It was just so *nice*. Bob's thing, his gift, his curse at the time, was that he saw details—so many details—and his mind adamantly insisted on assembling those details into a big picture, a coherent story. Ron picked up on this, said a master's thesis was just too small to contain what Bob had been doing. And Ron was right. It should have been easy, and yet it was taking *years*. So the student and the professor had the conversation, and not long after, Bob went back to Arizona, which was still an awful place, and the two of them turned his master's thesis into a doctoral dissertation. Dr. Pappalardo crossed the stage in 1994.

After a lengthy postdoctoral research position, Bob found a job as an assistant professor of planetary science in Boulder, Colorado, in 2001. The singular focus of his professional life had been the evolution and activities of "icy moons"—the frozen satellites circling Saturn and Jupiter and Uranus and Neptune, those planets so large as to host planetary systems of their own, in miniature. That's what he studied, that's what he taught.

But was he *satisfied* teaching? he was asked that day on the ninth floor of NASA headquarters.

He was, but maybe he wasn't. He was a good instructor, he thought. He liked the university and he loved living in Lyons, just north of Boulder, along the foothills, and lying in his hammock on the front porch, watching the sun sink behind Longs Peak—what climbers called the "flat-topped monarch" of Rocky Mountain National Park.[11] He had a girlfriend. He had a cat. He contra danced—Boulder was great for that because you could drive north to Fort Collins or south to Denver, both with thriving communities of light-footed locals—hands four from the top, gents on the left, ladies on the right, face your partner, do-si-do—and every weekend, he'd strap on his Volkswagen Passat, this silver thing, custom plates (ICY SATS)—a little too much car, he thought, but he had the cold weather package and it was fine, and he had snow tires—and he'd drive north or south to swing his partner round and round.

So he was content, but was that enough? Was that the same as satisfaction? Colleagues joked that Pappalardo was unhappy in front of the classroom because he took teaching too seriously. Being a good teacher was wearying work, especially at a big state school. You grade two hundred twenty papers, and they are brimming with pages of pilfered prose, and there are only so many times you can debate with undergrads the answer to a multiple-choice question before you start fantasizing about gasoline, matches, and applying both to oneself. His graduate students were great, though—curious, creative, conscientious—and he took care to advise and mentor them as Ron Greeley had advised and mentored him, to sustain that unbroken chain reaching back to Plato's Academy. Bob relished pairing, like Gregor Mendel in his garden of pea plants, the expertise of different grad students—his potential planetary scientists—to see what might grow from the couplings.[12]

Bob came to Colorado by way of Brown University in Providence, Rhode Island, where he'd put in six years as a postdoctoral researcher—years longer than most for that kind of position, but he was at the time analyzing data returned from the flagship Galileo, NASA's spacecraft at Jupiter. He was officially an "affiliate member of the solid-state imaging team," and who would be in a hurry to leave that? Every two months, Galileo would complete an orbit of Jupiter. The relentless machine made by human hands carried cameras, sensors, and mapping tools, and every second it spent circling Jove served some purpose. When it wasn't directing data to Earth, Galileo was studying Jupiter below or one of the two dozen natural satellites encountered along the way. (The known number of moons had grown since Bob pressed toothpicks into his painted polystyrene planets.)

Pappalardo's job was to plan the mission's "imaging campaigns": what the onboard camera would take pictures of, and when, and why. That imaging campaigns were even possible was a triumph. NASA made space exploration look easy, but it never was. In Galileo's case, scientists learned soon after launch that the spacecraft had a faulty high-gain antenna. Its collected data should have *surged* back to Earth, an Amazon River of zeroes and ones. Instead, it trickled home as though through a kinked garden hose. Galileo, consequently, could return only a fraction of the intended science. For members of the mission, then, it meant being methodical and selective. It wasn't enough to know what you didn't know; you had to know the *best things* no one knew, or even thought to ask, and then have the spacecraft collect the data to answer them. Observation planning was high-stakes work, made more so by the harsh reality of orbital mechanics. You miss your one shot at a particular moon, and that might be it, ever, for an entire generation of scientists. Galileo might never pass that moon in that configuration again, and it might be thirty years before NASA got another spacecraft to the Jovian system to fill the gap.

Galileo's project and camera leads, in addition to running a spacecraft, taught and managed university departments and advised NASA and generally pushed planetary science forward. Scholars such as Greeley and his counterpart at Brown, Jim Head, or Torrence Johnson, the Galileo science lead, didn't have time to plan which of the two blobs spotted on Europa might be more valuable to image, or how to image them, or which color filter to use, or what camera mode should be used, or what compression level should be applied thereto. But postdocs and graduate students *did* have that kind of time, and though Bob was relatively young, his work, and that of his fellow affiliate members of the solid-state imaging team (and the grad students thereon), were critical to mission success. It was tiring, tedious, taxing work. You had to stay on top of it. I mean, the spacecraft never stopped. The advisors consistently and unflinchingly reviewed the plans. Greeley in Arizona would sometimes ask of Bob or Louise Prockter, a graduate student at Brown, *Is that picture worth a million dollars? Because this costs a million dollars per picture.*[13] And you had to make the argument that it was, or you had to find something better to target, and it was back to zero. And you learned to argue the science. On this flyby, do we take images of Europa, or do we point the camera instead at Gilgamesh Basin on its fellow traveler Ganymede, Jupiter's largest moon? The Europa images would be the highest resolution ever taken. The Ganymede images would be among the worst. But those few feeble Ganymede grabs might settle some surface-age question that has long vexed scientists. Which do you choose? (They chose Ganymede.)

By the time Bob took the job in Boulder, if he wasn't yet the world's foremost expert on Jupiter's moons, and Europa and Ganymede in particular, even the world's foremost expert might think he was. He had managed, over the course of his college career and afterward, to read everything ever written about the Jovian worlds, and he possessed an unnerving ability to retain not only the literature but also the location of said literature—e.g., "I recall reading a paper on that in *Nature* in 1979. Turn three pages in and you'll find a chart that might be useful." He had a fine run of accepted papers on the geophysics of Galilean satellites and spent an awful lot of time writing reports summarizing where, exactly, scientists were in their thinking about the icy worlds. For planetary scientist Fran Bagenal's book *Jupiter: The Planets, Satellites and Magnetosphere* (literally "the book on Jupiter," as in "She wrote the . . ."), he led the Ganymede chapter and cowrote a good bit of the Europa chapter as well.

When Bob first came into the field, all the foremost experts taught—Greeley, Sagan, Head—and he had benefitted immeasurably. Who was he to do less than the founders of the field? But by 2004, it felt like all the best young scientists were moving to research institutions. How would that affect planetary science? Who would teach the basics to the next generation? Could Bob? Would he really be the best person to carry the torch if he became a burnt-out ice moon obsessive arguing with undergrads the merits of "D. All of the Above" every March during midterms?

Being pulled aside and presented with this overture from Jet Propulsion Laboratory—it was exhilarating. The lab, located in the San Gabriel Mountains of Pasadena, was an important place when it came to altitudes above the thermosphere. As an institution, it had long secured a future for the city, which was first given geography in 1771 as part of the Mission San Gabriel Arcángel's planting of Catholic flags across the lower territories.[14] (The Spanish Franciscans didn't really count the indigenous peoples' millennia of settlement and cultivation of the land. No, it definitely all started in 1771.) Next came colonists from the northeastern United States,[15] and by 1886 Pasadena had become something of a winter outpost for well-heeled New Englanders. The Second World War made Pasadena permanent when Southern California served as a staging area for the Pacific campaign. While the army was in town, it partnered with the California Institute of Technology, whose engineers were developing an American response to an ascendant German technology called jet-assisted takeoff rockets. Thus was born Jet Propulsion Laboratory. Still managed by Caltech, it was, since 1958, more famously a NASA center, having evolved into the primary robotic research and development arm of the agency. (The buildings belonged to NASA; the employees belonged to the university.) All the multibillion-dollar large strategic science missions—the flagships—flown beyond the asteroid belt had been designed, built, and flown from there: Voyagers 1 and 2, both still coursing through uncharted regions of space; Galileo, now vaporized in the Jovian interior, its mission completed; and Cassini, only four months now in orbit around Saturn. Of course, JIMO, too, was a lab effort, and evidence that not everything its engineers touched launched.

JIMO was part of Project Prometheus, a NASA headquarters-directed initiative to use nuclear reactors to power the propulsion and payload of spacefaring vessels. The program was a personal priority of Sean O'Keefe, the administrator of NASA, and was modeled in part after the U.S. Navy's fleet of nuclear submarines.[16]

(O'Keefe was a former secretary of the navy.) The idea was to send spacecraft on long-term, multiplanetary science missions in the farthest, least hospitable reaches of the deep outer solar system. JIMO would be its pathfinder, its ship of the line, its fission-fueled flagship—the first of what could be an armada of similar such vessels searching the cosmos. The Prometheus reactor would change everything. Power was king in space exploration. No matter what went wrong millions of miles from Earth, the mountain people of JPL could summon an ancestral sort of strident braininess, whiten blackboards with complex equations, and send signals clear across space bearing So Crazy It Might Work instructions to get a wayward spacecraft back on track. They could heat the spacecraft by boosting power to certain components or orient it to endure direct rays from an unfiltered sun. They could shake the spacecraft, spin it wildly, extend arms and swing them round and round. They could reprogram every byte of its onboard computer. But in order to do any of this, the spacecraft absolutely needed to maintain power. If the lights went out, it was Kobayashi Maru. It didn't even take a lot of power to keep things running. The New Horizons spacecraft set to launch to planet Pluto in 2006, a mere two years away, would travel three billion miles—to the very edge of the sun's influence—and run rigorous analyses on the unmapped world using two lightbulbs' worth of power: about two hundred watts.[17]

The Prometheus reactor, however, would produce up to two hundred *thousand* watts of power—a number so large that scientists had no context at all for how to use it.[18] If power was king, Prometheus would be the supreme and undisputed overlord of the solar system. So with the administrator's blessing, JPL engineers swung for the fences with JIMO. They settled on a spaceship that was as heavy as an eighteen-wheeler and longer than the *Millennium Falcon*. It would require three separate launches to get to space and would need to be assembled in orbit, the same way NASA was building the International Space Station. It would then fly to the Jovian system, enter orbit around one of Jupiter's moons, look around, study this area or that, fly to another moon, orbit it, and another, and another.

JIMO was ideally suited for studying Europa because the moon resided in the heart of the Jovian radiation belt: a pulsing, rippling, four-million-mile halo of death that surrounded the largest planet in the solar system. Electrons there zipped about at just under light-speed, and when those particles smashed into a lesser robot's brains, zeroes got flipped to ones, and the spacecraft might have a very bad day indeed. Maybe a vital image would be wiped from existence. Maybe the computer instruction that said *Absolutely do not do this* suddenly read *Go for it buddy—YOLO!* and the billion-dollar mission would be lost forever. A starship like JIMO, though, was no mere robot. It was Optimus Prime! It was an electronic Aeneas on a ce-lestial battlefield, radiation but a refreshing breeze tousling its hair. You want a flagship? asked JPL engineers. We'll give you a flagship.

But first you had to get that reactor into space.

It was the size of a trash can. Something so small couldn't cause a nuclear accident on the scale of Three Mile Island, or much dam-age at all, really, even if NASA put its scientists on that problem called "the Earth" and how best to make it go away. If Prometheus blew up on launch, someone might be killed from a chunk of metal hitting his or her head, but there would be no mushroom cloud, no documentaries thirty years later about where all the cows with two heads came from. But for some people, crowning a colossal missile with a uranium-powered, atom-splitting nuclear device and firing it into orbit . . . it was just a little too . . . doomsday? An awkward brush against the concept of an intercontinental ballistic missile? The thing wouldn't switch on until it was six hundred miles from Earth, but flying fissile fuel over Florida retirees . . . it was asking a lot.

The real killer, though, was JIMO's price tag: ten billion dollars.[19] No science mission had *ever* cost that much. The Hubble Space Telescope cost a third of that.[20] The shuttle *Endeavour*—the shuttle fleet then being the heart of human space exploration (NASA's raison d'être)—cost a quarter of that.[21] JIMO may as well have come in at a hundred trillion dollars. Ten billion? NASA headquarters would never keep that kind of cash flowing for a science mission. Everyone knew that JIMO would die the moment the NASA administrator retired—and all signs suggested that Sean O'Keefe, who pushed Prometheus prominently, would do just that very soon. But JPL wanted that money, so JPL needed a plan.

Enter Bob Pappalardo. It was regulation DC weather for late October that day: cool and overcast, with a light breeze.[22] Over lunch at NASA headquarters, only a few minutes' walk from the Smithsonian National Air and Space Museum, a manager from California explained to the assistant professor from Colorado that while Jet Propulsion Laboratory had the best spacecraft engineers in the world, project teams needed strong scientists on point. After all, engineers left to imagine what's possible without scientific guidance can come up with ideas that are . . . unorthodox? Unconstrained by reality? We can't keep up with Europa science the way you can, Bob, and JIMO isn't the only thing the lab has in the hopper for spacecraft concepts to get there. You've got to be ready. If Congress sends an extra billion dollars to NASA, the agency is going to ask for ideas. Let's go to Venus or Pluto or Neptune's moons, and you need something to slap onto the administrator's desk. *Glad you asked! We've been thinking about this one for a while!* We're all one big, happy space program, but it's every NASA affiliate for herself. If Jet Propulsion Laboratory isn't ready with the razzle-dazzle, Goddard Space Flight Center in Greenbelt, Maryland, or its neighbor, the Applied Physics Laboratory in Laurel, might get the gig, and then you've got a thousand Pasadena engineers on the payroll with nothing to design, build, or fly. And this we know: Once JIMO joins the choir invisible—it's a dead mission walking, Bob—NASA is going to try Europa again. They'll ask for a more manageable mission concept. And they want density, not volume. NASA headquarters knows science from science fiction.

Bob knew that a big part of the lab's Europa program *was* science fiction—and not just JIMO. By 2004, JPL had spent money on such Europa concepts as "melt probes," which would have required landing on an unmapped moon in a robot-broiling radiation environment and penetrating an ice shell harder than concrete and kilometers thicker than any hole ever drilled on Earth to reach an ocean that, technically, might not even exist.[23] Good luck with that.

Even superb studies of plausible mission concepts had been unable to find traction at headquarters, as Bob knew firsthand. He had consulted briefly on a JPL study in 1998 for a possible spacecraft called Europa Orbiter—an outgrowth of an ambitious program called Ice and Fire—and it was a real contender. The project gathered momentum when the spacecraft Galileo got a good look at Europa's tarnished crystal facade, and magnetometer measurements hinted at liquid water churning beneath its icy exterior. NASA convened a science definition team to establish the best science attainable with a small, sub-billion-dollar spacecraft. They determined that Europa Orbiter's goal would be to answer the water question: Was it real? Or something else? Then, if it was real, the orbiter would map the ocean in three dimensions and, lastly, figure out why Europa's surface looked like a cue ball scraped by a madman with a rusty nail.[24] You do those three things, and you're in good shape for a subsequent lander mission, a concept for which was already on the books. It was about the size of a pizza, the lander, but it would reveal an awful lot about the Jovian moon's surface—e.g., was it solid or slushy?

Based on Jet Propulsion Laboratory's promise of an inexpensive Europa mission able to overcome the historically ten-figure toll to cross the asteroid belt, in 1999 headquarters signed on and seeded the lab with fifty-eight million dollars to begin detailed development work.[25] And once that money was spent, lab leaders came back to headquarters, all smiles and with a plan in hand that was twice the quoted cost—but, hey, you're on board, right? And, hey, headquarters definitely was not, and Ed Weiler, the head of science missions for NASA, canceled it with prejudice.[26]

THIRTEEN POINT EIGHT billion years earlier—three minutes, in fact, after the universe began—hydrogen nuclei formed: good old atomic no. 1, the lightest element on the periodic table.[27] Until then, space itself had been bounding outward from a single point to the entire observable universe. It cooled into a quark soup, quarks came together to form baryons, and electrons were new in town and turning heads.[28] It was a busy three minutes. When hydrogen nuclei stepped onto the stage (though not into the spotlight—light as we see it didn't exist yet),[29] so too did those of helium, lithium, and beryllium (nos. 2, 3, and 4, respectively, though their parts were small indeed), and it took another four hundred thousand years of universal cooling before the nuclei could draw in those eligible electrons and form stable, bona fide atoms. Over time those atoms met, became gravitationally attracted to one another, and formed clouds in space called nebulae. A trillion galaxies or more formed from the clouds over the next nine billion years, and one of them was spiral-shaped and destined to be called the Milky Way.

Our nebula was not a particularly peaceful place, though it was stunning, from the outside far away and looking in: a celestial cloud of white, blue, beige, and burgundy, and it was very, very big—quadrillions of miles from end to end—just ridiculously large, really. All across the nebula, stars formed and exploded with un-nerving regularity, contaminating the cloud with smithereens of stardust from which other stars and systems would emerge and ex-pire and further enrich the ether.

It is how the elements beyond beryllium were born. A small star is a fusion-powered factory, its dense and powerful interior slowly squeezing together the nuclei of hydrogen atoms and turning out fresh helium. Fusion reactions let there be light. Larger stars do this on a scale commensurate with their size, and when their avail-able hydrogen is used up, fused fully into helium, they double down on the whole process and start fusing helium to beryllium and car-bon, and down the line to iron—our friend Fe, twenty-six protons now forced into a single atomic nucleus. And that's as far as a star gets before all that iron and heat and pressure destabilize it, and it finally says forget it and explodes in a cosmic cataclysm. The resul-tant forces then *really* get to work on the atoms at hand, chaotically fusing and forming everything up to (and including) uranium, with ninety-two protons stuffed in its nucleus: our atomic bombs, fueled by supernovae themselves. And by now, the bulk of the periodic table is forged and scattered across the cloud, again and again and again, stars forming and failing and feeding the fertile miasma.

Across ages, epochs, and eras, meanwhile, a paltry puff in the cloud that would become our own oozed fluidly and with a sono-rous turbulence. From its nascency, gravity really worked on our little parcel of nebula, drawing it ever inward on itself, slowly but inexorably, until at last, nearly five billion years ago, it buckled and collapsed. As it did, its erstwhile inner turbulence, those fluid mo-tions, caused our contracting cloud to exhibit a net sense of rota-tion. Our tiny wisp began to spin. The more material it drew in, the faster it spun, and though the sun today could hold more than one million Earths, it began like this, as a swelling union of dust and hydrogen, and it grew and grew and grew, its rapacious core inhaling everything available, growing ever denser, more massive and molten, one mote of dust at a time until nearly all of the cloud was consumed. It was becoming a protostar.

The spared, swirling fraction of a fraction of gas and dust on the protostar's fringes flattened all the while into a thin disc of vast diameter, the way pizza dough stretches and flattens as it is tossed. These particles, they *aspired* to such sizes as grains of sand, and perhaps one day grains of rice, but for now, they were but specks circling some crazy ball bursting from within. Over time they clustered by chance into particle pelotons behind which other specks could hide, drag diminished, energy saved, the groups growing bigger and bigger still. The clusters' comfortable wakes tempted more things yet, and in due course these growing balls of material reached sizes sufficient to start self-gravitating and attracting yet more stuff, until they themselves collapsed into solid celestial objects: the first asteroids. Onward they went, colliding with one another and growing larger and larger as they accumulated debris and other solid material. The farther from the swirl's center you were, the lower the temperature, and thus the more solids these planetesimals had handy, because a new building block was introduced to the material available: ice. This swarming supplementary matter allowed planetesimals to grow ten times larger than Earth, itself now forming from rock and metal nearer to the disc's interior. Planetesimals soon showed signs of becoming protoplanets with attendant superhot centers and such distinct, differentiated layers as crust, mantle, and core. All of this was happening at once: the protoplanetary disc, the protostar, and the whole thing still submerged in the thinning local nebula.

Then, the awakening. The heat and pressure at the disc's core, ever increasing and increasing, could increase no more in its present state, until fantastically, this enormous, round, burning-hot *thing*— almost the entire mass of what was once merely a haze of atoms—

reached eighteen million degrees Fahrenheit, and its pressure and heat were now so great that the nuclei of its constituent hydrogen atoms commenced fusing together, creating helium and releasing astounding amounts of energy. As the newborn star flickered to life, it blasted a wave of heat and plasma outward, concentrating the cloud at the orbits of the protoplanets Jupiter, Saturn, Uranus, and Neptune, each of which immediately went to work acquiring this sudden influx of hydrogen and helium. Jupiter, the hungriest of the four, was largest and best positioned to dominate, and within a million years, its atmosphere was as massive as its core: ten Earth masses of solid stuff surrounded by ten Earth masses of hydrogen and helium. Feeling now like a real master of the universe, it then dove headlong into an unstable, runaway phase of accretion, and in ten thousand years—on cosmic timescales, the firing of a brain's neuron—Jupiter inhaled three hundred Earth masses' worth of gas.

When a planet forms that fast and that heedlessly, weird things happen around and inside it. Jupiter migrated, first inward, truncating the disc of debris available to the protoplanets nearer to the sun, leaving Mars material enough only to grow twice the size of Earth's newborn moon. Saturn saved the solar system by pulling Jupiter again outward and away from the sun. The planetesimals now settled into place between Mars and Jupiter, and were so agitated by events and the giant world's gravity that they were unable to organize into proper planets. Thus was born the asteroid belt.

At the severe pressures of Jupiter's interior, hydrogen acts like a metal. Hydrogen is simple: it's basically just a proton with an electron going around it. Take a ball of hydrogen atoms and squeeze it tightly enough—say, a million times the pressure of Earth's atmosphere—and the atoms get so uncomfortably close to one another that the electrons stop caring which protons they're orbiting. As long as there's a proton nearby—any proton at all—the electrons are happy, and they will just start hopping around from atom to atom. The interior has, at this point, become a highly conductive metal fluid: *liquid metallic hydrogen*. The planet's core heats the liquid metallic hydrogen, causing it to rise, and once it reaches the outer layer, it cools and sinks, again and again and endlessly, generating in the process a massive magnetic field.

In Earth's night sky, Jupiter is not all that special: a flickerless pinhole of light in a dome of darkness. If its magnetic field were visible, however, it would be the size of three full moons in our sky.[30] When charged particles in space travel through that massive magnetic field, they get trapped and start zipping around at the speed of light. Space is a deep, dark, deadly domain, infinite in its dangers, but it is a boundless Switzerland compared with the wilderness of this, the Jovian radiation belt.

As for the rest of the solar system, what nebular material was not consumed by Jupiter, Saturn, Uranus, and Neptune was at last blown away by the stellar winds. It took five hundred million years for all of this to happen, from collapsed cloud to solar system, planets circling and a star to steer them by.

After four and a half billion years, the solar system more or less settled, with a thin layer of life having taken hold on Earth. There, an artist named Giusto Sustermans painted what would become the definitive portrait of physicist-philosopher Galileo Galilei.[31] The somber, sober subject was seventy by then, round, wrinkled, bearded lavishly, and posed touching a telescope. Draped in black, a white collar wraps around his neck, and higher up, silver brows furrow below a hairline struggling valiantly to hold on a little longer. The man in the painting seems wise, but more than anything else, he just looks tired.

But when Galileo discovered what he called the "stars of the Medici" orbiting Jupiter, he was preening, dynamic, forty-six, and famous. Already, he was a caustic seventeenth-century version of Carl Sagan. He wrote poetry. He loved wine and women. He sold science to the masses (his books were published in Italian rather than Latin), and in salons from Pisa to Padua, he took his ideas on tour, throwing down against fellow philosophers.[32] A graceful winner he was not (and win he did; he was Galileo, after all). He seemed to find a satisfaction only in the scholarly equivalent of Mortal Kombat finishing moves, extracting twitching spines and still-beating hearts from his debate opponents.

Galileo certainly had the CV for such high self-regard. At twenty, he had discovered the law of the pendulum: that the period of its swing is independent of its amplitude. (This had real ramifications for timekeeping, though he would be dead before they could be leveraged.) At twenty-five, he took a swing at Aristotle, asserting that two objects dropped from a great height would land simultaneously regardless of weight; density, he declared, was the deciding factor. He was right and, naturally, was not modest about it.

Cosimo II de' Medici, the grand duke of Tuscany, eventually made Galileo court mathematician. (Only the pope would have been a better or more powerful patron, and Galileo was friends with him, too.) In 1610 the acerbic Italian fixed a modified spyglass on planet Jupiter and discovered three tiny stars in its vicinity. Observations over several evenings found a fourth, and movement, and it didn't take long for Galileo to work out what he was looking at: objects in space circling another object. There was no end of implications to this, and no better scientist to make the discovery. Heliocentrism had been around for a while—the Greek astronomer Aristarchus of Samos advanced the idea in the third century BC, though his work was lost in the centuries to follow.[33] More recently, just before his death in 1543, the Polish astronomer Nicolaus Copernicus had published *On the Revolutions of the Heavenly Spheres*, positing that there might be multiple centers of motion in the universe: that the planets circle the sun, and the moon orbits Earth. Absent proof, however, everyone went right on thinking that Earth was center of all things because it was safer, made more sense, and had zero incompatibilities with Scripture and its armed enforcers. Galileo had no such qualms, however, and beat his drum—*hard*—upending cosmology itself and humanity's Very Special Place Indeed in the universe. He probably thought a lot about that while spending his last days under house arrest, having annoyed the Inquisition with this heliocentrism business.

The four stars he found were, in fact, moons—Io, Europa, Ganymede, and Callisto, now collectively called the Galilean moons. (They are named for the lovers of Zeus, king of the gods on Mount Olympus, though Galileo did not name them. A German astronomer named Simon Marius did the christening, having had the misfortune of discovering the moons *one day* after Galileo.)[34] So dominant was Jupiter in the whirling, collapsed cloud of cosmic dust and gas that formed the solar system, that it attracted a disc of its very own called the Jovian subnebula. It was a microcosm of the wider solar system, with Jupiter playing the role of the sun and its moons the planets.[35] The stuff closest to Jupiter formed a world of rock and metal—Io—and moving outward, as temperatures dropped, ice increased as a formative factor. Europa is made of a lot of rock and some ice. Farther out, Ganymede is icier still, and Callisto—the most distant of the lot—is the iciest of them all.

Europa is the smallest of the quartet. It is a little smaller than Earth's moon, with a little less gravity. "Small" in celestial objects is relative; one standing on Europa's surface would notice no curvature on its horizon. Its atmospheric pressure is about one-trillionth that of Earth, meaning, in effect, that to stand on Europa is to stand in space. Above, two major moons hang in the darkness: Ganymede, giant, its pale-bronze hammered surface sometimes smooth, sometimes sharp, scratched, and scaly, splotches of snow seemingly splattered at random; and Io, an unsettling, Gigeresque orb, yellow with stains of orange and brass. Farther out, Callisto, brown and speckled like amphibian skin. There are more than a hundred other moons that could be witnessed on the dome above if you brought a good pair of binoculars, but nothing unsettles the soul like Jupiter, twenty-four times larger than a full moon in Earth's sky, this looming leviathan, this aptly named and veritable god of planets, robed in bands of tans and reds—a spherical windstorm in space—its clouds of hydrogen and helium ever a slow churn driven by some unknown force from deep within its interior.

Looking down and to the horizon, an astronaut on Europa is casting her eyes across a postapocalyptic Antarctica: an endless tundra of gashed ice. In places, it is snowman white—the stuff of pure water. Elsewhere, it is sepia, seared and poisoned by the radiation belt into which Europa is submerged. Those gashes: in shadows they are cinnamon, scarlet, sienna, and they break up the landscape as though the whole world had been smashed on a marble floor and then reassembled haphazardly. There are steep cliffs and deep troughs and Grand Canyons of ice the color of prison cells. In places, the lacerations curl and meander like spaghetti. Some icebergs tower, some stoop in subjugation, and they meet chaotically across the expanse. It is three hundred degrees below zero Fahrenheit, and there is no weather, no wind, no rain, but there is ferocious radiation: Io-borne ions beating endlessly into the ice for billions of years, making some of its surface something almost like snow, depths unknown.[36]

Beneath the ice, the ocean, the seafloor, is Europa's thousand-mile-diameter core, which is made of iron. (Just like Earth's.) The mantle surrounding it is four hundred miles thick and made of silicate rock. (Just like Earth's.) A sixty-mile layer of liquid water covers the mantle—not some green alien goo that technically fits some Poindexter's definition of "water," or water with an asterisk, or water for extremely large values of x, but *liquid water*. A saltwater ocean. And the whole world is wrapped in an ice shell fifteen miles or so thick. (Unlike Earth.)

Here is why there is a liquid saltwater ocean on a small moon so far from the sun that its surface is three hundred below zero: 1. gravity, 2. mighty Jove, and 3. the odd interplay of cohort moons. In the time it takes Ganymede to make a single lap around Jupiter, Europa has circled it twice, and Io four times. As they orbit, they pull each other toward and away from Jupiter. Because of the clean numbers—4:2:1—their "orbital resonances" are stable. They'll go right on doing this forever. (Were they unstable, the forces at play would eventually rip the moons apart. This happened once at Saturn, and its breathtaking rings are a cemetery made of the moons' mortal remains.) An effect of these contra dancing orbs is that their orbits are not perfectly circular. Sometimes they are nearer Jupiter in an orbit; sometimes they are farther away. When a moon is close to Jupiter, the giant planet's astounding gravity, in effect, stretches it. When a moon is farther out, it gets respite. Closer, clenching. Farther, solace. It is like squeezing a tennis ball repeatedly. The more you do it, the warmer the ball gets.

If ever Io were an icy world, those days have long passed. No ice could survive the forces so near Jupiter, the heat, the insistent choking by a planet so immense. Io is the most violent body in the solar system; at any one time, there are more than a hundred volcanoes actively expelling hot rock into space. Under the force of Jupiter's gravity, Io is literally turning itself inside out. Here on Earth, the moon we see today is the same one the cavemen saw, and the dinosaurs before them. Things just haven't changed that much for eons. But in fifty years, the surface of Io—a moon just slightly larger than our own—will look totally different.

Europa has just enough space from Jupiter that it's able to hold on to its ice, but not enough distance to be unaffected by the situation in which it finds itself. Those cracks on Europa's surface are crushed ice under Jove's flexing fist. It heaves, this body!—the Europan tides rising and falling by a hundred feet or more over its three-and-a-half day revolution of Jupiter.[37] The friction caused by Europa's eccentric orbit becomes heat welling up from the constricted mantle, and though Europa's surface is six times colder than Antarctica in winter, twelve miles down, the ice becomes water in a flowing, relatively warm ocean.[38] And there is a lot of water in that warm, swirling, gushing, meandering, swelling sea: three times as much as there is on Earth, with global currents more powerful than Earth's as well.[39]

To create life as we know it, a world needs organic molecules—compounds with carbon, hydrogen, nitrogen, and oxygen, which most planets have. Such life requires water for processes such as ingestion, metabolism, and excretion, and while water is less common than organics, it is certainly found beyond the third planet. Life lastly craves chemical energy, which is the toughest of the three to come by. There is no sunlight in the Europan ocean, and thus no photosynthesis. It's ink black down there. But on the ocean floor, water touches rock, which is conducive to interesting chemistry. And what Europa lacks in sunlight, it makes up for with unbridled chemical reactions powered by something else. Though its seafloor is a wondrous mystery, all the same, planetary scientists have a pretty good hypothesis for what it looks like: it looks like Io.[40]

It's not quite as dramatic down there—Io is Mordor, and Europa's distance helps stave off Sauron's more malevolent designs. But the terrible forces causing Io to hemorrhage its entrails are at work on Europa as well, and, on its ocean floor, great geysers gush heat and chemical compounds from the mantle and into the water. This happens on Earth, too, but for different reasons: at the bottom of our ocean, hydrothermal vents ceaselessly billow brutal columns of scorching water, an endless supply of nutrients blasting from the bowels of Earth, and life teems there—life simple and complex—despite the total absence of sunlight.

Ganymede, meanwhile—the third of the three moons swinging about—is an icy world being squished by Jupiter. But because its shell is so thick, it manages to melt only in the middle, leaving its water sandwiched between ice layers. Its water is never afforded the chance to touch rock, which is a bad thing indeed if you want the kind of chemistry that likely yields fish.

No one knows how you go from lifeless material to living material. It is the eternal mystery: Where did I come from? What scientists *do* know is that it takes a long time to happen. A celestial object can have all the right ingredients—organic material, water, chemical energy—but if the pie hasn't had time to set, it doesn't have life on it. The best guess for life's baking time in the oven: five hundred million years.

Europa's ocean has had more than four *billion* years for life to get started.

The possibility of a Jovian moon growing things that swim hasn't escaped the notice of artists since Sustermans. Author Arthur C. Clarke and film director Stanley Kubrick developed the plot and themes of a story involving the Jovian system based loosely on a much older story by a poet named Homer. In 1964 Clarke went off and wrote his novel, *2001: A Space Odyssey,* while Kubrick drafted a screenplay of the same name, and along the way, the two compared notes on each other's works, incorporating the best stuff into each. The film premiered in 1968, two months before the book hit shelves, and though Clarke had written a masterpiece, Kubrick had created something superlative: a piece of artwork from which there was no turning back.

The pacing and storytelling of the two works track similarly. A magnetic anomaly pulsing from the lunar interior leads astronauts to drill a core sample of the area in order to work out what is happening there. Twenty feet down, when the drill is stopped cold by something unexpected and impenetrable, they begin to dig and soon call in a team with serious hardware to do a proper excavation.

They discover not far beneath the lunar surface a large slab—a monolith of perfect dimensions, 1:4:9—the squares of the first three integers: $1^2:2^2:3^2$. When the monolith is exposed—a sign that life on the planet below has reached some level of ability and sophistication—its first encounter with sunrise causes a piercing beam of radio energy to be blasted across the solar system. The plot of each work, novel and film, involves a human expedition to the target of said radio beam.

Here, however, the stories diverge. In the novel, the signal is aimed at a moon in the Saturnian system. To the extent that the spaceship of exploration, *Discovery One*, flies to Jupiter at all, it is only briefly—a single chapter in which the crew uses the planet's gravity as a slingshot to speed them along to Saturn. It is a harrowing episode, but Clarke's imagination outpaced the abilities of special effects artists of the era. No one could create a Saturn that Kubrick found convincing, so rather than render a set of third-rate rings, the perfectionist filmmaker simplified the story. The monolith's signal stopped now in the Jovian system, and all that followed did so in Jupiter's orbit.[41] The plots more or less converge there.

This small science-fiction factoid might have been consigned to the back of a Trivial Pursuit card were it not for the sequel, *2010: Odyssey Two*, written by Clarke and published in 1981. Here the author faced a dilemma. Should he pen the story as a sequel to his novel, or to the film, which was released first and known more broadly? Did Dave Bowman, lone survivor of the computer HAL 9000's psychotic break, abandon *Discovery One* in orbit around Saturn (at the moon Iapetus), or around Jupiter and one of its moons?

Clarke chose to write the novel as a sequel to the film. *Odyssey Two* opens with the Chinese spacecraft *Tsien* landing on Europa, its taikonauts trudging across the icy world on a reconnaissance mission for sources of water a vital commodity if humans are ever to settle space. Before they can complete their task, however, some sort of evolved sea life emerges from a crack in the Europan ice shell, destroys the landing vessel, and slays every taikonaut but one, whose final, frantic message to Earth is: ". . . relay this information to Earth. *Tsien* destroyed three hours ago. I'm only survivor. Using my suit radio—no idea if it has enough range, but it's the only chance. Please listen carefully. THERE IS LIFE ON EUROPA. I repeat: THERE IS LIFE ON EUROPA."[42]

The novel ends with a mysterious message to Earth from an unknown celestial entity:

> ALL THESE WORLDS ARE YOURS—
> EXCEPT EUROPA.
> ATTEMPT NO LANDINGS THERE.

THE PROSAIC PRESENT remains squidless. Planetary science is a plucky upstart as far as scientific disciplines go, belonging once exclusively to the field of physics, and then to astronomy, and, since the start of space exploration, to geology. (As spacecraft and their scientific instrument payloads—the tools they carry that collect the data—evolve in sophistication, the fields of chemistry and biology will grow increasingly vital to the discipline.) To do geology, you need access to, or images of, hard surfaces. You need to see rocks and ridges and rubble and regolith. You can work backward. The snake looks like an ancient riverbed. Those stones, smoothed, get that way when water rubs rock. That's a volcano, that's a gorge, that's a cliff. It's easy on Earth because you can just look down and start studying, but before the rise of rockets, Mars existed only through telescopes, a smear of fire with ice caps on each end, and Venus was veiled in teal and impenetrable clouds.

It was the Apollo program that really made geologists tilt their heads upward in unison and ply their trade to untrodden worlds. Down in Alabama, Wernher von Braun, the exfiltrated German genius of rocketry, built the biggest booster ever to lift Luna-bound star voyagers. But nobody was sure, exactly, what the moon was made of or how hard its surface might be. What if we looked up each night to a lunar surface not of solid rock but of soft dust accumulated over billions of years and almost entirely undisturbed? What if the *Eagle* landed, and just . . . kept landing! Just sank right into the moon as though its surface were a powdery snowbank. While precursor robotic reconnaissance spacecraft had touched down on luna firma, would every part of the moon be like that? Oh, but it was even worse. What if the *Eagle* didn't sink—what if astronauts set up Tranquility Base, walked around, planted the flag, and called the president, but when they climbed back into the lander, what if the moon dust was *flammable*? They'd pressurize the *Eagle,* and boom: Mare Tempestatis. NASA needed geologists to sharpen their pencils and solve the problem most ricky-tick—and to the great relief of Apollo astronauts, they did. The magnitude of the achievement of geologists to understand an orbiting ancient alabaster rock was reflected in the *second* thing Neil Armstrong said, after taking a giant leap for mankind:

"And the—the surface is fine and powdery. I can—I can pick it up loosely with my toe. It does adhere in fine layers like powdered charcoal to the sole and sides of my boots. I only go in a small fraction of an inch, maybe an eighth of an inch, but I can see the footprints of my boots and the treads in the fine, sandy particles."[43]

And planetary science was off to the races.

Bob Pappalardo entered that race in earnest twenty years after taking a desk in Carl Sagan's classroom, with Jet Propulsion Laboratory beckoning him: Come to California. Set up a Europa laboratory. Help us build a proper program on Europa. *Help find that life*. You won't get there from Colorado, Bob. Are you interested?

Yes, Bob said. I am interested.

Seventeen months of negotiations followed. They weren't tense, exactly, but they were tedious: title, lab facilities. He wanted a salary such that in Los Angeles, he could buy a house similar to the one he had in Lyons. That delayed things. I mean, what does a house in Colorado cost? *How much?* You're not getting a house that size for that money in L.A.! You're not getting a house that size in L.A. for *double* that price.[44] So there was a firm but gentle and sometimes terse but generally respectful back-and-forth. Ultimately, the numbers fell in Bob's favor, but there was one more thing he wanted.

During preliminary discussions, the lab teased the prospect of Bob being the project scientist of a Europa mission—its leader, in other words—should a mission go forward. This was so far outside the scope of the career Bob had imagined for himself that it veered into areas incomprehensible. Before the lab came calling, Assistant Professor Pappalardo's goal—he'd written it down and everything in one of those what-do-you-want-out-of-life? workbooks you find in the self-help section of bookstores—was to be the lead scientist on a camera carried by some spacecraft that might one day return to Jupiter. He wasn't even asking to be Spock; Gene Roddenberry would have listed him in the credits as Blue Shirt Crewman #3, maybe. It was still an enormously ambitious goal—any scientist's crowning professional achievement. But to be *project scientist*? Look, one flagship launches every decade. And there's only one person who gets to be project scientist. It would be far easier, statistically, to be an astronaut—NASA employed about a hundred of those, versus *maybe* five or so project scientists to launch a flagship mission to the outer solar system, and not "at the time," but *ever*. Part of it was because the outer planets were so aptly named—Jupiter (the nearest outer planet) was one wire past Mars on the classroom solar-system-around-the-yellow-lightbulb model, but took vastly longer to reach. Pregnancies lasted longer than a trip to Mars, but by the time a spacecraft reached Jupiter, the same child would not only be born, but would be old enough to play on a youth soccer team.

Which made Mars an enticing object of exploration—more so, even, because every robotic Mars mission could be framed as a precursor for human exploration. And though astronauts hadn't touched a celestial object beyond Earth since the last Apollo flight in 1972, NASA was thoroughly and indelibly an astronaut-led, astronaut-centric organization, and Mars its elusive but inevitable target post-Apollo.

Programmatically, the Red Planet was an enterprise unto itself at the agency and competed against only itself for flight projects. Every twenty-six months, as orbits aligned, *something* would launch for Mars. For every other object in outer space, however, it was urban warfare. Headquarters was, on a good day, apathetic to aspirations for an outer planets flagship (the outer planets as in *all* of them, and their hundreds of moons of fire, ice, rock, and metal) and, ordinarily, entirely antipathetic if not actively antithetical. As a result, the outer planets competed against not only themselves as worlds worthy of exploration (e.g., Do we go next to Io or Iapetus?), but also against every other object in the solar system. You wanted a flagship to Neptune, you had best be better than Venus. You wanted a Triton orbiter, and you're crossing swords with Mercury, Ceres, or Saturn. NASA selects a spacecraft to study an asteroid or comet, and it is never *Mars* that will want for hardware on the launch pad; it's Europa or Titan or even the moon. And if you were a non-Mars researcher, you couldn't help but feel some irritation? annoyance? envy? Look, the Mars community labored mightily for its success, and if you studied the outer planets or Venus or the giant asteroid Vesta, you probably, in fact, *worked* on one Mars mission or another, or did research on Mars data, because the grant money was there and your car payment was due next week. Despite an unbroken chain of NASA budget squeezes and shortfalls going back to the end of Apollo, everyone needed Mars to prosper because it was like social security for solar system scientists. The government might meddle, but it would never kill it. Places like NOT MARS, however, were always maybes, always if-we-can-afford-its.

Then again, you never knew. And if some sympathetic functionary at NASA headquarters got traction and convinced the right person to give Europa a chance, Bob wanted it *in writing* that he would be the project scientist of that mission.[45] JPL management personnel didn't say no, but they couldn't say yes on paper. They could, however, formally declare that he would be their top candidate for project scientist should a mission go forward. It was a sign, at least, that they were serious, and Bob signed on the dotted line. Three years, he figured, and NASA would want his mission. As the lone survivor of the *Tsien* could attest, however, Europa would not prove so hospitable to callers.

In the end, it would take seventeen years, six major studies, multiple missions approved, multiple missions abandoned, friendships formed and enmities established, funding raised and budgets lost, congressional hearings, unlikely alliances, technological breakthroughs, terrible losses, and stunning discoveries to get NASA to make it official.[46]

But much of that was yet to come. A contract signed, Bob Pappalardo packed his life and his cat and pointed his car westerly— one more northeasterner with his sights set on Pasadena. He had once missed a total solar eclipse, but that was OK. They could have the sun. In 2006 the child of *Trek* was California bound. This time, he was after a moon.

Situations Vacant

LOUISE PROCKTER FINISHED THE EMAIL AND SENT IT. Just like that. One sentence and a CV attachment, and it was out there now, three branches down the tree of life that sustained the American space program: Drafts to Outbox to Sent Items. She opened her planner to December 21, 2006, pressed pen to page (blue ink for Europa's turbulent ocean, currents coursing fathoms below), and crossed it from her list of things to do that day. Volunteer for new Europa mission study. Done. From her little office at the Applied Physics Laboratory of Johns Hopkins University in Maryland, Louise worked full-time on the MESSENGER mission to Mercury, part of the spacecraft's camera team. Her role on the project kept page after page and line after line full in her color-coded Levenger planner (Mercury in red for its intense heat), but she would fill the *margins* with blue ink, if necessary. I mean, this was *Europa*. She belonged on that study.

And why not? She'd paid her dues on the late JIMO's science definition team, where she co-led the geology and geochemistry group. She'd planned observation orbits of the spacecraft Galileo around Jupiter before that. (Poor radiation-poisoned Galileo, asleep at last.) She'd published a steady stream of papers on icy moons, and Europa in particular. She knew more about dark terrain on Ganymede than probably anyone else in the world and had written the dark terrain section for Fran Bagenal's book on Jupiter. The last twelve months alone, working on MESSENGER—a tortured but accurate acronym for the Mercury Surface, Space Environment, Geochemistry, and Ranging mission—had proved among the most exhilarating of Louise's life—and that spacecraft had another four years to go before entering orbit around the planet nearest to the sun.

Space exploration was not for the impatient. It took years to get a mission approved by NASA, years further to get it built and off the ground, and except for Mars (with its favorable celestial alignments with Earth), still more years yet for it to reach its destination. In the case of MESSENGER, rather than flying directly to Mercury, which would have been faster but bananas in its fuel requirements, like most spacecraft in NASA's fleet, it would instead pinball around the solar system using slingshot maneuvers called "gravity assists," swinging by this planet or that and leveraging the encountered planet's massive well of gravity and atmospheric friction to speed up, slow down, or make major trajectory changes. This allowed the vessel to carry less fuel at launch, which in turn meant NASA could use a smaller rocket to launch it from Earth, saving tens of millions of dollars.

MESSENGER's mission plan called for the spacecraft to lift off and oval the sun until it again met Earth for a course correction that would kick it inward to Venus and again around the sun for another, more harrowing Venusian encounter (a terrifyingly low two hundred ten miles above its surface), really reducing the vehicle's velocity now and angling it inward toward Mercury, circling the sun another dozen times, thrice buzzing the tiniest inner planet, easing, easing, easing ever so slightly into formation with Mercury, until finally, finally, *finally* the spacecraft might slide gingerly into the tiny planet's ethereal orbit. The distance from Earth to Mercury at their closest was about fifty million miles. The distance MES-SENGER would travel ultimately from launch to arrival: nearly five billion miles.[47]

So far, the vehicle had had its gravity assists from Earth and Venus. For much of the duration of these "cruise phase" operations, MESSENGER's science payload remained active, each instrument team working full-time, including Louise and the rest of the camera crew. During the Earth encounter, the spacecraft captured two of the most stunning shots of our azure orb that she had ever seen. There it was: South America, with Africa wrapped along the lower crest of Earth like some great giant's gentle hand holding up the world for inspection.[48] And while flying away, a parting image of the Galápagos Islands gilded with a glint of sunlight.

None of this just *happened*, of course: the images of Earth, the orbital adjustments, the golden glint of sun. (Well, that one just happened.) It had to be planned, all of it, and rigidly, every image, every video, every second of every sequence, taken by a spacecraft speeding six and a half miles for each of those seconds, the laws of physics alone keeping the spacecraft in precisely the right place at precisely the right time. The whole point of the exercise (beyond bagging opportunistic new photos of our home planet) was to work out the wrinkles before you got to Mercury: team dynamics, mechanical calibrations, image sequence plans, camera control software development. All of it was critical to ensuring that the prime mission at Mercury moved from day one as if on rails. The work did not stop during the years-long cruise phase, and it could be spellbinding, tedious, challenging, or all three of those at once, but even with its decadal timescales, there was an urgency inherent to space exploration that energized everyone, and especially Louise, who simply could not believe that this was her life, but who never stopped to think too hard about it because there was so much work yet to be done. Earth, Venus, Venus, Mercury, Mercury, Mercury, orbital insertion—you never punched out early. And the excitement of it all, you internalized it. It became you. And eight weeks after the first flyby of Venus, one year after seeing the sun-kissed Galápagos Islands,[49] you're writing a one-sentence email to Curt Niebur at NASA headquarters, CV attached, and you're telling the guy trying to get a Europa mission going, *Hi, Curt, I would like to submit my application for membership of the SDT for the new Europa flagship study,* and you add in parentheses—because, let's face it, who in the community *wouldn't* want to be on the study, and you are one hell of a Europa scientist but nothing is ever a sure thing—*or the Ganymede study—I believe I could also make a useful contribution to that study,* which you could, and, you add, again, because, look, you really are one of the best qualified Europa scholars out there, *but I think I could make a greater contribution to the Europa study.*[50]

And just like that, you send it and see how far you can ride this thing: in ten years, Louise had gone from undergrad to section supervisor of the Planetary Exploration Group at the Applied Physics Laboratory. So why not send that email? I mean, we were talking about Europa here. If there was a chance—any at all—of going back, she had to be part of it.

BEFORE SHE FLEW spacecraft for a living, Louise Prockter sold newspaper advertisements. To her parents' shock and horror, she announced at seventeen that she was absolutely not going to university, and that was that—settled, over—and she found a job at a local paper in London, where she addressed envelopes and made tea for her bosses. It was 1982. The move to sales was a big promotion. They placed her in charge of the paper's sits vac, or situations vacant—want ads for companies hiring—and she was good at her job and she had an active social life and she was well on her way, sits vac by day and antics by night and aspirin by morning. She wasn't, she recognized at the time, a particularly good person, but rather, just *a person*, a Londoner, and she ran with a certain crowd, and she was muddling through, making her way. It was life.

Not long after Louise entered her twenties, a competing paper hired her to run its sits vac, and she felt good about the work she was doing because sits vac in general charged double the price of regular ads, and this was her second go at it, and to have that kind of responsibility was like a stamp of approval from the Powers That Be that she was highly competent with a sharp eye for detail, no college needed.

All good things, though, and after a couple of years, she grew restless and wanted something more out of life and she found a job selling typewriters. Olympia-Werke, out of Germany, had London locations, and where better to work in those days than the office supply industry? But Louise realized quickly that she hated calling potential clients to sell them on typewriters, and she eventually came to hate when clients called her as well and solved that problem by not answering any ringing phones in reach. She had a company car, though, and worked alone, so she spent her days driving endlessly around southeast England (that part was a pleasure), and yet few Mastertype 120i typewriters deployed to British businesses and schools can thank Louise Prockter for their presence (though they were quite nice, she had to admit, with cathode-ray-tube screens and everything). She quit after a year, but wasn't quite finished with the office supply industry and found a job at a small manufacturer that made PVC ring binders. The job still entailed sales, though she was given charge of the company's marketing as well, and made pretty good money. Businesses needed binders, and she had them, and if she didn't have them, she worked at a *binder factory*, and so she could sufficiently solve any supply shortcoming. At twenty four, she felt things were looking up after the typewriter debacle.

Then Prockter and her boyfriend split up after four years of dating. The two shared all the same friends, and he got them in the split, so she was not only single now, but single without any friends, and thus without her previously active social life of antics and aspirin. By day she sold ring binders, which was exciting, and by night she was bored out of her mind, which was not. Louise searched for something to distract and fill the hours (you couldn't really work nights, even in the dog-eat-dog world of binder sales), and courses at Open University, England's highly regarded correspondence school, struck her as the most productive option available. On her application, she checked the box marked General Science. It was—well, when she was a kid, her family wasn't well-to-do, and London was expensive, but the museums were free. You could spend the day walking around galleries and exhibits, and nobody cared whether you had money or not. The Prockters spent a lot of time at the Natural History Museum. There was this giant skeleton of a brontosaurus on the other side of the museum entrance and a life-sized model of a blue whale alongside it for scale. (The brontosaurus-to-whale scale was roughly 1:1, excluding neck and legs.) But her scientific curiosity went beyond that. No question: whales were great, dinosaurs were great—giant lizards from millions of years ago, Fred Flintstone used one as a crane and, after work, a slide—but deeper within the museum was a geology area, and in her young mind, some part of the physical universe suddenly clarified. Rocks, realized Louise, had context. They belonged in certain places, and could *only* belong there, and told stories about the world they left behind. Less clear was the museum entrance itself. Louise, attired in her school uniform (black blazer, school badge on the pocket, white shirt, red tie), just couldn't wrap her brain around it. The facility had been built in one of those gorgeous

Victorian-era buildings that somehow survived World War II, and its entrance had been renovated to accommodate the public.

The doors at the museum opened on their own. You'd walk up to one, and it would slide open. Nobody controlled it. Each time Louise's family arrived at the museum, she would try to outsmart the door. She'd walk slowly when nobody was around, creeping up to it as though she were trying to snatch a squirrel from a park bench, and it would figure out what was going on and open. She would edge along the wall so as to avoid stepping on a suspicious plastic mat in front of the door (was it pressure sensitive?), and the door would somehow figure out that there was this little girl trying to get in, and it would open helpfully. The whole thing baffled her. It wasn't magic. It couldn't be. It had to be science. But how? Her father was a respected engineer with a zeal for astronomy and geology (on his office wall hung a map of Earth's tectonic plates—the large slabs of planet that slid along the molten upper mantle and compressed into mountains and subducted below for recycling—that were responsible for continental drift). He had earned a correspondence degree from the Open University, too, and throughout her childhood, Louise watched him toil on whatever assignment came in the mail, piles of papers, textbooks. She recalled her father once carving up a sheep's brain on the kitchen table—could still smell it, in fact. Her mother was a biology teacher, and their freezer was ever in possession of one bovine organ or another, eyeballs for the next day's class. So Louise's decision to enroll in the Open University made sense, and checking the General Science box, doubly so.

She had to take the usual classes—physics and chemistry and biology—but one of them, earth science, was new to her. What did it even mean? For the next year, the school mailed her modules that she completed at home; little kits with, for example, mineral samples for a geology unit—her first formal study of the subject—

and she would spend evenings identifying olivine and feldspar and such. To learn Newton's laws of motion, Louise had to rig a pendulum on the doorframes and bookshelves of the new apartment she had leased using the pretty good money she was still making at the PVC binder factory. After completing each module, she would return the homework to the school for a teacher to mark, and the next assignment, toted by mail carrier, would follow. Active evenings of experimentation bled into bleary-eyed weekends, when she would wake at ungodly predawn hours to watch televised lessons broadcast by the university.

That summer, she attended a residency on the Open University campus, where she met fellow students and spent time in an actual laboratory. She loved it, the whole experience, scholarship, having friends again. She also loved her good company car and her new apartment-slash-laboratory that her pretty good PVC binder factory salary afforded. But . . . So to everyone's shock and horror, she quit her job at age twenty-seven and enrolled full-time at Lancaster University as a "mature" student.

Little latitude was allowed in the courses chosen, or when one took them. University educations paid for by the state permitted no playing around with electives in fifteenth-century Russian folk stories, or whatever. Louise chose to study geophysics, and the modules handed down from above reflected that. Now you are taking an earth sciences module, and here are the classes you will have. Now you are taking mathematics. Now physics. The courses came at a relentless pace, with final exams given at the end of the three-year program. To Louise's surprise, she did very well.

During her final year, she was given her first real choice, coursewise. The two available electives were cosmology and something called planetary science, the latter taught by a professor who happened to be passing through, and, for whatever reason, that was the one she chose from the catalog. From the very first class, she didn't know-know, but she just *knew*—knew that this was what she wanted to do for the rest of her life. The students looked at images of planets—not even wild stuff, really, things she'd seen on TV and in magazines, but the teacher asked Louise to look closely, to ask questions about what she was seeing, and to hunt down some answers. One day he handed out a paper and assigned them to read it.[51] It had been published in the *Journal of Geophysical Research* and was titled "Atmospheric Effects on Ejecta Emplacement and Crater Formation on Venus from Magellan." This, he said, is everything a paper should be. It is one of the most astounding papers that—you, you should read it! This is how a paper should be written. This is great science![52]

The paper looked at the morphology—the shape and the texture and topography—of impact craters on Venus, and how the planet's atmosphere and gravity affected the meteor collisions that created them. The images it included had been produced from radar data taken by an orbiter called Magellan—those being the only type of data available for the Venusian surface—and they astonished in ways the customary images of Venus (i.e., fuzzy cloud tops as seen from space) could not. You didn't need a doctorate to tell that the craters were weird, but also, to Louise, beautiful. They looked living, somehow. Organic. Animate. The photographs might just as easily have been found in a medical textbook.

Across sixty-six pages of small print, complete with tables, graphs, and diagrams, the author concluded that the atmosphere of Venus shields the planet from small objects from space and preserves the effects and processes of impacts that might be lost on other heavenly bodies. *What a result!* her professor declared. *What a paper! The research that went into it!* He really could not get enough of it. Much of the science was beyond Louise, but what she grasped, she grasped tightly and held close to her heart.

That paper, written by Peter H. Schultz of Brown University, changed her life. She now knew for certain what she wanted to do, what good, bare-fisted science looked like. It wasn't some grand theory of everything, some *Eureka!* that solved the great mystery of the universe. It was something far more profound in implication: meticulous work conducted over a number of years to solve a small oddity on another world. How does gravity and atmosphere affect the formation of impact craters on the planet Venus? Now we knew, or, at least, had a plausible framework for understanding. The paper's references section listed eighty-one other papers written by scores of scholars over forty years. There was at work here a Confucian interaction of generations, nationalities, and specialties of knowledge, with everyone before, and surely everyone after, driven by his or her own simple need to know the world as it is. And now Louise Prockter knew, too.

She graduated in 1994 and formally stated her intention to study planetary science. A faculty advisor suggested she apply to graduate schools in the United States, and Louise thought that was just a terrible idea. She was suspicious of the currency and leery of the American sense of humor or lack thereof. Her advisor, unsympathetic, gave her a list of names of planetary scientists in the United States and suggested she write them letters. She did, and some actually responded, including a man named Jim Head at Brown University, a graybeard and force among the chosen few in the field, and who had also once been part of the most arresting and audacious achievement of the twentieth century, if not all of human history.

It happened twenty-six years earlier, when James W. Head was himself a newly minted Ph.D. in geology and in need of a job. While flipping through a job placement book produced by Brown, he came upon a full-page advertisement that read, OUR JOB IS TO THINK OUR WAY TO THE MOON AND BACK.[53] Well, who would say no to a problem like that? The page listed a DC telephone number, and he dialed, reaching a company called Bellcomm, a subsidiary of Bell Laboratories. It had been established in 1962 at the request of NASA to handle systems analysis for the agency's program called Apollo. Bellcomm hired Head straightaway and installed him at NASA headquarters, where he was effectively an agency employee. His job: *to help them think their way to the moon and back.* Putting a man on the moon was going to happen, the NASA people said, but they were stuck on the "and back" part. Their immediate problem involved figuring out where an astronaut could safely land on the lunar surface, the integrity of its uppermost layer questionable. Was it snow-like or as solid as steel? This was a serious problem. All they knew for sure was that the moon was a giant rock and they needed a rock expert. That was but one problem, however. Once the *Eagle* landed (gently but firmly), and the astronauts saluted a flag and telephoned the president, there were still two hours and twenty-nine minutes to kill on the shortest visit. On days-long expeditions, you'd need a good deal more for these guys to do, and if you're going to travel that far to visit a giant rock, the only science it made sense to do was geology. And so, once the lunar rigidity was worked out (provisional determination: the moon is solid), Dr. Head was told to teach aspiring Apollo astronauts geology. To his surprise, they proved the most enthusiastic student body imaginable. Give type-A personalities a job to do—tell them that *only they can do it*—and watch them hit the books and take up tiny rock hammers and study stones and emplace seismometers and practice pickax employment in austere alabaster environs . . . if the moon program lasted long enough, these guys would probably find dinosaur bones up there.

But it didn't and they didn't, and in 1994, when Louise Prockter wrote Jim Head, he was back at Brown and long entrenched there as a distinguished professor of geological sciences. He invited her to visit. She arrived that July, on what was (coincidentally) the twenty-fifth anniversary of the Apollo 11 moon landing. This seemed ancient to Louise, the Apollo program belonging somehow to the same age that produced the Wright Flyer or Amelia Earhart, but it was brought back to life for the occasion, on television and in the papers, and especially in the Lincoln Field Building at Brown, where walked the man who helped choose lunar landing spots, a literal rock star who once ran an astronaut academy. Compounding coincidence: The same week that Louise arrived, a comet called Shoemaker-Levy 9, which had been swinging around the solar system for more than four billion years, suddenly shattered into smaller pieces. A salvo of cometary fragments the size of football fields and shopping malls careened into Jupiter. It was the first time in history that anyone could directly witness the collision of two celestial objects.

The buildings of Brown University blend by design into Providence proper—little labyrinths of metal fences, manicured lawns, courtyards, and redbrick buildings—and in front of Lincoln Field the week of Louise Prockter's arrival, scientists set up rows of telescopes so that anyone walking by could stop and see the once-in-an-eon event: the pitiless bombardment of planet Jupiter. There were folding tables lined with pizza boxes and punch cups, and music—a carnival-like atmosphere and days of celebrations for Apollo and Jupiter. For an aspiring planetary scientist, Louise had to admit, it was as though the universe and the university had conspired to make clear to her that she needed to settle in. She was home now. Providence indeed.

In 1996 she completed her master's at Brown (thesis: "Axial Volcanic Ridge Architecture: Classification and Interpretation of Volcanic and Tectonic Features from High Resolution Sonar Images of the Mid-Atlantic Ridge [24°N–30°N]"). It wasn't so much a goal as a milestone; at Brown, you enrolled directly in the Ph.D. program and earned an M.S. along the way. At the time, the NASA spacecraft Galileo—newly established in orbit around Jupiter—was gearing up for its first encounter with a Galilean moon. Brown hired a researcher named Robert Pappalardo to help Jim Head and Geoff Collins, another doctoral student, plan the imaging campaign. Up until then, Louise had focused her studies almost entirely on the volcanoes of Venus and the ocean floor of Earth. Her specialty was surface geomorphology, which meant that she looked at images from planetary bodies and studied the textures of their facades. She did a lot of mapping, and her job was to tease from them a surface's history. She had an eye for structural geology as well, understanding surface tectonics, the fractures and faults. It was a skill that could be applied to any planetary body: Venus or Earth or Mercury or an asteroid or the moon, but once those first Galilean images arrived from five hundred million miles, the Jovian system had her henceforth.

But those images almost never happened. The spacecraft Galileo had launched from Cape Canaveral seven years earlier, and during its first eighteen months in flight, it talked to Earth using a tiny, low-gain antenna—perfect for transmitting telemetry tidbits and confirming course corrections, but not much more. It was a temporary thing, and once the spacecraft was sufficiently far from the sun and free from the attendant risk of thermal damage, engineers ordered it to open its massive, mighty main antenna dish, which worked and looked like a giant umbrella.

The robot responded that it could not.[54]

This alarmed engineers. This high-gain antenna was the only way the spacecraft could send data to Earth from Jupiter. It could still collect the science, but without a way of returning its findings, Galileo would be like a library that fills itself with books, locks the door, and self-incinerates.

Clues culled from spacecraft telemetry—a surge in a motor's current here, a slight wobble there, a slowdown in spin rate—revealed that three antenna "umbrella" ribs were jammed. The reason, soon determined: dry lubricant, which got that way while Galileo sat for years in a warehouse waiting to launch. (The spacecraft, which could have launched on any big rocket, had been tied politically to the space shuttle, which needed very important missions to justify its existence. When the shuttle *Challenger* blew up, those very important missions were forbidden from launching while NASA conducted a years-long investigation.)

The first idea from engineers was to do what you do when any umbrella doesn't open: close it and push it open again. The antenna actuator was not wired for closure capability, though, only opening. (On Earth, before liftoff, humans had to fold it into its stowaway position.) The next thought: spin the spacecraft sunward a few times and let the hot-cold cycle walk the ribs open. This "thermal expansion" technique was tried without success. The next notion: swing the low-gain antenna around a bit—really work it and shake up the spacecraft—and maybe motion might make the umbrella open. Six times they did this, and six times it failed. To jar the ribs loose, they then tried just jackhammering the thing open: sending spikes of electricity to its actuators. But even after thirteen thousand hammer strikes, the recalcitrant radio wave dish would not budge.[55]

But—and this is why mere salutatorians don't bother applying to Jet Propulsion Laboratory!—the engineers, their pencils now nubs, their blackboards white, found a way forward, and slapped a So Crazy It Might Work on the project manager's desk. The Galileo mission, they said, winding up for a pitch, had two major elements: 1. the spacecraft proper and 2. an attached atmospheric probe that would, on arrival at Jupiter, be launched like a torpedo into the gas giant. The probe's job was to study Jove's mysterious makeup and interior physics; it would live for less than an hour before being vaporized by the pressure of Jupiter's atmosphere. Meanwhile, the observing Galileo would receive and record the frantic and increasingly desperate reports from the probe, and later, leisurely transmit that recording back to Earth using the high-gain antenna. The spacecraft's digital recorder was an old reel-to-reel job like you might see in a police procedural. Mechanically, it worked like this: Rewind the tape. Record the data from the doomed probe. Rewind the tape and play back the data for scientists to study.

The engineers pointed out that everything on a spacecraft has a second vocation, and the tape recorder was no different. The high-gain antenna—it would never open, was vestigial now, as useful as a human appendix—was designed to deliver data to Earth at one hundred thirty-four kilobits per second—fast for 1986, when originally slated for launch, but not always fast enough.[56] On occasion, the spacecraft might be commanded to collect more data during an encounter than could be transmitted straightaway—like, say, from a high-resolution image. Likewise, when the spacecraft was on the far side of Jupiter and thus unable to see the Earth, it would not be able to return any data at all. To act as a buffer during bottlenecks and blockages, the spacecraft would fire up ye olde reel-to-reel and make temporary recordings to get the antenna over the hump. It was useful, a nice-to-have.

The project manager, of course, knew all of this.

Then the pitch: Rather than use the recorder as an occasional and brief buffering mechanism, said the engineers, why not use it to record everything, continuously? The tape had just enough length to store the totality of data collected during major encounters with Jupiter's moons (more or less). You could record an orbit, play it back to Earth, erase it, and repeat for each encounter. Rather than act as an intermediary, the tape deck could serve as a hard drive: the spinning heart of the spacecraft Galileo. As for the antenna, being limited to low gain was certainly a problem. Its present bandwidth—about one-ten-thousandth of what the high-gain could do—obviously wouldn't work;[57] it would take years for the data recorded during a *single* moon flyby to get back home, and the spacecraft had at least ten flybys planned in its first two years alone. The math didn't work. But computer scientists had devised a new storage methodology and compression algorithm, and could program them into Galileo from millions of miles away, just rebuild its brain completely, one zero and one one at a time, the way you might upgrade a home computer operating system. And by cranking up the sensitivity of the Deep Space Network, the global array of giant spacecraft antennae that listened to Galileo from Earth, engineers could get Galileo above the one-kilobit mark. Its new, improvised transmission speed would thus be one-one-hundredth of what was originally planned.[58] It wasn't pretty, and thirty percent of the planned science would be lost, but it was orders of magnitude better than would be achieved otherwise. It would work. The mission could be saved by the tape recorder.

NASA gave engineers the go-ahead. Crisis averted. Fives highed. Paychecks earned. Congressional inquiries and rolling heads forestalled.

Four and a half years later, on October 11, 1995—fifty-eight days before Galileo would arrive at last at Jupiter—a computer in the Space Flight Operations Facility at the Jet Propulsion Laboratory (mission control for robotic spacecraft) said something was amiss. The spacecraft had announced an anomaly.[59] A big one. A terrible one.[60] Twenty-two million miles out from Jupiter, the spacecraft had taken a family portrait of the planet and its major moons and, as instructed, rewound the recorder reels to playback and return the images to Earth. This should have taken twenty-six seconds. Fifteen hours later, according to the telemetry, it was still rewinding.

This was not a very good day at mission control.

The Galileo recorder was long obsolete—had been before launch, reel-to-reel practically papyrus by then—but engineers searched high and low and found a laboratory spare that had been built for the Magellan mission to Venus. They spooled up a fresh tape, fed it the same conditions as Galileo at Jupiter, and pressed Play . . . and the tape ripped right from the reel, *slap-slap-slap-slap-slap*—[61]

It couldn't end like this—it just couldn't—so they continued studying the problem, piecing together trajectory clues from Galileo and comparing them with the spare on Earth. Nine days later, having disassembled and inspected every recorder component and crevice, they had a suspicion and a plan of action. Perhaps the tape wasn't torn. The spools, it seemed, were perhaps stuck, snagged on debris from the natural wear of a new recorder never broken in, never meant for dedicated operation. If the reel wasn't severed, and if the recorder, in fact, still worked, they could theoretically just gun it over the hill to clear the debris.

They tried it.

It worked.

Afterward, the project declared the compromised stretch of tape off-limits, which meant that much less storage and that much less science, but rolling heads and congressional hearings had been avoided yet again.[62] And though crippled, the recorder was able to do its job, the data collected vastly exceeding expectations.

Some of Galileo's most earth-shattering results came from perhaps the least obtrusive of the scientific instruments in its payload: the magnetometer, proposed by Margaret Kivelson of UCLA to study how the Jovian magnetosphere (i.e., the region of space dominated by the planet's magnetic field) affected its moons. Magnetometers were generally afterthoughts on spacecraft—something you added because they didn't take up much space or power—and rarely returned results relevant to anyone outside of a small community of theoretical physicists. It was a "no-surprises" scientific tool, and yet when the magnetometer returned data on Europa, it found a startling surprise indeed: an intrinsic magnetic field, which should have been impossible. Europa was far too small to generate such a field, and, weirder yet, the ice moon's field was pointed the wrong way relative to Jupiter. The whole thing just didn't make sense, no matter how many blackboards physicists filled, but the data weren't lying. There it was, a magnetic field emanating from the Europan interior.

Krishan Khurana, a research geophysicist at UCLA, published a paper positing a reason: that the mysterious magnetic field flowing from Europa's insides might be *induced* by that of Jupiter, suggesting a subsurface conductor of some sort.[63] It was the same way an airport metal detector worked.[64] Despite the name, a metal detector doesn't detect metal. Instead, it produces high-frequency magnetic waves that pass through, say, the car key in a traveler's pocket, inducing a little magnetic field of its own. That *induced signal* is what metal detectors detect. In Europa's case, either its interior was made of copper (it wasn't), or there was an extant saltwater ocean down there—as had long been thought but was impossible to prove. But a quirk of Jupiter's physics presented an opportunity: the massive planet's magnetic field was tilted by ten degrees. If Galileo took measurements of Europa when the icy moon was on the *other* side of Jupiter's field, and Europa's field flipped, you had evidence of induction and, consequently, of an ocean. Kivelson, as instrument principal investigator, made the case to the Galileo project management for further measurements, and it was a hard one, because the spacecraft was by then limping along and radiation poisoned, flying on borrowed time. *We waste this observation, and we lose a textbook of information.* But she was insistent, the way only a genius seventy-one-year-old space physics pioneer could be, and she prevailed.

It was worth it. Until Galileo, though physics certainly suggested an ocean inside of Europa, it might well have frozen solid hundreds of millions of years earlier. In 2000, doubts were dispelled when Kivelson published a paper presenting the first direct evidence of Europa's subsurface ocean.[65] The magnetic field flipped, as hypothesized, which meant the ocean was extant, liquid.

And where there was water, there was life.

Six years after Kivelson's paper made manifest the need to explore Europa in earnest, Dr. Louise Prockter's outbox still hummed in her office at the Applied Physics Laboratory.

The next day, she received her reply from NASA headquarters.

From:	Niebur, Curt
Sent:	Friday, December 22, 2006 3:31 PM
To:	Prockter, Louise M.
Subject:	Re: SDT membership application
Importance:	High

Louise,

Are you interested in a joint appointment as a member of the Europa SDT and cochair of the Ganymede SDT? Dave Senske of JPL will chair with you.

Thanks,

Curt[66]

Replied Louise: "Yes please!"[67] And now it was her job to get NASA back to Jupiter.

CHAPTER 3

The Dark Ages

FOR PLANETARY SCIENTISTS, THE JIMMY CARTER–
Ronald Reagan years were in retrospect like the Dark Ages, and
they, the monks tending in enclaves to the embers of civilization.[68]
For a solid decade starting in late 1978,[69] NASA launched no planetary science missions, and pretty much the only space science data
trickling back to Earth came from the Voyager 1 and 2 flybys of
the farthest planets of the solar system, where you'd get three weeks
of data and then three to five years of silence—hardly enough to
sustain an entire field of scientific inquiry.[70] The Voyager findings
at Jupiter fueled a desire by the careworn planetary science community to return there, but that required Reagan to fund the spacecraft
Galileo—something his administration worked diligently to avoid
doing upon assuming power in 1981. The new president believed
he had a mandate to slash nondefense spending, and he was following through, and if you weren't building bombs, battleships, or
Black Hawk helicopters, your budget was up for grabs—and grab
they did. While NASA's top line fared well overall, that money was
directed largely to the space shuttle program, which had become

something of a flying Statue of Liberty in the public imagination.[71] Anyway, the shuttle had military applications, including the deployment of spy satellites[72] and, on paper at least, stealing satellites from foreign governments.[73] The supply-side marauders would still get their squeeze from the agency, however, and that meant science. Before the toner was dry on new presidential letterhead, the White House told NASA that of Galileo, the Hubble Space Telescope, and the joint NASA–European Space Agency International Solar Polar Mission to study the sun, it could keep two (for now). And just like that, Solar Polar was gone. The Europeans had invested in it more than one hundred million dollars, and America thanked them for the trouble by withdrawing without warning, leaving the Europeans seething.[74] The slaughter continued with the spacecraft VOIR, the Venus Orbiting Imaging Radar: vaporized. This cancelation, too, went over poorly. If the Solar Polar abandonment was an uninvited concupiscence thrust upon America's allies abroad, the Venus cancelation was at *least* a rude gesture suggesting the same to planetary scientists at home.

But that Galileo mission—how it vexed and annoyed the White House. How the administration wanted this half-billion-dollar monstrosity slain! This expedition to Jupiter . . . we—we'd *just been there* with Voyager! Why were we even talking about this? So the Office of Management and Budget zeroed out Galileo in its tentative plan for the agency.[75, 76] As for those twin spacecraft Voyager: What, exactly, was there to learn about planets past Saturn, anyway? Uranus! Neptune! Did it matter? I mean, come on![77] Just issue the shutdown command, and we could also switch off this devil-begotten Deep Space Network as well, those gigantic, twenty-story radio dishes required to talk to them. That's two hundred twenty-two million dollars saved overnight.[78, 79] Between Galileo and Voyager, we could cut costs by a half billion.

To somehow save what was becoming even to outsiders a

sinking ship, the public started pitching in. In one instance, Stan Kent, a California engineer, created what he called the Viking Fund—a private, pass-the-hat effort to cover costs for Deep Space Network downlink time for Viking 1, the last surviving spacecraft on the surface of Mars. Donate now to feed a starving robot—send checks to 3033 Moore Park Ave. #27, San Jose, CA 95128.[80] The Viking program had once been the zenith of NASA space science, the most ambitious agency endeavor since the Apollo program, and, when conceived, a prospective precursor to Apollo's obvious heir: human missions to planet Mars.

Between 1965 and 1976, NASA had sustained a steady sequence of sophisticated Mars probes. Mariner 4, a flyby in 1965, was humanity's first successful encounter with the Red Planet. Mariners 6 and 7 followed four years later, imaging up close the entire Martian disc, and those images, stitched together, revealed a real rotating planet—just like Earth.[81] Mariner 9 in 1971 was the first spacecraft to enter orbit around another planet, mapping Mars in high resolution and capturing dust storms and weather patterns. Like elapsing lines in the book of Genesis, each spacecraft in succession made Mars a world as real as our own. By the time the Viking landers left launch pads at Cape Canaveral in 1975, no hope remained for extant alien civilizations, but flora and fauna of some form were still on the table. And the question remained—the ultimate question— the same that had fueled fiction and stirred scientists for centuries: What did that Martian wildlife look like?

The American space program has always marched inexorably toward Mars. Before the *Eagle* landed—before the first naut, cosmo, taiko, or astro—before Sputnik—before even the formation of NASA itself—there was *Das Marsprojekt,* a work of speculative fiction by Wernher von Braun, the German rocket scientist relocated to the United States immediately after World War II. No mere thought experiment or flight of fancy—no ray guns, no saucermen—the plot was a thin veneer over How to Do It, and the author the person most likely to make it happen. Von Braun wrote *Das Marsprojekt* in 1948 after finishing work reconstructing for his new American hosts the V-2 rocket, a ballistic missile he helped develop during the war. The book was later stripped of its fictional elements, and much of it repurposed as a nine-page article in the April 30, 1954, issue of *Collier's Weekly,* then one of the most popular and prestigious magazines in the United States.[82] The first serious study of how to get to Mars, von Braun's plan involved a space station and a flotilla of reusable rockets and shuttles, and necessitated a crew seventy strong for a Martian stay exceeding one Earth year. Upon arrival, astronauts (well, "spacemen"—*astronauts* had not yet been invented) would enter orbit and scout suitable set-down sites for the human beachhead. (He didn't discuss robotic exploration because digital, programmable robots had not yet been invented, either.)

For von Braun, Mars was always the plan, the moon merely a waypoint, and fourteen years later, when Armstrong leapt from that bottom rung of the lunar lander ladder, it was von Braun's Saturn V rocket that got him there.[83] He (i.e., von Braun) was by then director of NASA Marshall Space Flight Center in Huntsville, Alabama, de facto "father of the American space program," and a minor celebrity. He had made multiple appearances years earlier on a 1950s television show called *Disneyland*—hosted by Walt himself—selling to forty million Americans the notion of robust, reliable rockets, moon shots, and Mars colonies.[84] When the shows aired, Yuri Gagarin was still an obscure pilot in the Soviet air force, and Alan Shepard a test pilot in Maryland. To the extent that Americans were even aware of U.S. space ambitions, it was von Braun soft selling Mars missions with Walt Disney. He had been working toward this for a very long time.

It was thus unsurprising that two weeks after American silicone soles pressed prints into fresh moondust, von Braun stepped into

Spiro Agnew's office and slapped onto the vice president's desk the next natural frontier for American exploration: the Red Planet.[85] The fifty-page presentation—the definitive plan to make mankind multiplanetary—represented the culmination of von Braun's life's work. His prescription involved many of the elements he had proposed decades earlier: the rockets, the shuttles, the station—even a nuclear-powered spaceship.

Unfortunately for von Braun, prevailing forces in Congress and the White House came quickly to see the Apollo program as the goal, rather than, as he had hoped, an early milestone of something much larger. You didn't build Hoover Dam and then . . . build more Hoover Dams downriver, said the politicians. We set a goal, and by God we did it. Why even *have* a NASA? wondered the White House aloud. By Apollo 15 in 1971, opinion polls pegged public support of space spending at about twenty-three percent, with sixty-six percent saying that spending was too high.[86] There would be no national political price for closing Cape Canaveral completely. Really, what were we doing up there?

Nevertheless, von Braun's sequence of space missions culminating in Mars exploration had so defined NASA that it was almost hardwired into the system. Nixon, having zero interest in the space program but even zeroer interest in being the one who ended it, entertained only the space shuttle element as viable because it 1. had those spy satellite applications and 2. could be a major construction project in Palmdale, California, keeping his home state in his column during the next presidential campaign. So the California-made, satellite-stealing space shuttle it was! NASA lived to flight another day.

Unlike the human adventure, which always sold itself, NASA's robotic program had a much tougher go of it, and that it survived the end of Apollo was miracle adjacent, especially since more than once it was nearly canceled entirely. Still, that twisting of Mars into the agency's genetic code led eventually to the billion-dollar Viking program: a pair of orbiters and landers that would answer the "life question."[87] Even by Apollo standards, Viking was gutsy and glorious. Mars, engineers and scientists had determined, was one of the most difficult places to land in the solar system. You needed parachutes to slow down, but because its atmosphere is limpid and ephemeral, you still needed thrusters to take you in—and you couldn't use both at once. Remote control was impossible because of the distances involved; computers would have to command any entry systems, and the descending spacecraft would be alive on the ground or vaporized in a hole by the time transmissions from Mars reached Earth—despite those signals traveling at the speed of light. No one even knew for sure the solidity of the Martian surface. One-third of scientists on the project thought it might have the consistency of shaving cream.[88] What if Viking landed and just . . . kept landing! All of which is why NASA launched two Viking surface probes: redundancy. One would probably crash, but the mission would go on. And yet despite the challenges and uncertainties, Vikings in 1976 would achieve their most undaunted landings since Erik the Red a thousand years earlier. Viking 1 set down softly near the Martian equator at Chryse Planitia (the Golden Plain), and Viking 2, halfway around the world, in the scalloped, higher latitudes of Utopia Planitia (the Paradise Plain).

To find life, the landers would each extend arms and scoop up soil samples for testing. (The Viking camera was a test of sorts as well: if something scurried or slithered on the surface of Mars, Viking would see it. Paw prints, snake trails—the cameras could capture them.) A scoop of soil would be tested for photosynthesis. Another test would look for markers of metabolic activity. A third would check for respiration. Any positive result would be tested and retested so that scientists might know conclusively.

For the first time since Apollo, the media were *absolutely into*

this NASA endeavor, which was great . . . except . . . well . . . scientists were being asked for instant analyses of data that might take
months, years, or entire careers to evaluate. Reporters pestered
them for instant, definitive statements in the affirmative or otherwise on the most consequential question in all of science, religion,
and philosophy: Are we alone?[89] Even on the easy stuff, scientists
hesitated to give yes/no responses—there's *always* an exception. But
with Viking, only an absolute answer would do! Is there life or isn't
there? To the dismay of editors around the world, Viking seemed
to produce only hedging headlines with terms such as *could*,[90] *hints
at*,[91] *still a puzzle*,[92] *inconclusive*,[93] *unsure*[94]—even weasel words
without scientific pretense, like *hope*[95] and *belief*.[96] Coverage in the
New York Times read as though written by coin toss. Two months
after the landing, the evidence for life was "beginning to mount
impressively."[97] Twelve days later, findings "did not bode well for
life-on-Mars theories."[98] Whom to believe! What were we *doing*
up there?

By the onset of the eighties, with White House budgeters picking off spacecraft like Special Forces snipers at a carnival shooting
gallery—*ding!*—*ding!*—the Viking answer to the life question was:
maybe, but probably not. And though Viking 1 still maintained a
heartbeat, still performed in situ science, it was elderly and infirm,
and a prime candidate for the purge.

Enter that Viking Fund, with ten thousand fans of far-off
worlds saying in unison, "Count me in!"[99] and contributing small
sums: five dollars, maybe fifteen. Robert Heinlein himself dropped
into the jar ten crisp one-hundred-dollar bills.[100] The fund's very
existence spread word-of-mouth, geek to geek, a science-fiction
convention here, a Dungeons & Dragons gathering there. It was
moving, but it was dismaying. American space science had been
reduced to a Beautify the Park initiative. Stan Kent dutifully sent
to NASA the sixty thou he had collected . . . and the agency had no

idea what to do with it or how to handle it, because federal offices were forbidden by law from accepting private donations for specific projects. Of course, this wasn't a bribe, exactly. It was just normal people sending money to a federal agency? Because they believed in it? Well, it was unheard of and completely crosswise with the rising Reaganite zeitgeist, but, look, things were bad, so NASA lawyers found a way and *cashed that check*.[101] It amounted to the first private funding of a deep space mission, and the initiative ultimately raised one hundred thousand dollars.[102] But the bake sale model of space exploration was neither ideal nor sustainable, and somehow things only got worse from there.

Emboldened by its progress and politically invincible, in December 1981 the White House Office of Management and Budget pushed in its pile of chips, proposing the termination of the entire planetary program at NASA. If you were a space scientist, it was just comic-book-caliber villainy. It was like canceling biology or math! But the president's science advisor concurred, as did the deputy administrator of NASA, who supported astrophysics, but that was about it (he was a physicist). He certainly didn't see a future for Jet Propulsion Laboratory and had only a few years earlier written a paper declaring the results from space science in our solar system thus far to have been of no major consequence.[103] The science return simply wasn't commensurate with investment. What's the rush to visit Jupiter? It'll be around in ten years. The Venus orbiter was just going to return higher-resolution data of something we had already. And it's not like there would be a political price to pay. The reptilians on Venus didn't vote and were unlikely to stage a counteroffensive. At the time, every single planetary scientist in the world could fit on a single international flight.[104] You could probably *win* votes by killing planetary. Selling off the Deep Space Network, running the bulldozers across Jet Propulsion Laboratory—or better yet, giving it to the U.S. Defense Department—that was prime real estate in the San Gabriel Mountains (Rose Bowl Stadium was three miles away)—it was an easy dollars-and-cents decision.

The California Institute of Technology, which managed JPL, fought back successfully. First, it authorized increasing the lab's intelligence and defense work to about thirty percent, as defense spending was sacrosanct in Reagan's DC, and better to make satellites for the war machine than to clean out our desks. Next: Caltech's powerful, well-heeled, and well-connected trustees took a sort of Chesty Puller attitude toward the problem: So, we're surrounded? Good—now we can fire in any direction! The lab was a California institution, and California was Reagan country. There were direct lines of influence that reached the Oval Office, and the trustees tapped into them. Through correspondence and cocktail party offensives, they began a systematic petitioning of the government across the nation's capital. They struck pay dirt finally with Howard Baker, the Republican Senate majority leader from Tennessee, who intervened personally on behalf of the Galileo mission, extending a political shield that made it invincible. From that point on, one by one, the dominoes stood themselves back upright. Preserving the mission meant keeping the lab and necessitated the Deep Space Network which in turn saved the Voyagers.

Salvation did not all systems forever go. The planetary program wasn't on solid footing, but having narrowly averted an apocalypse, scientists could circle the wagons and take stock of the situation. If their field was to survive, everyone knew, it would have to find a new way of doing business.

THE FIRST TRY was called Planetary Observer. The idea was to take flight-proven, Earth-orbiting satellites and refit them for cheap, focused missions around other rocky worlds. Rather than build billion-dollar battlestars that did everything (including attracting too much attention from appropriators and budgeteers), why not build a lot of low-cost, "no-miracle" spacecraft that each did one thing really well? We could actually afford to build those and could launch them in regular cadence.

Programmatically, the idea with Planetary Observer was to replicate an even older mission model called Explorers, which managed small satellites that could be built and launched in a short amount of time. These were single-issue science affairs, live hard and die young. One might map the magnetosphere near Earth and, eight months later, fall from orbit. Unlike larger lunar and planetary missions, their budgets were practically rounding errors in congressional appropriations. Accordingly, Explorers program managers didn't have to go out and seek permission to get a new mission going. The Planetary Observer program, the idea was, would work the same way. They would keep costs low by using only heritage hardware and off-the-shelf parts. Earth satellites were cheap and plentiful, and worked well. There was no reason that you couldn't repurpose one for, say, the planet Mercury or the asteroid belt. The spacecraft's science payload would be small and sharply defined. A mission might study only atmospheric conditions. Later, another precision expedition would work on magnetic fields. The next might do composition or imagery. Each would build on the one before, would keep the science coming in, would keep the cadence going. The program, by design, would control costs by way of a fixed budget. And because everyone in the community could be confident that there would always be another orbiter just around the bend, you didn't get a "Christmas tree" effect, with everyone trying to hang his or her science instrument on the spacecraft. Instead, you waited for the next chance in two years and submitted your spectrometer specifications then.

Right away and on cue, the White House had a problem. Fine, NASA could start up an office to manage this new "heritage spacecraft" paradigm—knock yourselves out—but we weren't about to give you a fixed budget. You want to fly a mission, you still go through us. So that's what happened. The first of its line to get the go-ahead was the Mars Geoscience/Climatology Orbiter.

The last of the Vikings had finally died. The Red Planet looked lifeless, but to know for sure, you would need to study its soil up close with big binocular microscopes on long black tables on Earth. The holy grail was thus a sample return mission, but if you were going to take samples, the more context you had for them, the better the science you might be able to do. So planetary scientists rolled up their sleeves and proposed a panoply of potential Mars missions. The orbiter, designed by Jet Propulsion Laboratory, emerged as best of the lot. It was a sensible first step toward an eventual sample return expedition that might be mounted, if all went well, in the late nineties. It was approved in 1984 for the following fiscal year, and it would cost less than three hundred million, total, top to bottom, start to finish.[105] That, everyone agreed, was a practical price tag in these tough times. In keeping with the spirit of simplification, the lab shortened the mission's name to Mars Observer.[106]

It was doomed from the start.

Like everything else, NASA needed to tie this thing to the space shuttle. America's "space truck" had been intended primarily to build space stations, but space stations were not being built, partly because there was no money left after building the space shuttles. And maintaining the fleet was proving to be a colossal (and expensive) headache.[107] But we couldn't just give up! We have four of them to use, said headquarters, and *we will use four of them*. So planetary launches would work like this: Mars Observer would be carried to orbit as a shuttle passenger, the payload bay doors would open, and the spacecraft would launch from there to Mars.

It was that final step that threw things off. NASA headquarters couldn't reach consensus on the rocket element that would blast

Mars Observer from shuttle to target. You could build the booster into the spacecraft itself (a common practice among Earth satellites) or use what NASA was proposing to call a "transfer orbit stage": an expendable rocket that would shoot the spacecraft from the shuttle and then separate, letting gravity do the rest. The advantage of a transfer orbit stage was that the design could be used again and again for future planetary missions, saving money down the line. The disadvantage was that . . . it didn't exist. And the whole point of the Planetary Observer program was to use cheap, off-the-shelf parts: no miracles.

The result was a comedically complicated spacecraft design process. NASA pressed the Mars Observer team to allow private industry to submit spacecraft proposals for either method of propulsion, and if a transfer orbit stage spacecraft *happened* to win the competition, Marshall Space Flight Center in Huntsville would build it. This slowed spacecraft selection, so to make up for lost time, the agency announced an opportunity for labs across the country to submit science instrument designs . . . before the spacecraft bus had been chosen. That was new! Not all spacecraft were created equal. Some could hold one instrument, which might be a camera or mass spectrometer or magnetometer; some could hold a dozen. Size, mass, power constraints—those things varied—and nobody knew which spacecraft would win or how many instruments it might be able to accommodate, so the selection process now had to account for multiple payload configurations. Ultimately, an eight-instrument payload was chosen by NASA, and it would ride on a transfer stage spacecraft capable of holding . . . seven. Moreover, the instruments turned out not to be the off-the-shelf parts previously planned for. Because the Planetary Observer program was already so hobbled, the mission was being treated by scientists as the Last Mars Mission Ever—because with Reagan at the wheel, it might well have been. Proposed instruments, therefore, pushed the technology envelope and promised spine-tingling new data, which meant an increased risk to budgets and schedule. (The original plan for Mars Observer didn't include a camera. But the public loves space pictures, and since taxpayers were footing the bill, they'd get their postcards from Mars. Add it.[108])

If the problems stopped there, the spacecraft might still have gone over budget, but respectably so. But they didn't! Here is the problem, it was later learned, with repurposing reliable, flight-proven Earth satellites for interplanetary exploration: "reliable" exists on a sliding scale. The Earth spacecraft technology chosen for Mars Observer was used frequently by the Pentagon. That was a major selling point. But—and this wasn't factored into the decision—if one of those Pentagon satellites circling Earth glitched—say, its avionics or thermal protection were wonky, or it simply conked out and crashed back to terra firma—the military would just go to the warehouse out back, wheel out another one, slap it on a rocket, and launch it to space. No one would lose a lot of sleep on the deal, and from that standpoint, yeah, the spacecraft were *totally* reliable. But take the same spacecraft, load it with delicate, one-of-a-kind, state-of-the-art scientific instruments, send it to Mars, and it glitches . . . then that's it! Years and hundreds of millions of dollars are lost, no spares waiting in the warehouse.

Moreover, launches local versus those interplanetary work differently. An Earth satellite goes up and all the dirty work basically happens at once: the trauma of launch, the engine firings to place it in proper orbit. By the time the satellite blooms like a flower, unfurls its solar panels and instruments and arms and booms, it is in a Zen-like state of total serenity, at peace with the universe and in position in orbit. The whole process takes hours. Not so for a spacecraft to Mars. You repurpose an Earth satellite for a multiyear voyage to the Red Planet, and, to take but one example, you're saddled with joints that weren't made to endure that type of torque. Another:

Earth-orbiting spacecraft have instantaneous and constant communication with Earth. The same luxury does not exist at Mars; it takes minutes for a signal to send, and minutes more for messages to return. Further frustrating things: we are not always listening. The Deep Space Network is shared among multiple spacecraft. It'll point to Voyager 1, listen, send commands, and then Voyager 2 (same), and then Pioneer 12 and so on, meaning that spacecraft in deep space can count on communications only for an hour or two each day. They have to be smart and semiautonomous, because if something bad happens, they can't wait twenty-two hours for instructions from Earth. They need to react immediately. All of this would thus have to be designed, built, tested, and integrated into the spacecraft.

The upshot is that Earth satellites, chosen to reduce cost and complexity, increased both. But not as much as the shuttle would! (Again!) In 1986 the space shuttle *Challenger* and all hands aboard were lost at launch. It was a national tragedy, and to prevent such horrors from ever happening again, NASA grounded the fleet for a two-and-a-half-year inspection and refit. (Galileo's antenna lubricant dried during this hiatus.) Mars Observer, tied to the shuttle for launch like everything else, was a relatively low-priority mission. Even once shuttle operations resumed, the Mars team was told there would be a long wait ahead. That turned out to be correct, for several reasons: because contracts with industry would have to be renegotiated; because the launch vehicle changed (after all that, the spacecraft was given a Titan III rocket—a blessing, really, as it allowed more mass than the shuttle did); because NASA was strapped for cash; and because Mars missions were tied to celestial dynamics. Ultimately, the fast, cheap mission first funded in 1985 didn't launch until 1992. Its original cost: three hundred million. Its cost by the time it left the launch pad: eight hundred million. Which might have been worth it, had the spacecraft not vanished without a trace.

What possibly went wrong was this. Mars Observer used pyrotechnic valves to pressurize its fuel system before beginning its orbital insertion. But its communications system, built for Earth operations, wasn't rated to remain active during the jolting, percussive pressurization sequence. To mitigate the possibility of damage or crashed communications, upon arrival at its destination, the spacecraft would switch off its radio, perform the Mars maneuver, and switch it back on.

Here is what NASA now knows for sure: the spacecraft successfully switched off its communications system.

It was never heard from again.

Mars Observer was the first total failure of a Jet Propulsion Laboratory spacecraft in decades. A big part of the lab workforce didn't know a spacecraft *could* fail, so successful had they been for so long. And no emotional closure would be forthcoming because there was no telemetry transmitted to perform an autopsy. The loss of Mars Observer devastated lab morale, embarrassed the cash-strapped agency, and set science back by years.

If it was any consolation, however, the lab's Mars program would not only recover, but become the dominant force in space exploration for the next twenty-five years. Here is how it happened.

MAYBE REAGAN WAS playing four-dimensional chess, though more likely it was just a happy accident, but while he was cutting planetary science to the bone, and then shaving bone to the marrow, and then just sort of idly needling around in there to see which bits of gooey tissue NASA could do without, the president was also funding a secret space program that would eventually pay Apollo-level technological dividends and enable planetary exploration for the next two decades.[109] It was called the Strategic Defense Initiative and, as envisioned, would be a missile defense shield designed to deflect or destroy an incoming Soviet nuclear strike.[110] The idea was to nullify the doctrine of Mutually Assured Destruction, which asserted that if Ivan launched nukes at Uncle Sam (or vice versa), it would be suicide because the other side would retaliate with everything it's got.[111, 112] The Strategic Defense Initiative, went the argument, would deter the Soviets from attacking because we would survive the onslaught unscathed. But for this new type of deterrence to be effective without destabilizing global security, the missile defense shield would have to perform flawlessly one hundred percent of the time—be able to knock out an endless, simultaneous barrage of thirty-five thousand nuclear warheads[113] delivered from launch silos, submarines, bombers, and mobile launch platforms.[114] Its strategic folly was that by its very existence, if the balloon went up, the Soviets would now have a legitimate *reason* to launch those thirty-five thousand nuclear warheads at once. You had to get through that shield! And if, say, just two or three got through, and Washington and New York City and Los Angeles were vaporized, then the shield wouldn't have done its *one job*.[115] So there was a downside.

Reagan set a former NASA administrator on the task of working this thing out, and the California Space Institute, a research group at the University of California San Diego, held a workshop to bring together experts on everything from military strategy to space science. In the end, the institute returned a report on how to do it, and the researchers didn't hold back. This is not easy stuff you want, they emphasized. It will be expensive and it will not happen overnight, Mr. President, because the things you want haven't been invented yet, and the research is disparate, unfocused, and abstruse. But, they asserted, by engaging NASA, national laboratories, academia, and industry, it could be done.[116]

To defend the United States from missile attack, the report explained, the United States would need to deploy space-based interceptors (a concept labeled derisively by Senator Edward Kennedy as "Star Wars"[117]). But there was more to it than mounting lasers to cheap communications satellites and calling it a day. If the Soviets were willing to push the button and wipe out humankind, they wouldn't worry much about the treaty violations incurred by blowing up our satellites, which wouldn't be hard in any event, because spacecraft were designed to be as lightweight as possible; every gram increased the cost of launch. Because something as small as an errant flake of paint, flying at tens of thousands of miles per hour, could destroy an orbital satellite, the fleet of spacecraft comprising a missile shield would need to be armored—otherwise the Commies would use a fusillade of impactors, particle beams, heat rays—even nuclear blasts—to target them immediately as a precursor to doomsday. So in order to do this right, the shield would have to be *built to last*, the totality of necessary satellite armor enormous—three times heavier than the Brooklyn Bridge. Forty thousand metric tons.

The good news, according to the researchers, was that the United States did not need to launch three Brooklyn Bridges into space. Everything necessary to build the armor was already up there. When it came to aluminum, iron, graphite, ceramics, and glasses, Earth wasn't the only game in town. Rather than mine it here and fight gravity to bring it to space, we could just *mine it in space* and bring it to Earth's orbit. Look, hull plating is simple technology: it's just sheets of metal. We establish resource utilization facilities in space, and it's practically free, relatively speaking, to manufacture not only armor but also a whole host of things. Shielding, propellants, explosives, structural foundations (e.g., beams, container tanks), projectiles, decoys—none of this is hypothetical. We've been running trade studies and feasibility assessments on it since Apollo. Moreover, the same infrastructure for pulling iron out of asteroids can also mine such things as platinum, chromium, manganese—

these are *strategic materials* that the Defense Department requires to maintain a state of warfighting readiness here on the ground—and such treasures are not readily available in the continental United States. Asteroids and the moon, though, are teeming with the stuff. And guess who knows how to go to the moon, Mr. President? The research and technology development for something like this has long been underway on an ad hoc basis at NASA, universities, research institutions, the Defense Department, and the Department of Energy—but there's never been a unifying purpose bringing them together. And now there is.

So the Strategic Defense Initiative went forward.[118] It became, by far, the single largest research and development program in the Defense Department's budget, which was saying something.[119] It wasn't an overnight thing, never looked like the time lapse of a battleship being built. It was more like the world's largest science fair, with foundational technologies explored, designed, prototyped, tested: adaptive optics allowing laser weapons to mitigate atmospheric blurring; miniaturization and weight reduction of spacecraft components; radiation hardening of computer processors that were now also an order of magnitude faster; artificial neural networks; guidance systems and control units; propulsive technologies. Five-pound, hundred-thousand-dollar inertial measuring units used for guidance were reduced to four ounces in weight and five grand in dollars. High-speed communications. Reusable rockets—still in development, but the Initiative successfully launched one a hundred fifty feet in the air, flew it three hundred fifty feet sideways, and landed it vertically, softly, successfully. Techniques such as aerobraking, to slow spacecraft down upon arrival at an atmosphere. The Initiative built and launched a spacecraft called Clementine to map moon resources, and they did it in less than two years and for eighty million dollars—an order of magnitude less than what NASA was managing for Mars Observer, which took seven years to launch.[120] It wasn't apples to apples, but there were some salient lessons there.

All in all, SDI was the most aggressive aerospace research and development project since Apollo. Congress appropriated one billion dollars to the Initiative in 1984 and increased appropriations each year through 1989—1.4 (billion), 2.7 (billion!), 3.2 (billion!!), 3.6 (*billion*!!!) (each year!), peaking finally at 4.1 billion.[121] The program ran out of steam under George H. W. Bush, however, with appropriations diminishing to a mere 3.6 (billion, *sigh*) and then 2.9 (billion, grumble grumble), and essentially ended under Bill Clinton. It changed names over time: the Ballistic Missile Defense Organization, the Missile Defense Agency. Had it continued as initially conceived, the program anticipated the need for crewed and robotic missions, a lunar colony, an orbital transfer vehicle— essentially a barge and tugboat to travel to and from the moon, the shuttle (because we had to find *something* for it to do), upgraded launch capabilities (something cheaper than the shuttle, because I mean, come on), and a space station.

By the time Mars Observer disappeared on August 21, 1993, NASA had a new administrator named Dan Goldin, who had previously worked on the Initiative while an aerospace executive, and before that was a mechanical engineer at NASA Lewis Research Center in Cleveland. When Goldin took the helm of the agency in 1992, it was one of the least stable jobs in government; NASA administrators served at the pleasure of the president, and President George H. W. Bush was cruising toward electoral defeat. But the likely incoming vice president, Senator Al Gore of Tennessee, a space enthusiast, liked him, and Clinton—who had no real connection to the program, emotional or otherwise—didn't bother replacing Goldin once elected.

The consolation was cold, however. Under the new administration, Goldin had one real directive: prepare for budget cuts, because

they were coming, and on the order of twenty percent.[122] The loss
of Mars Observer was to the agency what dropping your lunch tray
was to a grade school student: a crushing embarrassment, but for
eight hundred million dollars' worth of food in front of five billion
people, to say nothing of the decades of science now likely lost and
the careers placed suddenly in jeopardy. Goldin, who brought with
him a reputation for being . . . mercurial (at his last job, they called
him Captain Crazy), and whose soon evident quick temper bore
this out, could have stopped this Mars madness once and for all.[123]

But he didn't.

Goldin, born in 1940, a young man during Mercury, Gemini,
and Apollo, was a Mars Guy—wanted his whole life to see an astronaut set foot on the Red Planet, had joined the agency in the
sixties specifically to be part of the inevitable human Mars mission. Far from losing his cool in the din of disaster, he rallied and
inspired the troops. So we lost Observer! It was a mess, foisted on
you—the whole program. It's over now.[124] Cancel the Mars program? Are we not explorers? We will grieve later; right now we
have a Red Planet to invade.

The chaotic aftermath of Mars Observer allowed Goldin to
implement a new development philosophy that came to be called
Faster-Better-Cheaper. It worked like this: all new planetary
missions would be limited to the small Delta II rocket, budgets
were capped at one hundred fifty million, and you had a three-
year limit from conception to launch. Go over mass, money, or
months, and the plug would be pulled—just like that. This would
have been impossible, but Faster-Better-Cheaper had something
going for it that no previous program had: that massive thirty-
billion-dollar infusion of Strategic Defense Initiative research
and development, which had finally filtered down to the civilian
sector. All the heavy lifting for technology maturation was done,
and from NASA's perspective, for free. From thrusters to star

trackers, the miniaturization, the radiation-resistant tech—it was sitting on a shelf or file server waiting for some special spacecraft to put it in play.

Faster-Better-Cheaper would be run under a nascent competitive mission program called Discovery, which had been nurtured by Wes Huntress, the new head of space science under Goldin.[125] The first Discovery mission to Mars would be a small lander and miniature rover that had been proposed originally by Scott Hubbard, a physicist at Ames Research Center in Northern California.[126] The idea from the outset was that if it worked—if this tiny lander and its Tonka truck passenger survived, set down on Mars, and *actually worked*—it could be built in large numbers and deployed in force to the four corners of Mars. Huntress adopted the proposal, nurtured it, and when Ames passed on putting it into development, the job went to Jet Propulsion Laboratory.

Project Pathfinder, as the mission became known, carried no transformative scientific instruments; this was no Viking. Its subtle aim was to remain just under the radar so as not to attract cancelation or Christmas tree ornaments, and to test what NASA could do. Miraculously, relative to JPL's previous Mars efforts, it launched on time and on budget in December 1996 and landed successfully on Mars seven months later.

Pathfinder reenergized NASA unlike anything since Apollo, its landing received by the public as though the Viking landings had never happened. Maybe it was because an entire generation had missed the Apollo-Viking window, had never seen a Mars landing before, had no idea that such landings were even possible. (Alas, Viking.) Orbiters were exciting, but landers? It was the first successful set-down on Mars in twenty years, the first unambiguously successful mission in ages, and, magnifying the achievement, the first real planetary science mission of the Internet Age, giving it publicity aplenty. (Poor Galileo, lapping Jupiter, was decrepit before launch with its faulty, radiation-poisoned reel-to-reel recorder.) Those images! Papaya Martian horizons and that little rover rolling around—the newly wired global community ate it up. And at one fifty mil, a bargain.[127] NASA had its mojo back.

Then came the next two Mars missions, which were . . . less successful. Both launched in 1998. The first, Mars Climate Orbiter, arrived at the Red Planet and swung behind it during orbital insertion, losing radio contact, as expected. The spacecraft should have emerged from the other side twenty-one minutes later. The Space Flight Operations Facility at Jet Propulsion Laboratory waited patiently for a signal to reach the Deep Space Network, and then, suddenly impatient, prompted the spacecraft anxiously for a reply, but response came there none. The orbiter either disintegrated in the Martian atmosphere or skipped off it, a stone on a pond's surface, forever lost in space. In any event, engineers determined later that a software bug introduced in a single subroutine had caused the catastrophic failure.

But the Faster-Better-Cheaper program *expected* some level of loss. Failure *was* an option! That was the whole idea! We keep flying these missions, we launch a veritable invasion armada to Mars in two-year cycles, each time our orbits align, and, well . . . there will be lost spacecraft. Yes, one hundred twenty million dollars was a lot to pay for a shooting star that no one would see, but think of the one that cost eight hundred million! We were practically making money on this deal.

Three months later, Mars Polar Lander arrived, and forty meters above the planet's surface, the spacecraft's descent engines switched off. The computer thought the lander had landed. It had not. But then it did, and hard, making a nice, one-hundred-ten-million-dollar hole in the ground. And suddenly—seemingly as soon as it had arrived—the goodwill garnered from Pathfinder evaporated, and headquarters had to come up with a plan, and quick, and it did, and very.

ED WEILER, NASA'S new head of science missions, was in Pasadena for the auguring, faced the press with everyone else, and then flew back home, exhausted. Over the next few days, there would be four opportunities to try talking to the lander. Maybe it survived, simply set down like a ballerina, but had, say, a radio problem. Hope faded, however, with each passing day, and on December 7 the mission was pronounced dead.

He was sitting in gymnasium bleachers when his BlackBerry's ringtone joined the din of bouncing basketballs and squeaking sneakers. The caller ID read DAN GOLDIN. He knew what was coming. Ed was at his son's basketball team practice. The hours you put in at the agency, you didn't miss family time when you got it. You couldn't hear anything in the gym, so he stepped outside to answer. It felt freezing that night, the wind lightly rustling the leaves, brown and desiccated on nearby trees, and Ed pressed Answer.

You have twenty-four hours to fix the Mars program, said his boss.[128]

Ed told Goldin he'd get back to him tomorrow.

Despite his boss's reputation for acerbity and ire and the legion of egos bruised agency-wide, Ed loved Dan—the two had always clicked, respected each other, were honest with each other—and the next day Ed walked into Goldin's office with a plan for Mars.

We're going to cancel the entire Mars program, he said.[129]

What do you mean? asked Dan, and not serenely.

We're going to cancel the program and start from scratch because clearly we are doing something really badly here.

Ed never wanted to run NASA science. He had spent the last nineteen years as head of the Hubble Space Telescope, which, for

an astronomer, was like being mayor of Disneyland. And how far
he had come. Born in blue-collar Chicago, his mother managing
the home, his father a steelworker and meat cutter; they scrimped,
his parents, saved and sacrificed to put Ed through Jesuit prep
school, to give him that better lot in life that every parent wants for
his or her child, and Ed didn't waste the favorable prospects they
bestowed upon him. He worked his way through college at North-
western University (disappointing the priests of his youth, who
called it a pagan school), studied astrophysics, was no slacker—did
a clean sweep: baccalaureate, master's, doctorate. There was never
any doubt about what he wanted to do with his life. When he was
twelve, his dad had found and ordered a four-dollar mail-order
cardboard telescope for Ed, and that was pretty cool—you could
point it at the moon, see the craters along its crescent, the terminat-
ing line of the moon where met the lighted and unlighted regions
of the disc, every feature along it cast into relief, and through his
telescope you could make out mountains and ridges—the moon
was no cue ball, no polished pallid orb, however it looked with
the naked eye, and though teachers could tell you there are mountains up there, to see them? With your own eyes? Mountains on
the moon? This stoked something inside of him—this fire he didn't
even know he had—and his dad saw this and went out and found
his boy a two-inch Japanese refractor telescope—a Tasco—and in
retrospect it was one step above a toy, but it *really* set young Ed's
imagination ablaze, and he joined the Adler Planetarium's Junior
Astronomical Society, which offered a *telescope-making class,* and
there you are, a teenager, a kid, a lover of the stars, living in a
working-class Chicago apartment, and now you are learning how
to grind your own telescope mirror, and he built—built!—young
Ed!—a six-inch, two-hundred-pound Newtonian reflector. It was
the early sixties, and he had decided already that he wanted to
be an astronomer, wanted to work for the National Aeronautics

and Space Administration. His entire life: meticulously mapped at thirteen years of age.

Ed didn't know that he would one day launch and lead the Hubble Space Telescope, inarguably the most successful science program in NASA's—if not the nation's—history. Like so many others, the plan initially involved being an astronaut, and he applied for the first class he saw advertised to the general public. There were thousands of serious applicants, and he made it through the first couple of cuts, but the numbers were not in his favor and Weiler never left the planet Earth.

After graduate school, there weren't a lot of jobs out there for astrophysicists, and young Dr. Weiler was set to sign, of all places, at American Hospital Supply, where he would work as a computer programmer. While doctorates in physics were scarcely useful in the labor market, a side effect of doing the discipline's advanced mathematics was an unrivaled ability to write complex computer software. In those days, that meant Fortran on mainframes, and American Hospital was right: Dr. Ed Weiler could play a CDC 6400 mainframe like Handel at the organ. Before he committed to a lifetime of writing subroutines for company accounting, however, he heard from his doctoral advisor that Princeton University was looking for someone to work on its small orbiting astronomical telescope, Copernicus, which NASA had launched four years earlier, in 1972. The principal investigator was a guy named Lyman Spitzer. Ed interviewed for the position, presented well enough, and was hired as a staff astronomer, although his job kept him mostly at NASA Goddard Space Flight Center, which ran space telescope operations. It was a case of on-the-job training. At the time, there were only a dozen or so "space astronomers" in the world—everything was ground based. So the community was small but eager. Copernicus, a one-meter telescope limited to spectroscopy (studying ultraviolet emissions from stars, for example), wasn't that powerful, but it was a science-making machine, and while working on the project, Ed met some of the greatest astronomers in the world, including Spitzer, who was his boss. He (i.e., Spitzer) talked frequently about another telescope he hoped to launch: a powerful orbital observatory free of the distortions and interferences of Earth's atmosphere. He had first proposed it in 1947—ten years before Sputnik—in a paper titled "Astronomical Advantages of an Extra-Terrestrial Observatory." Spitzer had spent his entire career lobbying for it, arguing the idea of it to the community, and eventually selling it successfully to Congress.[130]

Ed had been three years on Princeton's payroll when one day someone knocked on his office door, and there stood Nancy Roman, the chief of astronomy at NASA headquarters. Ed didn't know her—I mean he knew *of* her, but Ed didn't exactly warrant the chief of astronomy's time—and he invited her into his phone-booth-sized "Princeton office" at Goddard. In addition to running NASA astronomy, Roman was chief scientist of the space telescope that Lyman Spitzer had championed for so long. (It was still a study at the time.) She made Ed an offer: How would you like to leave Princeton and come work for the government? You would be a staff astronomer and my assistant on the space telescope we want to build.

Ed said yes. Maybe it was a subconscious concern about the longevity of Copernicus. Maybe it was those years of animated discussion with Spitzer about this space telescope and how it would revolutionize astronomy. Maybe—and this one required a deep look inward—it was that he worked for *Lyman Spitzer,* universally recognized as one of the greatest astronomers to ever live. Spitzer had literally written the book on interstellar mediums—the same book Ed had used in graduate school. You work for a guy like that and you're offered a job at NASA headquarters, you ask yourself: If I stay in academia, will I ever be as good an astronomer as a guy like Spitzer? Or could I affect the field of astronomy in other ways, enabling new space missions—things like this orbital telescope— for the rest of the community, for the future Spitzers out there? It took some humility, but he knew the answer. He could best help space science advance by going to NASA, and that's what he did.

One year later, Nancy Roman retired and Weiler was hired for her position. Thus, in 1979—three years after finishing school—he became NASA's chief of astronomy and the chief scientist on what would be called the Hubble Space Telescope. In the early days of the project, his job was to advocate for its science and protect it from engineers and project managers who might make Hubble less than it could be. He considered it his moral responsibility to defend the science, to build the community around it. He saw Hubble through to the launch pad in 1990, and when it reached space— over budget by a stunning order of magnitude, but surely capable of miracles—he and astronomers the world over awaited eagerly its very first image, and it came down, and something was TERRIBLY WRONG. The images were . . . blurry.[131] Astronomers realized eventually that there was a defect in Hubble's mirror; its perimeter was too flat by one-fiftieth the thickness of a single human hair, but in a precision instrument, that was enough.[132]

And how the world laughed. If you were a stand-up comic, the Hubble Space Telescope provided an endless supply of easy jokes at NASA's expense. Had Mr. Magoo designed the thing? Jay Leno: "What sound does a space turkey make? Hubble, Hubble, Hubble."[133] Bumper stickers appeared on cars: IF YOU CAN READ THIS YOU ARE NOT THE HUBBLE SPACE TELESCOPE. And so on. The good news, scientists soon realized, was that shoddy craftsmanship was not to blame; the mirror was made absolutely flawlessly. It was just flawless to the wrong specification, which meant that you could correct the problem simply by crafting "eyeglasses" for it. That's exactly what happened. In 1993, after three years of ridicule, astronauts on the shuttle *Endeavour* went up, installed the fix, and the problem vanished. (Finally, the shuttle had a purpose!) Hubble went on to become as transformative to the field of astronomy as Galileo's telescope.

Five years later, Wes Huntress was ready to leave his position as head of NASA science and recommended Ed for the job. Dan Goldin made the offer:

DAN: Ed, do you want the job?[134]
ED: No thanks!

He was happy running the Origins program, which concerned itself with the very beginnings of life, the universe, and everything. He *loved* working on Hubble. And, I don't know, he didn't feel *qualified* for the job. He'd never felt qualified for any job he had ever held, and this one, it didn't seem like a burden, exactly, but who was he to do things like meet regularly with Congress? Did he even want to do something like that? He recognized the opportunity there, but just—I don't know.

DAN: Well can you at least help me recruit somebody to do it?

Ed agreed to do that much. After two months of searching, though, he had changed his mind. He saw the experience level of the candidates who had lined up for the job, and one night he sat down, dug deep, and made some decisions: I'm one of the most senior people in the science program. If Dan brings in one of these rookies who's never worked at headquarters, they're going to ask me to be a mentor, to hold his or her hand. Do I really want to spend months of my life holding the hand of some academic who's never been in the government?

No, he decided. He may as well do it himself.

So Ed called the ninth floor of NASA headquarters, but Dan was on a NASA jet to California. Turned out that Dan had a phone on the plane, though, so Ed called it.

ED: Dan, I changed my mind. You still interested?
DAN: Let me think about it.
[Click]

When the plane landed, Ed received a call.

DAN: You've got the job.

Canceling the Mars program was one of Ed's first major acts as agency science mission lead, and over the next year, he helped put a new plan in place. Scott Hubbard—he of Pathfinder—was brought to headquarters as the first director of a new formalized NASA Mars program. NASA, the Mars science community, and Jet Propulsion Laboratory then put their heads together to come up with a reasonable program of missions, starting with a water-divining orbiter, followed by twin rovers in the vein of Viking and a high-resolution reconnaissance satellite. Two missions would next be submitted for competitive review in a program called Mars Scouts. Afterward, a rolling science laboratory would be sent to seek signs of organic material. All of this would culminate in a sample return mission, the details of which were still fuzzy, but they had decades to figure them out. These new missions wouldn't have quite the constraints of Faster-Better-Cheaper, though the goal remained to hit every launch window, so "faster" was a perpetual demand. (Variants of each of these missions had been studied extensively and developed internally at Jet Propulsion Laboratory for years; they weren't simply being dreamed up and launched in twenty-six-month boxes.)

What followed was perhaps the greatest unbroken string of successes in NASA's history, and considering the challenge of Mars, certainly the greatest at Jet Propulsion Laboratory. In 2001 the agency launched the orbiter Mars Odyssey, a water finder. In 2003 the rovers Spirit and Opportunity launched to get good looks at Martian soil and rocks in order to better understand the water processes at work. By now, Ed had managed to double the space science budget on such achievements before taking a job as director of Goddard Space Flight Center, a lateral move that placed him closer to his beloved Hubble and its successor in development: the James Webb Space Telescope.[135]

The Mars milestones, meanwhile, didn't stop. In 2005 the Mars Reconnaissance Orbiter launched from Cape Canaveral with a mandate to map Mars in greater detail than ever before and to give planetary scientists a global view of Martian weather conditions.[136] Each mission arrived successfully at the Red Planet. The two rovers landed without a scratch. Mars was everywhere, and the data coming back painted not an abstract world, cold in the mind's eye, but a place as real as Earth, with every little detail being teased from rovers, landers, and orbiters alike. The next NASA lander, Phoenix, was in development, set finally to right the scientific loss of the Mars Polar Lander.[137] Never before had a planet other than Earth been such a hive of activity.

The price tags for such successes grew by hundreds of millions of dollars with each successive mission. Which was great if you studied Mars, but across the planetary science community, the number of unhappy faces proliferated. The Martian momentum was nice, but were we not neglecting the wider solar system?

Still, these cries found few allies, especially as the George W. Bush administration went all in on human exploration. The space shuttle program was at last drawing to a close, giving urgency to NASA's need for a new grand plan. A space station now orbited

Earth, as God and Wernher von Braun intended,[138] and the next move in the late von Braun's playbook called for lunar development as a stepping-stone to Mars. As author of the agency's genetic code, one way or another, von Braun was going to have his way.

They called it the Constellation Program. NASA would build gargantuan "heavy-lift" rockets, moon landers, habitats—all of it. We'd go to the moon and take the lessons learned there to the Red Planet. It was finally happening: a true successor to Apollo. And with all this Mars mania, robotic exploration was placed on a pedestal, Mars robots as human precursor missions. Since every dollar spent learning more about Mars meant mitigating risk to astronauts down the road, these missions were worth every dime to the astronaut-centric agency. This gave Jet Propulsion Laboratory greater latitude in planning its most ambitious Mars rover yet: the Mars Science Laboratory.

The vehicle was about the size of a Jeep, weighed just shy of a ton, was a six-wheeled beast relative to the three previous rovers, and was born to amass data and mint new Ph.D.s. It would go farther than any vehicle ever landed on another world, would do more, had lasers, cameras, a full soil-analysis suite, and a mandate: figure out if Mars was once habitable. It was the final rover in the pipeline before the Mars people reached for the brass ring: sample return.

PLANETARY SCIENCE RUNS on ten-year timelines. The community's driving document is called the Decadal Survey. You don't even have to specify in casual conversation, just, *the Decadal,* and everybody knows what you mean. The very first planetary Decadal was commissioned in 2001 at the behest of Weiler, who took the concept from the astrophysics community, which had, for fifty years, organized itself around a similar concept.[139] With each survey, the top minds in the field would come together and decide:

What do we know? What have we yet to learn? How do we learn it? Where should we focus our energies, and in what order? Astronomy is expensive, and their decadal reports were a way for the field to focus on big questions and the big things necessary to answer them. It's how astronomers came to support so effectively the great telescopes of the twentieth century—the Very Large Array, Hubble, Spitzer, Chandra—and the cornerstone of the twenty-first, the James Webb Space Telescope, which, in addition to luring Ed Weiler to Goddard Space Flight Center, had by 2006 nonupled in price, five hundred million to four-point-five billion, threatening to engulf the agency's entire space science budget.[140] When spreadsheet cells stretched ever rightward like that, the Decadals were how you kept White Houses and Congresses from losing their nerve.[141]

The planetary science community had no such document, and its unceasing political problems reflected this. NASA's science office built its exploration road map from a jumble of reports from scholarly bodies both internal and ex. Accordingly, there was no unity, no clear community consensus. Weiler wanted to change that, and he asked the Space Studies Board of the National Research Council of the National Academies of the United States—itself an independent, nonprofit, nongovernmental scientific body—to appoint a committee to steer what would be called, formally, a Solar System Exploration Survey. Its goals: to paint a nuanced portrait of where the field was and why it existed; describe where it needed to go in terms of exploration and technology development; prescribe the care and feeding of the community regarding things such as research funding and retention; and present priorities for the next ten years that might enable it to achieve those goals. The "survey" side of the name, in other words, referred not to some sort of ballot box vote, or group of graduate students holding clipboards and asking anyone to rate Venus-as-Object-of-Exploration on a scale of one to ten, but rather, "survey" in the way Shackleton, far inland at Antarctica, took measure of what he faced, pointed a finger and said, "That way."

This was, in all, a yearlong effort:[142] July–July, 2001–2002, with the solar system split into general areas of inquiry: the inner planets (Mercury, Venus, the moon, and Mars); the giant planets (Jupiter, Saturn, Uranus, and Neptune); the primitive bodies (asteroids, comets, and Pluto); and "large satellites" (the massive, mind-blowing moons of the solar system such as Europa and Titan). The steering committee later added an astrobiology panel to the mix. For each of these areas, it commissioned a subpanel of subject matter experts who would meet multiple times to determine the state of knowledge of its subfield and how best to improve it given possible spacecraft on the drawing board. Panels sorted prospective projects by mission size and prioritized them: the small Discovery-class missions to launch every two years; medium-sized missions called New Frontiers to launch at least every three; and multibillion-dollar missions called flagships, one per decade. The steering committee would then take each panel's endorsement and come to consensus on an integrated list of mission priorities. (The giant planets group might recommend a Neptune flagship, and the inner planets team might push for a flagship Venus lander. The steering committee would decide which of those had higher priority this decade.)

Mars, as ever, was the exception to all this. At the time, the National Research Council was already actively doing something similar to the Decadal Survey, but specific to the Red Planet. Rather than double the effort or present conflicting recommendations, the Decadal steering committee essentially absorbed that working group as a stand-alone panel—Mars—and instructed the inner planets panel to ignore Mars entirely. The upshot is that everyone competed, but Mars competed with no one, and when it came to mission recommendations, Mars was essentially treated as its own space program with its own funding and its own small, medium, and large mission classes, none of which were integrated into the report's recommendations. Mars would do whatever Mars wanted.

The first planetary science Decadal was published in July 2002 and rounded out at two hundred fifty pages. NASA's planetary science division, though not legally obliged to follow its recommendations, moved to adopt them all the same and came quickly to expect that any agency mission proposal give Decadal justifications for its science goals in the way that a judge might expect specific case citations in litigation. In the final analysis, the Solar System Exploration Survey endorsed as its top flagship a Europa Geophysical Explorer. This recommendation was unambiguous: "One Flagship mission is recommended for this decade—the Europa Geophysical Explorer."[143]

In 2003 it fell to Curt Niebur, a green, almost entirely anonymous agency functionary at NASA headquarters, to make the recommendation a reality. Three weeks after he had been hired by the agency, Curt's boss asked him if he would like to be program scientist of the Europa flagship, called JIMO—short for Jupiter Icy Moons Orbiter. Program scientists were something like intermediaries: they represented NASA headquarters to mission science teams, and represented the mission science teams to NASA headquarters. It involved a very careful give-and-take, and there was always a good chance of annoying everyone on every side. JIMO, an agencywide initiative supported by the administrator himself, had a big audience. Curt was too young and naive to know that he should have been terrified by the very notion of the project, let alone the project itself.

When JIMO died (officially, it was "indefinitely deferred"[144]) in 2005, Curt couldn't let Europa go. He had been given this moon—the moon the community wanted—the moon deemed the field's *supreme scientific concern*—and in meeting after meeting, he continued clutching a copy of the Decadal and jabbing a finger at its flagship endorsement, asking how headquarters planned to proceed.

But it was like shouting at the ocean. After JIMO, headquarters leadership cheerfully consigned Europa to oblivion. With Mars hysteria moving the agency's needle in the press and on Capitol Hill, there was effectively zero support for flying further flagships to any of the outer planets, but Europa in particular was *luna non grata*. That moon was hard—marinated in a spacecraft-killing radioactive morass—and the shielding and power requirements necessary for survival were too great to overcome in a time of austerity. We had spent tens of millions of dollars failing to get a mission going and had given the Decadal its due. It was time to move on, Curt, and your continued heretical braying about that Jovian ice ball wasn't doing you any favors here.

But Curt would not relent. Practically and professionally, surrender would have been easier, and certainly smarter. He could have just opened his agenda and drawn a line clean through the icy moon and moved on to the next thing, because in his portfolio of missions, he now also held the Mars rovers Spirit and Opportunity. He was on the winning team! But the solar system stretched beyond the Red Planet. The Mars community framed its exploration as the search for water and thus life, but Europa was by far the wetter and more hospitable of the two. Mars might have had microbes eons ago, but Europa might have fish today. What more reason did you need? Without even getting into the Decadal endorsement—which it did, right on page 196—this sort of frontier science was the whole reason he had come to headquarters, and he wasn't ready to yield. So from his tiny, glass-walled interior office at NASA headquarters, Curt closed the blinds and hatched a plan. Leadership didn't want to go to Europa? Fine. He would give them what they wanted: Not Europa.

Which icy moons in the outer solar system were the consensus community targets for exploration?

There was Titan at Saturn.

Enceladus, also at Saturn.

And, yes, Europa at Jupiter. (I mean, we can't ignore the Decadal.)

Curt proposed forming full teams to develop, over the course of one year, competing missions to each of those worlds. Then— unlike any such early-stage flagship concept study that had ever come before—NASA would run those missions through complete, independent reviews of their technical, management, and cost feasibilities, as well as their science rationales and implementations— everything. The agency would treat each prospective mission as though it were going to fly. Once the independent review results came in, the leaders at headquarters could take a look, and *they* could choose the best mission to launch, the best moon to explore. Maybe it would be Europa. Maybe it wouldn't. But at least something in the outer solar system would see a flagship.

It was the only thing he could think of to shift some of the attention away from Mars.

This kind of contest had never before been considered, let alone attempted. That's just not How Things Worked Around Here. NASA didn't decide flagship destinations by competition. Multibillion-dollar missions were directed from above and developed down below (i.e., *Hear ye, scientists and engineers: Send us your spacecraft concepts, for as you have long pined, we are going to Saturn. NASA has spoken*). Sure, the smaller, focused missions competed for flight—the five-hundred-million-dollar jobs—but never the flagship targets. Headquarters gave guidelines—place and price, science sought—and an institution such as Jet Propulsion Laboratory would take it from there, develop a nominal mission profile—more of a vision, really, though coincidentally, I mean, since you happen to ask, we've been working on this problem for quite a while! And the institution would go to headquarters, drop a three-inch stack of full-color-printed pages on the associate administrator's desk, wind up, and pitch:

"This spacecraft is a steal. You fly this thing off the launch pad and it's going to *appreciate* in value. I don't—look, I like you, but if you take this deal, I'm pretty sure you don't like me! [*slaps desk, laughter*] You seem like the kind of person who knows a good price when you see it, so I'm OK with you taking the money out of my pocket here, only . . . [*sotto voce*] don't tell my manager. Look, this orbiter, it's an investment, really. I can—OK, let me ask you this. You have scientists on this mission, yes? And what do scientists like to do? I bet they like to use ion and neutral mass spectrometers. They do, don't they? I could tell the minute I said it. See this spacecraft? See this spot right here? Perfect place for an ion and neutral mass spectrometer. Right? Look, I'd never speak ill of another NASA center, and we could make a *lot* more by selling you a spacecraft just like the ones they make, but [*leans in closer, almost a whisper*] you really don't want their thrusters on your spacecraft. I'm not saying they're bad—I'm not—saying—they're—bad. But they don't thrust. [*raises hands, eyebrows*] That's what I heard, honest to Charlie. And let me ask you this: You look at their spacecraft—go on, I encourage it!—and you know what you *won't* see? Space for an ion and neutral mass spectrometer. Not *this* one, anyway. Not comfortably, at least. But you drive a hard bargain! I'll be right back. [*stands, absconds to breakroom for full three minutes*] I'm sorry about that. My manager wanted to talk to me about the price. Look I'll be frank: she's not happy about this price I've put on the table. I knew better, but this is between friends and I can take a little dressing down if it means helping out a friend. So what do you say? [*dangles keys*] You want these? You want these. You *do*, don't you. I know you do. [*extends hand*] So do we have a deal? [*lets outstretched* *hand hang, says <u>nothing</u>*] We have a deal! Whoa, firm handshake you have there. I'm going to need to get my wrist checked out later! OK, sign here, and I'll have our people draw up the paperwork. You'll be sending me Christmas cards for the next twenty years for this one, mark my words."

The other reason no one had ever tried putting flagship destinations into competition was the consequence of failure. By putting the outer planets in opposition, Curt wasn't risking only Europa; he was risking the entire outer planets program. If every target came back over budget, headquarters could write off everything beyond the asteroid belt as needing technology development and delay exploration indefinitely. Mars was easier, a launch window opening every two years, and astronauts would one day walk there. It took six, seven, eight years *or longer* to reach Jupiter and beyond—and that's after the eight, sometimes ten years spent designing and building the thing. Which meant most of a manager's career, literally, might be spent fighting for the flagship, defending its funding across multiple Congresses with different parties in power, different White Houses, and new NASA overseers, who might be scientists but who might be political attachés. And that lowly, beleaguered manager would be fighting alone because the dominant human spaceflight half of the agency certainly wasn't going to pony up coin *to explore Titan*, three hundred degrees below zero—so cold a human's skin would burn off, just slough away if unprotected. Curt knew that if the three full-fledged flagship studies came back with three impractical, overpriced proposals, three unworkable schemes, agency administrators could take hold of the lovingly prepared plans, each in a binder three inches thick, and bludgeon him with them for the next ten years.

Complicating Curt's plan was that he wanted Jet Propulsion Laboratory to lead the Europa study. Headquarters leadership had zero faith that the lab could keep Europa in the cost box, and wouldn't want to give it yet more money to fritter away on yet another study. But of the three moons, Curt was adamant that Europa had to go to JPL. The lab had the best engineers in the world—women and men so smart that when you set foot on campus, you risked developing a nosebleed from all the brainwaves. Yes, as NASA leadership noted pointedly and in colorful metaphors, those very engineers had been working headquarters-funded Europa concepts for eight years now, starting with the Europa Orbiter in 1998 (which asked: What is the smallest Europa mission we can do?) and ending with the recently departed JIMO (which asked: What is the *biggest* possible Europa mission imaginable?). In both cases, the answer was something very, very expensive. Still, Curt knew that those dollars spent and lessons learned had yielded solutions to technical obstacles and sharpened the science objectives of an expedition there. And if the Europa people returned another battlestar? Then we'd go somewhere else—go with the winning proposal—Titan or Enceladus. Europa might have to enter stasis, but the outer planets would go on.

So Curt presented his competed flagship proposal to NASA leadership and received, generously speaking, a Not No. Emboldened, he pressed the issue again and again, suggesting study structures and timelines and—look, Curt, this isn't working—OK, Curt . . . OK . . . OK . . . again with this?—and conversations could come just short of name calling, and more than once Curt walked out of a meeting wondering if he would be told to pack his desk at the end of the day. But this was his job! And if we weren't going to follow the Decadal, then what were we even doing here?

And then the federal budget made the decision for everyone: NASA was simply tapped out. It would have died right there, Curt's gambit, were it not for the hiring in August 2006 of Jim Green, who was 1. the new head of planetary science and 2. Curt's new boss.

Jim was an outsider not only in the division but also in the entire field. He was a scientist, but he wasn't a *planetary* scientist per se. He was a space physicist, and consequently came to headquarters carrying none of the baggage that a partisan of a particular planet (e.g., of Mars) might pack. Green was a pragmatist, saw the value of a balanced exploration portfolio, and understood immediately the allure of the outer planets and what Curt was attempting to do with the studies. Green even suggested adding a fourth moon to the mix: Jupiter's Ganymede, the largest moon in the solar system and the only one in possession of a powerful magnetosphere. This made what had then been a trio of studies quad, and collectively, the mission investigations would become known informally at headquarters as the Quad Studies.

The four flagships would be spread across the agency: two at Jet Propulsion Laboratory, one at the Applied Physics Laboratory, and one at Goddard Space Flight Center. Each would have a study lead and science team co-leads. Curt got his wish, and the Europa study would be conducted by JPL. An engineer named Karla Clark, who had been running Europa studies since 1997, would take lead, with Bob Pappalardo and Ron Greeley running science.

The Ganymede study, similarly, would be helmed by veterans of JIMO and Galileo: Louise Prockter at APL and Dave Senske at JPL, with the latter lab leading the overall effort. If successful, the mission, with the broader aim of studying the entire Jovian system, would also get a good glimpse of Callisto—the largest moon in the solar system that was not differentiated (i.e., it had no crust-mantle-core makeup; it was just this big ball in space).

Titan would be run from the Applied Physics Laboratory at Johns Hopkins University. Whereas Jet Propulsion Laboratory (its rival of sorts) was a master of the Big Expensive Missions, the So Crazies It Might Work, the Applied Physics Laboratory was famed for cost control and smart design. APL had recently launched the spacecraft New Horizons to Pluto—the smaller lab's first outer planets mission—and Titan would be another good opportunity to take a swing at the solar system beyond Mars. Ralph Lorenz and Hunter Waite, both of the orbiter Cassini's Titan probe, Huygens, would lead the science.

If Europa had the advantage in terms of mission refinement and well-honed science goals, Titan was by far the beguiling new thing. Cassini, now two years into its mission, found something new with every Titan flyby: standing liquid lakes of methane! Gullies and riverbeds! Weather, rain, a cool breeze in a dense atmosphere. (Denser, even, than that of Earth. When Cassini released Huygens, it took two and a half hours for the probe to touch Titan's surface. The lander had to switch to a smaller parachute halfway down to speed things up and get on with it.)[145]

Huygens, indeed, had discovered a primordial Earth. Titan's tantalizing atmosphere and winsome weather meant you could explore it as pioneers once explored our world: on wheels, wing, and watercraft (well, liquid-methanecraft). And unlike Europa, there was no belt of blistering radiation for a spacecraft to endure, no miles and miles of granite-hard ice to penetrate. Its liquid was easily accessible, and if it had life . . . well, it would be like nothing encountered ever before. It wouldn't be based on water, as on Earth. It would be exotic, truly alien: methane-based biology.

The Enceladus study, based out of Goddard, was a long shot because so little was known about the Saturnian moon. Before Cassini arrived at Saturn, its team had allotted three tepid orbits to observe the drab, airless ice ball. Enceladus was too far from the sun for an active surface and too small to retain a warm interior. It was a checkbox moon, interesting because it orbited inside of Saturn's hazy E-ring—a sort of ring around the rings—and that maybe something about that would make a fine footnote to a more interesting paper.

Then Cassini arrived, encountered Enceladus, and it was as though Gaspard Marsy himself had carved the white marble moon with hammer and chisel.[146] Its southern surface was mystifyingly young and, though veiled in darkness, was warmer somehow than its sunlit equator. Cassini found four stunning stripes on that pole, like some great galactic beast had slashed its claws in anger across its surface, leaving gashes parallel and preternatural. And finally, the impossible: from those claw marks—tiger stripes, they were dubbed—blades of water vapor were being blasted hundreds of kilometers into space. The moon wasn't dead—it wasn't even an ice ball—there was liquid water down there! Enceladus, this teeny, tiny ball circling Saturn, was venting a subsurface ocean into space, and despite the heat and velocity of the water being blindly blasted into the ether, some of that water came back down to the Enceladan surface *as snow!*[147, 148] Which was an arresting idea, but wasn't even the exciting part. The majority of water vapor that managed to escape was going to Saturn, where it *formed* the E-ring. It was all so phenomenal, so preposterously magical—and it rewrote the scientific understanding of active, geologically living ocean worlds. For much of the history of planetary science, the consensus was that everything in the solar system bar Earth was dead: a scattered cluster of spinning rocks circling a giant fusion reactor. That was *especially* true of the outer moons, formed and frozen for billions of years, and orbiting where it was simply too cold for anything interesting to happen.

But there was Enceladus! It soon seemed clear to planetary scientists that beneath its icy crust, below its liquid, interior ocean, was ongoing hydrothermal activity. Areas around such hydrothermal vents on Earth—our planet's plumbing system—invariably teemed with life. (Earth scientists had observed it directly in our oceans, along the midocean ridges where tectonic plates were spreading apart.) Moreover, because water touched rock on the Enceladan ocean floor, the ocean was conducive to chemistry. If Earth was any indication, an energy source plus chemistry meant the possibility of genesis around Saturn, too.

And *unlike Europa,* at Enceladus, you didn't have to drill through miles and miles of ice to get to the ocean: *the ocean was coming to you.* This was like a wheat thresher rolling across an entire field of science! Even if a focused Enceladus mission study didn't prevail, it would certainly sharpen the objectives of some future expedition. The brain power behind the study, Amy Simon-Miller of the spacecraft Cassini and John Spencer of New Horizons to Pluto, promised something special . . . and you never knew.

Green found four million dollars to make this thing happen— one million per corner of the quad—and that wasn't easy. The planetary science budget was eviscerated by twenty-two percent between 2005 and 2007—the Iraq war wouldn't pay for itself, after all—and for an agency already living paycheck to paycheck, the threat was mortal. Jim searched sofa cushions and spreadsheet cells to find funds.[149]

Curt had watched worriedly, meanwhile, as flagging funding for research reduced the roster of scientists and engineers able to sustain careers—especially those who worked in the outer planets. When Galileo plunged into Jupiter years earlier, never to return, research resources receded and withered away. Most of the scientists working with the data collected by Galileo weren't NASA employees— some survived only on soft money (i.e., they lived off grants). They had mortgages to pay, children to support. So Niebur pushed for an Outer Planets Research Program—a NASA funding line for research grants—and a Cassini Data Analysis Program (same).

Which is one reason NASA needed to get another flagship mission going: because big missions meant money for grants, the lifeblood of many a scientist's career. Feed a starving scientist. The oil business (Earth was a planet, too) paid royally for genius geologists.

You lose a first-rate Titan scientist to Exxon Mobil, and she will quickly come to count on those extra zeros on a paycheck—a paycheck guaranteed to deposit every two weeks—to say nothing of the generous health plan, four weeks of vacation, and the 401(k) that Big Oil could promise. To do soft-money science was to have googled the nearest homeless shelter at least once . . . just in case. So every chance that Curt could get, he sent money their way. It would do no good to get an outer planets mission off the ground if there were no scientists left to study the data returned.

Engineers needed scratch as well. Enter headhunters for the defense contractors. If a woman could land a rover on Mars, she could probably land a missile on some nebulous enemy in some as yet undeclared war. So when NASA needed an engineering study conducted on, say, flinging a brick beyond the asteroid belt, Curt would look at his list and assemble a team of engineers who could then feed their families for a few months more. (And to double his investment, Curt would tell them to blast that brick to, oh, I don't know, *Europa*.) He wasn't some space saint saving scientists, and it wasn't welfare. He was a program scientist at NASA headquarters who was just doing his job, impeding a repeat of Europa Orbiter's catastrophic, study-canceling cost estimate. He was building a foundation of knowledge and eliminating uncertainties, buying down risk on future mission studies.

During this time, Congress came to the notion that "flagship-adjacent" outer planets missions could be launched for less than one billion dollars. Curt knew this wasn't possible, but as a matter of due diligence, he announced the "Billion Dollar Box" studies: Everyone, please plan a mission, if you would, to, *oh, I don't know*, Titan, and another to Enceladus, and keep it within a billion-dollar cost cap.[150] How much science can you do for the money? (The answer eventually returned: not enough!)

During all this, the years spent working to keep an outer planets

program alive at NASA, Niebur saw daily just how highly organized and thus highly effective the Mars community was in comparison. They marched in lockstep. When they disagreed among
themselves, they settled it among themselves, and when they
emerged from meetings, they had a plan: something to rally around
and, more important, something to show NASA headquarters that
they had rallied around. Orbiter, orbiter, orbiter, orbiter, or: lander!
lander! lander! lander! or: twin rovers, twin rovers, twin rovers,
twin rovers. They spoke in a single shout, the volume of which
surmounted the din of discussion by scientists recommending the
rest of the solar system. They were like the U.S. Army in Panama,
blaring Van Halen outside of Manuel Noriega's hideout until he
surrendered from sheer auditory exhaustion. Mars roared. NASA
listened. None of this just happened. It was made to happen. The
community of scientists studying the Red Planet was organized by
the Mars Exploration Program Analysis Group, a cross-discipline
science committee chartered by NASA to help set Martian exploration priorities.

Why didn't the icy moons of Jupiter and Saturn have something
similar? And Uranus and Neptune. Pluto. The outer planets community, Curt came to see, needed a group of its own to organize,
push for missions, and prepare for spacecraft arrival when they were
flown. NASA obviously wasn't willing to just throw a bone to the
outer planets. Yes, four flagships had flown thus far, but each took
decades of hard work to get to space, and the first—the two Voyagers of the 1970s—the proofs of concept for large strategic science
missions—were beneficiaries of specific circumstances that would
not be repeated for some time: an alignment of planets that came
only once every hundred and seventy-six years. You could launch
from Earth and quickly hit Jupiter, Saturn, Uranus, and Neptune
in a single go.

"The last time the planets were lined up like that," said the

NASA administrator to Richard Nixon, "President Jefferson was sitting at your desk—and he blew it."

"Do two of them," replied Nixon.[151, 152]

And they did. Later, the spacecraft Galileo to Jupiter and Cassini to Saturn were conceived and launched during what amounted to a fourteen-year hiatus in Mars exploration following the failure of the Viking landers to find life there. But those days were done, and Mars was back, a crimson Death Star fully armed and operational; no mission or moon would stand in its way. So for the outer solar system, missions of opportunity were past tense. Further exploration would require a concerted effort and the force of intellect. The outer planets community needed a group to coordinate and take full measure of its scientific strengths and shortcomings.

To lead this prospective organization, Curt approached the University of Colorado's Fran Bagenal, the British planetary scientist and lead editor of the magnum opus *Jupiter: The Planet, Satellites and Magnetosphere*. Because each chapter had been written by a separate set of scientists, and those authors needed to be nurtured, encouraged, cajoled, and occasionally threatened with having their throats slit, Curt figured she would be perfect for calling cadence for the outer planets and making everyone march in the same direction.

He staged his ambush at a planetary science conference in Kentucky in 2004. On the Sunday evening preceding proceedings, the American Astronomical Society held a reception at the Louisville Slugger Museum & Factory. The two, Curt and Fran, had never met, and before walking in, Curt had to google what she looked like. He spent half the night wandering around the museum, trying fervently to hold in his head a mental image of her and matching it to everyone he encountered.

Fran? he asked a woman he thought was her. I'm Curt Niebur.[153]

Curt who? she asked in her distinctive Dorset accent, having only the vaguest notion of who this NASA nobody was or what he wanted.

How would you like to be the first chair of OPAG? he asked.

What's . . . *oh*-pag?

Curt explained what it was: the Outer Planets Assessment Group. Like the Mars group, MEPAG, but . . . for outer planets. He went on to list the reasons why he thought she would be perfect for the job . . . and she rejected his overtures outright. The Jupiter book had been a grueling effort, Curt!—she was wounded!—needed time to recover, to give her psychological scars time to heal.[154] Niebur, predictably, did not give up so easily. Surrounded by display boxes of baseball bats and wall murals of sports icons, the conversation continued, and the more they talked, the more assertive she became about how such a group could achieve its goals, how necessary such a group was.

It has to be you, Fran, he said. Just do it for a year.

And the once reluctant leader agreed, her mind steadied now, her scars healed as if on command. But if Bagenal was going to steer this ship, she was going to do it correctly. She didn't want it run from Jet Propulsion Laboratory, as MEPAG was. The Mars group had essentially become a lobbying arm of the lab, and she wanted OPAG to have total independence. MEPAG, with human spaceflight on its side, was a fierce competitor; therefore, OPAG would have to be aggressive in making the case for the outer solar system. The community would have to come together, Curt, to write the twenty-page document—the three-page document—the one-page document—the movie clip—the three-bullet slide—the elevator speech. We had to be ready to sell our science in any circumstance, and not just to the bureaucrats at the NASA centers. Mars had "Follow the water." That was their mantra. Where there's water, there's life. The public loved it. Drank it up. It was something that made sense whether you held a doctorate in planetary science or not. Follow the water indeed. On Mars? Ha! Nothing had wiggled on Mars in a hundred million years. So . . . OPAG could *follow the whales!* By conversation's end, she was exhilarated by the nascent organization's prospects.

There was nothing personal driving this intensity. Indeed, a pivotal adolescent event involving the Red Planet had led Fran to planetary science in the first place.

It was, of course, a Carl Sagan Moment.

Bagenal's parents were musicians. Her father made medieval musical instruments. Her mother taught violin. They lived in a small house in Dorchester first built in 1666. In her youth, she stayed up to watch the Apollo landings. Her parents didn't. That was fine. After the Apollo missions, there was talk of humans going to Mars—Mars!—and that interested her. More terrestrially, the study and understanding of plate tectonics had hit a high, and Fran was interested in that, too, this whole business of understanding how the Earth was put together.

When she was sixteen, she learned through her older brother that Sagan was visiting Cambridge University, three-plus hours away. Well, she had to see him with her own eyes and made the trip northeast. His talk was about Mars. It fascinated her to see how effective Sagan was at arguing with stodgy British academics—his parries, ripostes, *prises de fer*. That stayed with her, and years later, when it came time to set the course of her own life, she studied physics and geophysics at Lancaster University. In Britain at the time, however, there were no opportunities to study planetary science—the field didn't really exist yet beyond a few enclaves in American academia—so Fran went to graduate school at the Massachusetts Institute of Technology.

It was her earlier work, however, that would involve her directly in the space program. Her senior project at Lancaster had been on space plasmas—gases in space so hot that their constituent atoms begin to break—and it turned into a paper published in the renowned *Journal of Geophysical Research,* a century-old, peer-reviewed scientific journal. She needed a job after coming to the United States, and while attending a one-off mixer at Harvard University for women in science, someone suggested, after learning that Fran did space plasmas, that she drop by the Center for Space Physics at Boston University. Scientists there were part of a NASA mission set to launch to Jupiter, a planet that, because of its massive magnetosphere, was to plasma what Paris was to writers. Fran didn't realize at the time how historic the spacecraft would prove. Nobody did. When she joined the science team of Voyagers the First and Second, she became part of an expedition that would give humankind Io, Europa, Ganymede, Callisto—didn't discover them, but made them real—and then Enceladus, Titan, Iapetus, Tethys, Dione, Rhea. Here was Jupiter, its crazy clouds aswirl, and here was Saturn, its rings suddenly so big, so close, so detailed, and more stunning than even the most fevered dreams of Galileo Galilei, Christiaan Huygens, or Arthur C. Clarke. And at once, we had to go back to both planetary systems to understand any of it.

Yeah, Fran was up for this.

IN LATE 2006, with the planetary science division at NASA headquarters fully behind the Quad Studies, Curt focused with Jim Green on a road map. Once a mission concept was selected, NASA would need to find three billion dollars. The basic budget for the agency's planetary science division covered a spectrum of standard-sized spacecraft endeavors, but there was nowhere near enough money to stand up another large, strategic mission such as Galileo or Cassini, or a would-be one to Europa (or wherever). No, getting a flagship out of PowerPoint and onto a launch pad would require what was called a New Start: a dedicated, long-term funding wedge in the NASA budget—something that didn't happen often. New Start planning would have to begin straightaway.

It was one thing to send something hundreds of millions of miles to distant worlds of frost and fire. If you wanted a real challenge, try sending something ten, twenty, thirty feet above Curt Niebur's office to higher floors at NASA headquarters and higher links on the NASA chain. Then try and get it a mile and a half down the road to the White House Office of Management and Budget, and from there, another mile to Capitol Hill as part of the annual budget request, and then back and forth from House to Senate, three hundred feet at a go—"I'm just a bill . . ." —and then a final mile to the Oval Office for the president's signature.

Deep space had nothing on DC. The American government was a taffy machine of forces competing and in opposition even when working toward common ends, itself not often the case. A New Start proposal could be (and generally was) killed as it made its way across desks in four separate buildings. It had to survive the division director, the associate administrator, the administrator, the NASA chief financial officer, the White House budgeteer, the president's science advisor, the appropriations subcommittees of the House and Senate, the full appropriations committees, the two houses of Congress, and, finally, the president of the United States. Things got even messier when a spacecraft or mission concept originated not at NASA headquarters but in Congress or the White House. NASA headquarters would have no choice but to salute the president and do as it was told—the wisdom of a mission or lack thereof irrelevant.

Niebur knew long journeys, having taken one of his own to the nation's capital from Breese, Illinois (pop. 3,000), all cow pastures and cornfields. It was as idyllic a place as you might find on the Earth planet, and Breese was the *big town* in his part of the windy prairie state. One birthday as a boy, he received a little three-inch refracting telescope. To the extent that it had a brand, it was "the white one," and that very first night, he hauled it to the backyard and aimed it expectantly at a pristine pitch of rural American sky. Curt got his celestial bearings, pressed expectant eye to eyepiece, and suddenly these points of distant light inflated into worlds as real as Breese. And each was so different! Mars, the color of rust and capped with white. Jupiter and its watchful red eye and tan and brown bands of clouds—and moons! It had moons! Four of them! AND YOU COULD SEE THEM. There was Saturn, rings and all, just as the textbooks promised! You're in sixth grade learning about this stuff and it's a few photos in science class and that light-bulb solar system circled by bead planets on wire arms, but then you strap on that telescope for the first time and all at once you're not dealing in the theoretical. You're a space traveler. Science suddenly isn't a class you take in school, something that begins and ends with fill-in-the-blanks and multiple-choice questions—it is a thing that is absolutely, metaphysically happening and *I have seen it* and there's no going back from that.

He majored in aerospace engineering at the Georgia Institute of Technology, figuring that spacecraft design was the closest he could come to exploring other worlds up close. As years elapsed, however, Curt's *Trek*-fueled fantasies faded, and statements like "let's be real, here" and "imagine the competition" suffused his thoughts, and he soon set sights on what would be realistic. Spacecraft design yielded to aircraft design—a good gig if you could get it, after all, and those firms paid good money.

Then he met Susan Mahan.

Susan was an Alpha Gamma Delta, Curt was a Delta Upsilon, and her sorority was just down the street. A mutual friend introduced them, and he asked her to the football game—go Jackets!—and she said yes. The school allotted each student a free pass to invite one nonstudent guest. Curt brought his dad. Susan used her

pass to bring his mom. It was a great date, she said. He was so taken
with her that afterward he had no idea who won the game. And I
mean Susan had it all. She studied physics, her path fixed, laid with
iron rails: she was going to work for NASA. She made the plan at
three years of age, when her family visited Johnson Space Center
in Houston. She played on an old Saturn V booster—saw mission
control—the works—and at the end of the tour there was a visitors'
center where stood *actual astronaut spacesuits,* and I mean, that was
just something that a three-year-old sees and never forgets, and
lining the wall, there were framed photographs of every astronaut,
and that was something else a three-year-old sees and never forgets,
because she noticed a conspicuous oversight.

Where were the women astronauts?[155]

Well, that just wouldn't do at all, and from then on, Susan was
committed to solving that particular problem. During her youth
and adolescence in Jackson, Mississippi, she put the full measure of
her energies into following the "astronaut track": a literal checklist
of things you needed to study and be able to do in order to qual-
ify for astronaut training. (She considered also becoming a NASA
physicist, though the agency issued no such list for scientists; she'd
just have to nose her way through on that one.)[156] She followed the
list with the devotion of Saint Sebastian, and it led her to Georgia
Tech, and here was this guy, this Curt Niebur of Breese, Illinois,
and he once harbored hopes of exploring the cosmos, too? But no
longer? It made no sense—how could you lose your faith in *space*?

What about planetary science? Susan asked Curt.[157]

Curt had never heard of it. There's such a thing?

There was indeed! He could go to graduate school, get a doctorate in it. Sure, there were engineers bending metal and building the
spacecraft, but *why limit yourself*? You become a planetary scientist
with a background in engineering, and you could do it all.

So that's what he did. It was a revelation, and Susan, his deliverance. They ended up together at Washington University in St. Louis, Susan studying physics, and Curt, planetary science. His graduate advisor was Ray Arvidson, the deputy lead on proposals for a pair of Mars rovers—the same robots that Curt would one day add to his portfolio as program scientist at NASA headquarters. Ray expected his students to get real-world experience in addition to doing research. (You were going to do research.) Thus spake Ray: Your day job is your research. Your night job is teaching assistant. Your sleep job is Mars rover operations.

So to give him that experience, Arvidson and his colleagues would say, *Hey, Curt, we need you to look over this part of the Mars proposal and tell us what you think.* And Curt would do his best and offer his suggestions. Could he have really swayed the opinions of Jedi masters like Arvidson or the mission lead, Steve Squyres? Unlikely. But they were giving him the chance to try and exposure to a planetary mission at its earliest iterations: PowerPoint, paperwork, and prayer. One of Curt's tasks involved developing different operational scenarios. Here's your rover. It's on Mars. You only have so much power and so much room on your hard drive. When the rover rolls, it takes this many resources. When it drills, it takes that many. They dumped those constraints on him and said, Curt, figure out the first ninety days of the mission. How many times are we going to drive? How far? How many drill sites will we do? How much data can we collect? When do we recharge? When do we return the science? It was a complicated simulation, and *they gave it to him,* this guy from Breese whose white telescope first brought Mars into his universe, the same Mars he was now planning space missions for. Curt, fortunately (in keeping with his character), was too young and naive to realize how intimidated he should have been, and when the proposal was selected by NASA for flight, he took it as a great compliment when one day he was told by a JPL engineer that the very complicated spreadsheet he wrote tracking a kazillion details would now be a part of operations planning—that the professionals at the lab would take it from here, expanding upon it, modifying it, making it more realistic. Curt was now some small part of space exploration. Just like that! Four years earlier, if not for Susan—who was herself plowing through a physics doctorate—his life's work might have been building bolts for Airbus fuselages.

They married in 1999. By graduation, Susan still had her sights on joining the agency. She wanted to work at headquarters, and few goals could have been more preposterous. NASA, everyone explained, just didn't work that way. If you were under fifty, to send a résumé there was to waste a stamp. So she attacked from the side, earned her way into the prestigious Presidential Management Fellows program, an internship operation run by the U.S. Office of Personnel and Management. Through it she managed to get that headquarters assignment.

The fellowship didn't pay much, so to make it work, Curt took a job at Raytheon, a defense contractor that paid pretty well. His work was hodgepodge: they just wanted someone with an advanced degree to take a look at different programs from a broad perspective and to help pull pieces together. Later, he transferred to a nonmilitary division, and his job was to encourage nonscientists—farmers, for example—to use Earth satellite data to do things like monitor the health of their crops.

It was good work, but after two years of Susan coming home each evening beaming with joy at the Best Job in the World, he was primed to move on. A position at NASA headquarters opened, and it was a long shot because—look, Susan aside, *there were no young people at headquarters*—none! But Susan, having kicked open the door as a fellow and now leading the entire Discovery mission program at headquarters, demonstrated to the AARP crowd that you didn't have to be sixty to do good work. And so when this Curt kid, this scamp, applied—Hey, is he related to Susan?—he was hired.

Three weeks after Curt started, his boss sent Susan an email asking if her husband would like to take over the Jupiter Icy Moons Orbiter as program scientist. You had to actually memorize email addresses back then, and everyone knew Susan's and no one knew Curt's, so that's just how people did things. He and his wife were a team. For his part, Curt had only a passing familiarity with Europa (he knew it was a moon of Jupiter) and knew next to nothing about this thing called JIMO. What little he did had come recently from a guy named Dave Senske at Jet Propulsion Laboratory. Still, replied Curt, resplendent with that new car smell: "I'm definitely interested!"

It became clear to Curt early on that JIMO was one of those big ideas that just didn't translate to the real world. But as long as Sean O'Keefe was NASA administrator, JIMO and Project Prometheus were untouchable, and it was Curt's job to make it a mission worth flying. He worked closely with the members of the JIMO science definition team: the ad hoc beaker of scientists whose job it was to determine the detailed scientific concerns of the mission. It's where Curt met Ron Greeley, the maestro conducting this symphony, and Bob Pappalardo and Louise Prockter. It's where he came really to understand the complexities of NASA and How Things Worked. It's also where he first learned to appreciate the dangers that an exploration gap would pose for worlds beyond Mars.

NASA began exploring the outer planets in 1973, when the spacecraft Pioneer 10 became the first vessel to cross the asteroid belt. Astronomers had a pretty good handle on the big things in the belt; there were more than a million asteroids out there one kilometer or larger, but it wasn't like *The Empire Strikes Back* or anything. You'd have a much harder time trying to *hit* an asteroid than miss it. (Space is big.) No, the *tiny* rocks were the problem. They were impossible to spot telescopically, and because Pioneer 10 was coursing along at thirty thousand miles per hour, an asteroid the size of a rice grain would pass through it like an armor-piercing shell. Before launch, scientists gave poor Pioneer 10 a name-appropriate ten-percent chance of being destroyed.[158] But it wasn't, and once it crossed the barrier impenetrable, NASA never stopped.

One by one, the outer planets of the solar system flickered to life. Jupiter was first, lit by Pioneer 10. The same year, Pioneer 11 launched for Saturn. Voyagers 1 and 2 followed, sweeping through the Jovian system just as Pioneer 11 reached the ringed world. The Voyagers each in turn would maintain the light, with Voyager 2 continuing to Uranus. After it illuminated that inscrutable ice giant, Galileo launched, its target: Jupiter, to stay. Voyager 2, meanwhile, cast its beam on Neptune, and Cassini launched, cruising along as Galileo, safely in orbit, made Jupiter's frozen and fiery moons real, and, while there, found an ocean at Europa. Just after Galileo's eight-year residency ended, Cassini commenced its tenure at Saturn. Only one planet remained: Pluto, and New Horizons would get there soon enough.

In all, it took seventeen years to go from inching past the asteroid belt to exploring every planet within three billion miles. From then on, humankind established a sort of permanent residency in the outer solar system with the orbiters Galileo and Cassini. But this was only possible because of a rigid exploration cadence: while one spacecraft was encountering a world, another was cruising to some other target. While one was cruising, another was being built. You break the chain—launch nothing new during another spacecraft's encounter—and you would end up, at some point, with an exploration desert. The outer planets would go dark. It took seven years for Cassini to reach Saturn. Seven! Galileo took six years to reach Jupiter. Voyager 2 took twelve to turn up at Neptune. Had NASA, for example, waited until Galileo ended in 2003 to begin building Cassini, the outer planets would have gone dark for a full decade—if researchers were lucky. Nothing being explored meant no new data coming in meant fewer grants meant fewer scientists in the field meant fewer pushing for exploration in the first place, and repeat, the outer planets ensnared in a system of exponential decay.

So the Quad Studies had to work. Something had to launch. Cassini was still fresh at Saturn, but it wouldn't fly forever. Something needed to leave Earth, and soon. The field was counting on it. For that reason, Louise Prockter as the science co-lead for the Ganymede mission felt like an increasingly inspired choice. Curt knew that she and Dave Senske would devise a mission almost irresistible—Galileo on steroids—and he knew that the Europa leads knew that as well. Her presence would force the Europa study to produce the most astounding, affordable, and capable mission possible, and even if they did not, Louise and Dave would still get something to Jupiter.

CHAPTER 4

The Center of the Universe

IN FEBRUARY 2005 JOHN CULBERSON SQUEEZED HIS copy of the NASA budget request, all three hundred eighty-nine pages of it, blood draining from his knuckles and rising to his face as he read each elapsing line.[159] There it was, in twelve-point Helvetica: "An investigation of Jupiter's icy moons will not be the first demonstration for Prometheus Nuclear Systems and Technology."[160] JIMO had been "deferred," and would soon be declared, formally, "indefinitely deferred," and there was nothing he could do to save it. No one at NASA headquarters had even bothered telling him, a lowly congressman—from Houston, yes, but *west* Houston—a dilettante, then, who represented neither NASA nor its employees. He'd adopted JIMO as his thing, and he'd promised Jet Propulsion Laboratory the money, and it had appeared, and yet JIMO did not.[161]

The insolence! Thanks for the check, Congress, but here at headquarters, we have some better ideas for the money. The shortsightedness! JIMO is too expensive, and we have more pressing needs elsewhere. Elsewhere? The parochialism! It was textbook

NASA headquarters: no vision, no concept of its best interests, and a smug willingness to bite the hand. Inch when you should yard. Punt when you should throw long bombs. Not that there weren't other worthy NASA programs out there, but JIMO was *the will of Congress*. If the agency wanted more money for moon rockets and space capsules (where the funds were now flowing), it could simply have asked him for it.[162] Sure, NASA was technically allowed to cannibalize the money directed by Congress for JIMO, but this was *can* versus *should*. Congress appropriated in lump sums, and a concurrent conference committee report explained the breakdown: what the money was for and why. But while the budget was law, the report was not. Agencies of the federal government, including NASA, customarily followed the guidance of the committee report because it was just bad form to annoy the representative-attached hand that was writing the check. Clearly, however, this is how it would be with the new NASA administrator.

Culberson wondered: Did headquarters not see that the Europa mission had it all? Did they not care? Finding life on Europa would be a turning point in human history and shift attitudes on a civilizational scale. John could see it, this magical and transformational moment, the announcement of life by the president on a stage stacked with scientists, scores of men and women wearing sobersided expressions commensurate with their roles in having changed everything evermore. He saw the stories the next day, the polls in the weeks that followed, reflecting the immediate invigoration of the instinctive love every American has for NASA. The agency wanted more money? Too easy. Find a fish in an alien ocean and watch the world rally. Even the gormless bureaucrats at headquarters had to know that. The public would demand deeper exploration yet, and JIMO's Prometheus propulsion, flight proven, would power those newly urgent and energized missions beyond the asteroid belt.

It was all set: the future of the agency, science, society, and spacecraft. And it was now "deferred," with NASA rejecting not only Europa but also its best chance ever to do away with chemical propulsion—a really great thing in 1926, when Robert Goddard invented the liquid-fueled rocket, and eighty years was a good run, but I mean *come on*.[163] NASA needed nuclear reactors in space, and knew it, and here was Congress, checkbook open, pen twirling between its fingers, *clicky click*, asking how much Mike Griffin, Sean O'Keefe's successor, needed. There was only one answer to that, and it wasn't *Thanks for the offer of Gene Roddenberry, but we're happy with Patrick O'Brian*. Also: *We're keeping the money*.

Whether headquarters realized it or not, it just *didn't work that way*. John had been doing this since he was a second-year student in law school. He was pushing fifty now. Had headquarters any idea what a personal affront this was? How heedless and self-destructive? Had they any idea what retaliation might mean if he were to affix their thumbs to a vise and tighten the screw? John Culberson knew how to make it very tight indeed.

He first entered elected office in 1986 by way of an open seat in the Texas House of Representatives. He was boyish in those days, clean-cut, single, eyeglasses then thick as a windshield, and nobody expected the seat to open in the first place, but without warning, the incumbent decided instead to mount a quixotic run for railroad commissioner, though not before endorsing an heir: Herman Botard, Texan, a real man of the people, a proven quantity.[164] Which was nice for Herman, but there was no way a seat in the House of Representatives of the Republic of Texas was going to open up without a battle, and five other would-be Texas state representatives stepped forward, including a couple of multimillionaires, and John Culberson, who was, on happy occasion, a hundredaire.[165] And don't think they didn't notice! A student without experience against men of ability, of achievement, of standing. And those glasses—a bookworm child against Proven Quantities? Well, it was going to be ugly but lessons had to be taught.

In his favor: John knew forward and back the rules of politics. He understood the commonsense issues that mattered to his fellow Lone Star State residents: Taxes. Traffic. Trade. He was a good, God-fearing Christian, went to church on Sundays, and fit the spirit, if not necessarily the image, of a Texas Republican. He learned political survival and success at the feet of his father, Jim Culberson, a photographer turned campaign consultant, who believed before most how a new form of media would win elections. Television advertising was practically a dark art in those days, but Jim was a real artist who knew the right incantations: how to point a camera, focus a lens, compose a shot, and tell a story. He went all in on televised campaign commercials and dispelled any doubts about radio's replacement by reelecting John Tower to the U.S. Senate in 1966. Tower, previously a Democrat, had slipped into the job by way of a special election when Lyndon Johnson left the Senate to become vice president. Tower was one of just two Republicans representing the South in the Senate and was Texas's first GOP senator since Reconstruction. The Democrats, who otherwise had the region in a hammerlock, should have picked off Tower with ease. But Jim had delivered the Republicans a W, and after that, for the family Culberson, it was T-minus-five and counting. And all the while, there was young John, hauling his dad's cameras around, watching how the game was played.

That's where he learned to turn this thing around, his done-for campaign for the Texas House in 1986, learned that the way to win is to sell your weakness as your strength, so that when they attack you, it paradoxically helps. In the end, the eyeglasses saved his campaign. Unlike his millionaire, middle-aged adversaries, John was still young enough to have books such as Xenophon's *Anabasis* on his nightstand, and in those pages, the solution to his problem.

He read and relished the story of a Spartan army of mercenaries moving to seize the Persian throne. He learned how that army—the "marching republic of ten thousand"—was beaten, its generals slain. He saw how, being built of good Greeks, the leaderless army did what good Greeks do: they held an election. Who would lead them on a desperate drive from the Fertile Crescent to the Black Sea? No winner emerged. Again they voted, and then again, until finally, young Xenophon, watching from the periphery, stepped up, threw his laurels into the ring, and said, "If you bid me lead you, my age shall be no excuse to stand between me and your orders."[166]

Well, who wouldn't vote for that sort of pluck and vigor? *You need a leader, and I won't hide behind my youth to avoid the burdens of the job.* And there it was! John's campaign strategy, written by the designers of democracy itself. It had certainly been validated, and he pushed his freshness and willingness all the way to the voting booth. It was the first of a string of decisive Culberson wins to the Texas legislature until the turn of the millennium, when George H. W.'s old seat in the U.S. House of Representatives opened up. (It had been filled for the previous thirty years by Bill Archer, but I mean it would always be George Herbert Walker's.)

John's opponent spent three and a half million dollars against him, while John spent dimes in comparison (he had no millions to spend) and he kissed the grandbabies of constituents who first voted him into office all those years ago, shook hands, and posed for pictures, his boyishness still there, even as crow's-feet crawled toward his temples—and in the end, John's handshaking and back-slapping and *Hey, I haven't seen you in years! I wanna fix the freeway and I need your vote. How's the little one? What? In college now? Last time I saw him, he was on the T-ball team! Traffic's a mess, and I want to solve it* won the day.[167]

John Culberson was sworn into the U.S. House of Representatives on January 20, 2001, all of this perhaps a footnote in Amerlean political history, this changing of letterheads but not of the guard, this sliding of now-dominant Texas Republicans across the political chessboard—as inevitable in retrospect as the tides in nature—if not for the intervention of Tom DeLay, the majority whip in the House. The two men had much in common: both were native Texans, both represented Houston (a quietly massive city, fourth largest in the United States, behind Chicago, Los Angeles, and New York City), and both were whips, charged with keeping party members in line, John having once served in that role in the Texas House.

Members of Congress are placed on committees, the work of government parceled between the lot, with some working on, say, military procurement, some working on education, and so on down the line, all the business of a functioning democracy.[168] In 2003 DeLay called Culberson into his office and made him an offer: he wanted John on the appropriations committee, and John turned it down cold. He represented west Houston, Tom. He was already on transportation, and if there was going to be a committee change, he needed to get on energy (oil and gas being to his constituents what computers are to workers in Silicon Valley).[169] Appropriations? Try telling a district of Republicans that you spend the government's money for a living.

DeLay was insistent, though, in the way that a man called the Hammer can be. See, said DeLay, you get on the energy committee or armed services or agriculture or education, and you can pump out legislation all day long, day or night—exhaust the nation's supply of ink and toner if you'd like—and every single bill you write can be killed or left to die by the full House. That's politics. But appropriations, John? The House cannot ignore it. The Senate cannot ignore it. The *president* cannot ignore it. There are only twelve bills that Congress absolutely must pass each year, and those are the twelve appropriations bills—one for each committee. The appropriations committee is the only one with jurisdiction over these twelve bills that *must be signed into law*. Every year. Do you see what I'm getting at, John? It's the power of the purse. Appropriators dole dollars to the entire apparatus of government. If an office has a single employee, spends a thin dime on paperclips, *the appropriations committee can touch them*. But wait, hear me out, just so there's no ambiguity here: even if the whole of the Congress decides to skip work for a year, those twelve bills absolutely must pass no matter what, and—hold on, John, because it gets better—in addition to designating money, you can make small, permanent law changes within an appropriations bill. You can do this to ensure that the people's money is spent wisely.

Well, who would pass up that kind of power? John agreed under the condition that DeLay accept that his (i.e., John's) starting answer would be no on everything, that he was a Jeffersonian Republican, didn't like spending *borrowed* money. And DeLay was OK with that because we're all Republicans for crying out loud! And John said he'd only do it if he could be on the appropriations subcommittee with jurisdiction over NASA. He'd grown up loving space, wanted desperately as a boy to be an astronaut, lived a few miles from Rice Stadium, where, in September 1961, President Kennedy vowed to the world that the United States would put a man on the moon and then bring him back safely before the end of the decade. NASA's aspirations had fallen a bit since JFK, and if John could touch NASA as a junior appropriator, what could he do once he had real authority?

All of this coincided with a multiyear reorganization by the House and Senate of their respective appropriations subcommittees, to account for a new Department of Homeland Security and, later, to reduce the number of subcommittees from thirteen to ten.[170] NASA—now without a home—was moved to the Subcommittee on Commerce, Justice, State, and the Judiciary, which was renamed the Subcommittee on Science, State, Justice and Commerce, and Related Agencies. (State would soon be pulled from that pile.) Aside from making sentences veritable battlefields littered with commas, here is the upshot: in 2003, during a particularly grim stretch of fighting and loss in Iraq and Afghanistan, NASA no longer had to compete with its former committee-mate, the Department of Veterans Affairs, for the same pile of money. The following year, John Culberson learned about JIMO.

BEFORE THE UNION Pacific Railroad came to town in 1895, Katy was little more than a rest stop for stagecoaches running between Houston and San Antonio.[171] In those days, it wasn't yet called Katy, and Texas was still the Old West as seen on film and in faded photographs, sunbaked clay on every horizon and stern, sunbitten men, their mustaches, muskets, and hat stars. It was older, perhaps, even than the Old West as we imagine it. The cowboy hat had been around for only thirty years, and John Stetson still made them, personally. Santa Anna had been dead for less than twenty years, and the rallying cry "Remember the Alamo!" might well have been a question, because men and women still walked the Earth who were there, and actually did.

Katy's big break came on May 10, 1869, when railroad magnate Leland Stanford—later founder of the university—drove a golden spike through a steel rail in Utah, at last hammering together the Pacific and Atlantic Oceans with the skinny, sun-warmed lines of the Transcontinental Railroad. Over time railways spiderwebbed outward from the main line to junction towns, and here, just west of Houston, this sleepy stagecoach stopover with its verdant fields of farmland suddenly became one such interchange, connecting Missouri, Kansas, and Texas by way of a new junction: the MKT, but what locals and passers-through called simply the KT.[172] The town built a train depot, and the name stuck, but spelled phonetically.

A place like that, connecting continental companies, cargo, and cattle, is going to bring people, and Katy grew as the pages of calendars fluttered away. The oil and natural gas helped, and there were crops, too—citrus and cotton and rice—but if you want to give someone a reason to pull up stakes and move to Nowhere, Texas, just plant a natural energy well on an open patch of ground and watch what happens. Workers drifted in first like bees to columbine and soon swarmed like hornets around a hive.

Katy wasn't the only place with rising population figures. Houston to the east had more than doubled from 1920 to 1930, reaching two hundred ninety-two thousand, after nearly doubling in size the decade before and the decade before that.[173] This growth didn't stop at people, but also the things they carried, and the things that carried them, specifically: the automobile, now affordable to the masses and eventually to eclipse the passenger rail once vital to Katy's success. Then came the traffic.

It didn't affect Katy at first, though, as Houston sprawled ever outward. The city of Katy was a nice place to live, but Houston was a nice place to work—all the big companies were there, anyway—and getting from one to the other briskly became nothing short of a traffic tribulation. Highways helped in the 1940s, and the Interstate Highway System promised relief in the 1950s (when Houston doubled in size yet again). Little Katy got its own stretch of black in the 1960s, part of Interstate 10 but called, locally, the Energy Corridor or the Katy Freeway. But being Katy—I mean, it was nice and all but it was mostly cows and oil wells from there to west Houston—it garnered a mere six lanes of highway on the outset and two service roads.[174] And yet Houston grew and Katy grew and traffic grew and gridlock grew—everything grew but freeway, which needed to grow most of all.

But that traffic. Something had to be done. The problem was this. You don't just *expand* a highway, even one as highly congested and generally unpleasant as the Katy Freeway (which, by the 1970s, had become one of the worst in the United States, built to accommodate a maximum of one hundred twenty thousand cars, but carrying twice that).[175, 176] When it comes to major roadway construction, repair, or renovation, the Texas Department of Transportation operates unapologetically on a pay-as-you-go principle: save up your money and develop a plan of attack.[177] What do you need to do the project? How do you handle existing traffic? How do you connect myriad sections of highway and byway old and new, above and below? How do you establish quality assurance? How much is this really going to cost? Then you save up more money until you can afford to finish the design. Then you save up money to relocate water, gas, electrical, copper, and fiber-optic utilities along the "right of way"—that is, the vast stretches of land, houses, factories, used car lots, lodging, and local businesses that will have to be leveled to put in the new roadways.[178] It's never just a matter of moving a telephone pole a few feet over: every node of the grid is affected, making it akin to relocating a spiderweb one mooring thread at a time *without disturbing the spider.* Then you need more money to actually buy that right of way, and these being Texans, don't think *that's* not a negotiation. You get everyone on board—and there are a lot of everyones, from environmentalists to bus drivers—and only then can you get final approval from the federal government, assuming it approves, and you get started on construction.

But no matter how you approach the problem, the first step is to get the money, and in the case of the Katy Freeway expansion, there was none, and no avenue by which money might be attained. Real Texans don't generally vote for tax increases, and certainly not the genuine articles of west Houston.

Years of John Culberson's life had been lost on that horrible freeway. He was born and raised in West University Place on the western front of Houston (easterly of Katy), an idyllic American hometown of freshly mown lawns, the *tsst–tsst–tsst–tsst* of early-morning sprinklers, and neighbors you still felt comfortable asking for a cup of sugar. Like all suburbs, however, it was the place you left when you wanted to go *somewhere,* and where you returned after doing *something.* And in proper suburban fashion, it existed only because of the automobile.

Long before John saw himself as Congressman Culberson, he determined that if ever he were in a position to change anything in the Great State of Texas, his beloved state that you absolutely Do Not Mess With, where stars at night are big and bright, he would do whatever it took to open up that freeway and fix it. He made that youthful vow official when he committed his name to a ballot. Once elected to Congress in 2000, locked into the promise and still angry at years lost in Katy traffic, Culberson had no choice but to find a way to fix the problem, and in his first year, he did. It required what he called "creative financing," and it was nothing if not that.[179] First the funding. The only obvious way to pay for the thing without raising taxes would be with tollways, which federal law forbade on interstate highways. Except John found, in subsection b of section 1216 on page 105 of a practically forgotten federal law from 1998 called the Transportation Equity Act for the 21st Century, an allowance for a pilot program in which three toll facilities *total* might be tested in places along the forty-seven thousand miles of Interstate Highway System in all the United States. The reason: "reconstructing and rehabilitating Interstate highway corridors that could not otherwise be adequately maintained or functionally improved without the collection of tolls."[180]

Well, why not Katy? The law, as *he* read it, let local governments test taking tolls on special "managed lanes" in the middle of a highway. Such lanes, as opposed to the wild and wooly free-range species of unmanaged roadway, would be actively monitored and controlled in response to changing traffic conditions, e.g., the occasional opening of high-occupancy vehicle lanes for all traffic. Moreover, bumper-to-bumper traffic would make sparse lanes quite the commodity, and the toll could thus be higher. Conversely, open roads in every direction would make toll lanes pretty pointless, and by the same logic, free (or practically free) to enter and exit. As intended by law, such an arrangement could reimburse construction costs, roadway repairs, and renovations.

This had not been done before anywhere in the country.

Culberson sharpened his pencil and got to work. He liked the concept of an all-new toll-oriented Katy Freeway operating on free market principles. That, therefore, was definitely going to happen. No federal earmarks, no federal management—Thomas Jefferson would have loved it. The Harris County Toll Road Authority, in whose jurisdiction fell Katy, would run the lanes, and tolls would take care of it all. And those tolls—the congressman really rubbed his hands together at this—wouldn't merely double the number of existing lanes, or even triple it. If his estimates were correct, they could raise enough to cover a freeway expansion from six lanes to— another first—*twenty*-six lanes, consisting of three managed lanes, four feeder lanes, and six main lanes *in each direction*. It would be the widest highway not only in the history of the United States, but in the history of the world! It would be a road to make even the ancient Romans envious.

Before anything could happen, though, he would have to get the county and state on board, and while it was, in John's view, *clearly* in the best interest of both, they fought back, forcing him to bird-dog it. They disagreed on design. Some lamented the possibility of sprawl, and of air and noise pollution, and how it would harm Harris County businesses in the megahighway's expansion path. The Harris County Toll Road Authority wasn't even sure it wanted to run what would be the first interstate tollway in the country. The office built toll roads from the ground up and operated them, maintained revenues, and paid back bonds. It was tried and true. But these would be controlled-access toll lanes within an existing, non-tolled freeway. More worrisome yet, the roads would use all-electronic tolling, which didn't exist in Harris County. Estimates or not, they didn't know for sure how much revenue they might generate, and whether it would cover the cost of construction and allow the authority to pay back bonds. Could carpools go free, for example, and how would that affect income? What about busses? What if everyone carpooled and took busses? How could you even measure that?

John held public hearings along the corridor. Listened to environmental concerns, business concerns. He went back to everyone involved and negotiated. He cajoled. He took the Gila monster approach: he ran in heedlessly and bit hard, and getting smacked only made him bite harder. He fast-tracked every single avenue of the bureaucratic process. He pushed successfully to get design work done while simultaneously getting right of way clearance and drafting relocation plans for utilities. This concurrent activity, too, had never before been done. The cash promised by his toll plan was a battering ram like no other, and the Harris County Toll Road Authority agreed to bring two hundred fifty million to the table (appropriated previously for Katy maintenance) to get things rolling. Culberson next went to the officials at the Federal Highway Administration. No way, they said. The Katy project would set a bad precedent, they said. It had never been done, they said. It shouldn't be attempted, they said.

John could not believe that people actually thought this way. The congressman called Mary Peters, the U.S. secretary of transportation. They went back and forth, and in 2003 she approved the permit. The Texas Department of Transportation, the Harris

County Toll Road Authority, and the now forced-to-be-flexible Federal Highway Administration signed a tripartite agreement in March, and by June, Williams Brothers Construction, which won the contract, broke ground.

In preparation for the expansion, the Union Pacific had already pulled up its ancient railroad tracks along Old Katy Road. The Metropolitan Transit Authority, which wasn't part of the agreement because it initially brought no money to the table, got free busses in the toll lane, and so it kicked in ten million to strengthen support structures for a future light-rail line, which it most certainly did not get.

So what by all assumptions would never happen became a project that should have taken a decade to complete. In the end, however, it took only five years, three months, thanks to John Culberson. And no matter if it was day, night, rain, or shine, if Congress was in recess and he was in town, construction workers in engineering trailers beneath freeway overpasses could count on a knock on the door without warning, and there Congressman Culberson would be, hard hat in hand or on head. Were you getting everything you needed? Yes, sir. On schedule? Nope, ahead of it. (Financial incentives built into the contract helped that part along.)

In truth, John just plain loved meeting the guys who made this stuff. He'd drive out west to the Beltway 8 overpass, where construction crews used cranes to carry these giant steel support structures—it was unreal, these things, enormous, the crossbeam skeletons of the overpass, each a single, prefabricated piece of metal (made in America!), with predrilled holes for bolts. And there was this one crane operator—the only guy at Williams Brothers Construction who could do it—who would hoist these gargantuan beams, carry them slowly to the support structures, and set them into place so that the bolts and the predrilled bolt holes lined up perfectly every time. John even brought along his brother to see this guy in action. It was fascinating, rousing. It was nothing less than the American republic as the founders intended it.

By the time the Katy expansion was completed in 2008, John Culberson had a new tool in his box now, shiny and blunt, forged in the fires of the legislative process.[181] The one thing he knew for sure about the tool: it definitely worked. And if it worked once, maybe it could work again.

THE FIRST TIME John set foot on the grounds of Jet Propulsion Laboratory was January 25, 2004. The lab was like a cross between a sprawling college campus and a military fortress. Surrounded by, dotted on, and nestled among Southern California's San Gabriel Mountains, classrooms and cafeterias and conference rooms were stacked in buildings six stories high, and interspersed were courtyards and warehouses with clean rooms where spacecraft were designed and built and battle tested for the rigors of space exploration. The sheer futurism of the place—everything from its neomodernist architectural aesthetic to its resolutely optimistic staff—suggested an academic oasis bringing forth a world only promised in Golden Age science fiction. I mean, they built spaceships for a living.

He was there that day for the landing of the Opportunity rover on Mars. There was no such thing as a routine landing on planet four; historically, most Mars surface craft crashed or vanished, including the European Space Agency's Beagle 2, which failed one month earlier. The signal travel time between Earth and Mars made the moment more harrowing yet; once landing operations commenced, seven minutes would pass before you received a successful spacecraft status signal. Engineers at the lab called this the "seven minutes of terror."

The problem with landing on Mars was that it didn't work like the old *Looney Tunes* animated shorts: a rocket ship slowing and landing upright, a little ramp extending, a little car rolling out. Opportunity would plow into the Martian atmosphere at high-hypersonic speeds, heat shield down, looking a lot like a flying saucer, the letters JPL emblazoned on one side in red. Once through the upper atmosphere but still high in the Martian skies, a parachute striped in white and orange would fire from the spacecraft's top, slowing the saucer to speeds more manageable, but slightly. Once slower (though not slow), the heat shield—the bottom of the flying saucer—would pop off and drop away, and a padded pyramid within would fall, tethered to the speeding saucer. The whole contraption would still be coursing toward the Martian surface like an incoming missile, but then the trademark JPL So Crazy It Might Work: eight seconds before slamming into the ground, the pyramid, still dangling and whipping away, would explode outward like an enormous, lumpy bag of instantaneously popped popcorn. Massive airbags covering every angle of the pyramid would deploy and inflate in a split second, simultaneously, protecting the package inside. Two seconds later, the saucer above would fire these giant retro rockets, blasting like hell to help out the parachute, to slow things down, to stop the landing from becoming an auguring. And then the impossible: three seconds before impact, slow and low enough to prevent a human-made billion-dollar crater on an alien world, the speeding saucer would cut loose the giant bag of popcorn, and it (i.e., the bag of popcorn) would smack into Mars . . . and then bounce! Like a fifteen-foot basketball![182] Just collide with the planet and bounce fifty feet into the air, and then come back down, and bounce again, and again, and again, twenty-six times total, bounding along the Martian surface for entire football fields, a giant's plaything on a desolate red world.[183]

Once exhausted of energy, the fight taken out of it, inertia would keep the airbag-enveloped apparatus rolling along, and rolling and rolling and rolling across Mars and down hills and slopes, over jagged rocks and wind-worn topography, this comical cluster of inflated Vectran gamboling along another seven hundred feet.[184] And when the ball would tumble not one inch farther, momentum depleted, the whole thing would stop, take a brief breather, and deflate, just melt away right there in the sun, a tawny sky above and penny-pigmented soil below. Beneath the material of the deflated airbag: the outline of the pyramid, still safely intact, about the size of a riding lawn mower. It would open slowly, the three petals of the pyramid peeling outward, revealing, what? A robot from space. A car. A robot space car! It was ridiculous, the *Looney Tunes* landing eminently sane by comparison. And the robot space car would itself then open, servos unfolding it like a Transformer, origami in reverse, its solar panels spreading, presenting a glistening back of black glass to a generous sun god. This creature is born, hatched and clawed from a tessellated metallic egg, wings now spread one planet removed from where it was conceived.

Well, if you were a Martian and saw all of this, it would have scared the hazy daylights out of you. The robot space car's neck rising—the thing had a head! A face, and four eyes like a spider, two big ones, two small ones. And if you were a representative from west Houston who had gone to watch this unfold with the four eyes of your own from Jet Propulsion Laboratory's mission control, where telemetry lines on computer consoles explained to edgy engineers what was happening in zeroes and ones, and cheers erupted—Opportunity had survived!—you were already a space enthusiast, but now born again, from that moment and evermore. You—one of only two members of Congress to even bother showing up for the event—had witnessed the impossible made manifest at the JPL Space Flight Operations Facility, in a room that engineers called not modestly (but not unearned) the Center of the Universe.[185]

Everyone wants to be an astronaut when he or she grows up, but John, he *really* wanted it. He knew vast swaths of that Kennedy speech from memory, the Rice moon speech, and long before he wore the pin of a congressman, he had the pluck of a star voyager. His parents fueled it. You want to make an astronaut of a young Texan, give him, on his twelfth birthday, a Celestron telescope powerful enough to see Jupiter's moons—really see them, these tiny white dots, the actual moons: Ganymede, Callisto, Europa, and Io. Let him swing the scope over and catch the craters and permanently shadowed regions of the moon. You want him to be an astronaut, take him to Cape Canaveral in Florida and claim the closest spot to the launches you can get without riding on the rocket itself. Watch Apollo 15 lift into space, screwed to the top of a Saturn V, all three hundred sixty-three feet of it, this gleaming white column, taller than the Statue of Liberty, pedestal and all. Experience the launch—it will give you no choice—you don't watch, you *experience*—the seven and a half million pounds of force suddenly blasted into the ground below, and outward. Feel the Earth tremble beneath the might of this colossus, its force leaving spectators' flesh flapping as though they're all on the back of some great motorcycle. The pennies rattling in their pockets. Higher, higher, higher it rises—but slowly! No bottle rocket, this is more like an inverted volcano.

Not everyone gets to be an astronaut, however, and that includes a little boy from Houston who wasn't good enough at math, had feet too flat to run fast, and wore spectacles like windshields. If a boy like that wants to save the space program, he'd better get elected to Congress—be given signature authority for NASA's checkbook—view a Mars landing from Jet Propulsion Laboratory, where spaceships are built and textbooks written about worlds previously unseen by human eyes, and where rovers are made that explode into bags of microwave popcorn, collide with other planets, survive, settle, and roll away.

Which is why later that day, the lab's Powers That Be ushered the gimlet-eyed John across campus, up and down the rolling hills of Saint Gabe, and through alleyways between hangars and office space. They filed, the group of them, into a building that could have been any building on any college campus in America, into a briefing room that might well have been retrofitted from a classroom, lots of rolling chairs, but also desks and desktops that folded across laps. Whiteboards and projection screens lined the room's walls, the phantom residuals of erased doodles in some corners, impenetrable physics equations on others, and John sat and the scientists sat and the lab managers sat when appropriate, while others perched and crowded in the room's rear. Here, everyone settled in and introductions were made, starting with John Casani, engineer of legend—an icon—one of the four pillars of Jet Propulsion Laboratory—that's what they called them, the "four pillars"—the other three (though not all present) being Kane Casani, John's brother and a former lab manager, now retired; Gentry Lee, chief engineer of Galileo and who oversaw all engineering of the rover that had just landed; and Tom Gavin, responsible for the lab's flight projects. Space exploration was young, and these guys were old. They had been at the lab . . . forever, had done . . . everything. Also in the room was Charles Elachi, the director of the lab, who offered similarly impressive qualifications. Down the line they went like this, one pioneer after another, John Culberson casting eyes on explorers he had been reading about in *Aviation Week* and the journal *Nature* for decades now. Then someone dimmed the lights of the lecture hall and presented to the congressman the past, present, and future of American space exploration, with the proviso that only he, a member of appropriations—a man who *understood now what wonders were possible*—could make that future happen.

Lab leadership walked him through everything way up there or soon to head way up there: Opportunity, the Mars Global Surveyor and Mars Odyssey satellites, the Mars Reconnaissance Orbiter.

Over in Building 179—the Spacecraft Assembly Facility the Mars Science Laboratory rover would soon be under construction. Congressman, number six is the charm: Mars Science Laboratory, the most ambitious effort ever attempted to understand another planet. When it launched (only a few more years!), it would determine at last the habitability of Mars. The Red Planet, they assured him, was covered.

They updated him on the Cassini mission to Saturn, launched in 1997 and seven years later settling in at last to reach its destination. Its companion probe, Huygens, would likewise land soon on Titan, a world virtually impenetrable by earthly telescopes because of its dense atmosphere. Who knew what wonders waited on the Titanian surface? There was likely liquid there, and the Huygens probe was designed to float. Cassini itself had made some impressive findings already, and its prime science mission had yet to begin. Among other things, they discovered that the radiation environment surrounding Jupiter—scanned along the way—was way worse than previously thought: it was a real spacecraft computer killer for future long-duration missions.[186] Sure, Congressman, Cassini had gone way over budget during development, but it was *all payoff from here on out*. This thing would be a textbook shredder, a bargain at twice the price.

Then the presentation turned to missions in the making, and the septuagenarian John Casani stood up carefully to present a project that he had been developing with other engineers, and he slapped onto the board something called the Jupiter Icy Moon Orbiter, or JIMO, part of a project called Prometheus. And before Casani could even begin explaining JIMO, Congressman Culberson just knew. It was a revelation! It was beautiful, this thing—it looked from the artist's illustration like the spaceship *Discovery One* from *2001: A Space Odyssey*. It could go anywhere, Casani explained, this vessel of exploration—Prometheus-1, it was called—but was designed especially for the Galilean moons and one called Europa in particular.

John Culberson had heard of Europa. Had seen it countless times through his Celestron. But what they were describing here . . . They discussed the ice. They discussed the ocean beneath it, how the physics facilitated it—the tidal forces from Jupiter—and how it had been proven with the Galileo magnetometer. The water was warm down there, or anyway, warm enough. More research was needed, but there would be much more liquid water on Europa than on Earth. And as had driven the Mars program, the congressman knew, where there's water, there might be life. They weren't promising anything, the presenters. More research was needed. But at the bottom of Europa, water touched rock, which meant chemistry of some sort was taking place. More research was needed, but if that water and chemistry somehow encountered energy—hydrothermal vents, say, as on the bottom of Earth's oceans (where life teemed), or—here was an idea—if Jupiter's harsh radiation environment produced oxidants and simple organics on Europa's surface, and if that material somehow made its way through the ice shell and into the ocean . . . Well, more research was needed. What were the engineering challenges of JIMO, really, next to what waited beneath the ice of this mysterious moon? And the congressman resolved immediately that JIMO, and a lander element, too, to scratch Europa's surface, were goals worth meeting no matter the difficulty, whether engineering or legislative. If more research was needed, then he would find the money to make it happen, because . . . it just fit. All of it. JIMO's place in the American space program and Europa's place in space, infinite space, wondrous to behold and seeded by God with life. The more he learned about the origins of the universe, its growth and expansion and evolution—looking at the Prometheus-1, looking at the diagram of Europa and the descriptions of its chaotic, cracked crust—you could see the perfection and genius of God's creation there, just as you could see it in every direction in the night sky, and those images from Hubble were snapshots of breathtaking arrangements of the same periodic table of elements found here on Earth, and the same fundamental rules of gravity, of thermodynamics, of physics, of mathematics. If life was here, it was there, in those other galaxies, and on the ice moon of Jupiter. Had to be. John was certain of it—perhaps more certain now than even the men giving the talks. But we needed to get there, which meant JIMO needed a patron—someone to push it firmly but gently through the legislative process. Congressman Culberson was resolved to do his part. And I mean, it was a sure thing! The Prometheus-1 was being built by the man who designed the Ranger and Mariner series of spacecraft, which, in the sixties and seventies, inaugurated an era of American exploration dominance of the inner solar system. The Rangers were instrumental in mapping the moon and enabling the Apollo landings; the Mariners mopped up the other planetary bodies. Mariner 2 was the first spacecraft to successfully encounter another planet (Venus), in 1962. Mariner 4 was the first spacecraft to successfully encounter Mars, in 1965. Mariner 10, first to Mercury, in 1974. Any single mission would have been an explorer's crowning achievement. Casani did them all. Then he launched Voyager 1, was the project manager on the spacecraft Galileo to Jupiter, and then Cassini at launch. He was an engineer who knew how to get past the hardest part of any mission. Three billion miles? That was easy. The first inch off the launch pad? That was hard. But he wanted to do it again, on this thing called JIMO that might—though John had resolved already, *would*—find extant life elsewhere in our solar system. There was a second Garden of Eden beneath that ice shell, and its discovery—no, its *confirmation*—would outdo even the Apollo moon landings in the public imagination.

By the end of the meeting, John Culberson had seen enough,

had seen his future and the future of humankind. He'd watched what the lab could do on Mars, knew what they'd done previously, seen the engineers who'd done it, and was given a glimpse of what they were yet to do, wanting only for a benefactor. He needed now only to insert funding into the federal budget, tell NASA to send it to the lab, and watch what happened next.

And he did and did, and what happened next, of course, was that NASA took the money and rather than send it to JIMO and Project Prometheus, the agency sent it to a Mars program where it would be used to draft rockets on drawing boards and design crew vehicles that might never travel beyond the International Space Station, if they were ever built at all. What happened was that NASA headquarters had decided to tug on Superman's cape. What happened, John would never let happen again.

CHAPTER 5

Station

TODD MAY GOT HIS FIRST JOB AT FOURTEEN AND FROM then on never went more than a week without working one place or another. The gas station job wasn't the first time he *worked*, of course; during summers as a boy he had mowed lawns in his Fairhope, Alabama, neighborhood and built up a pretty good business: one or two yards a day. This was just his first official job, and he made two-fifteen an hour checking oil, pumping gas, cleaning windshields. His mom and dad were big on instilling a work ethic in their boy, and in high school his football coach was inflexible on the subject: you worked during the summer. The gas station job didn't last forever, and he next found work at a farm, where he threw bushels of new potatoes onto trucks for fourteen hours a day, five days a week, side by side with migrant workers. He painted houses. He worked in a cement plant. He worked for a fiberglass pipe company in Biloxi. He worked in a chicken processing facility. He was the only white guy there, and he sat there all day long grabbing chickens as they passed by and hanging them up on hooks. So it was that before he was old enough to buy cigarettes, let alone beer, Todd had learned that he was a lunch pail guy. How to appreciate hard work and the people who did it. And he learned also that he was a pack mule, that he could work from sunup to sundown, too, sweat, do manual labor, and not think twice about it. And all that—the hanging of chickens, the hauling of cement sacks, the loading of potatoes, the scrubbing fingers free of fiberglass shards—almost prepared him to work for Alan Stern.

The two men moved to NASA headquarters in 2007. Alan had been hired to run the agency's Science Mission Directorate, which meant if it went to space and did science, Associate Administrator Alan could touch it—and he had a long list of things he wanted to touch. A prolific scientist who spent his career trying to send a spacecraft to Pluto, Stern saw a science program flat on its back: a beleaguered bureaucracy bereft of innovation, inattentive to researchers, and flying far fewer missions than it could. The cause: cost overruns. The James Webb Space Telescope, which would succeed Hubble, was proposed as a low-cost five-hundred-million-dollar project, leveraging technology developed during the Strategic Defense Initiative.[187, 188] The telescope was now flirting with five billion dollars.[189] The Mars Science Laboratory, meanwhile, endorsed in the 2003 Decadal Survey as the top medium-class Mars mission (i.e., it would cost less than six hundred fifty million dollars), had, by 2007, ballooned by one billion and counting.[190, 191] The new rover bothered Alan especially because it never had a prayer at coming in on budget. It would weigh five times more than Spirit or Opportunity—tilted the scales at nearly one ton—which meant everything from actuator lubricants to avionics software had to be redesigned and tested.[192] And the schemes to land the wheeled beast on martis firma were evolving into increasingly elaborate Rube Goldberg devices. Airbags alone or retrorockets wouldn't work. Parachutes wouldn't work. No combination, it seemed, of those proven landing technologies would work. Simulation after simulation created crater after crater. What they finally proposed went way, *way* beyond the lab's trademark So Crazy It Might Work. They designed an autonomous landing platform that would be delivered with parachute and retrorockets. Twenty-five feet above the Martian surface, it would stop, float in midair like a flying saucer scoping out a cow, and from there become a sky crane, lowering the rover onto the surface before cutting its tether and blasting away. This was certainly ambitious, but it was never "medium." And Jet Propulsion Laboratory had really boxed headquarters into a corner on this one, because once any mission— but especially one to Mars—crossed the billion-dollar boundary, the world started watching. Failure, in short, would be untenable, would set the program back by decades. So it had to work.

But whether you wanted to stare at stars or roll tracks across rusty alien dirt—difficult endeavors, both—Stern wasn't taking excuses today. Setbacks may not have been your fault, but they were your responsibility. You were engineers, so solve them without help from the Bank of NASA. After all, the American space program was a Nice to Have. Was the future of humanity to live on other worlds? Almost certainly, but in the present and on this one, there was a Great Recession and a housing bubble beginning to burst, which meant when NASA went over budget, it had to do without, or rob from other internal initiatives. In space science, that meant scuttling disciplined missions to reward profligate ones.

Alan didn't like that way of doing business—he'd led missions that fell victim to the wastefulness of other people—and he intended to fix the problem. Not the national celestial malaise, or the fractured hand of Adam Smith, but the agency's acquiescence to projects with swelling price tags. To get more missions to space, he would bludgeon overbudget projects into submission: this is the funding you agreed to, launch on it, or don't launch at all, but the coffers are hereby closed. He had succeeded with New Horizons, his Pluto mission, which launched a year earlier. Why couldn't fiscal discipline work across the Science Mission Directorate?[193]

But Stern was an outsider when chosen for the job, and he knew it. Though he helped lift from Cape Canaveral one of NASA's most prominent, promising missions, he had never actually worked for the agency. (He ran the Pluto mission from the Boulder branch of the Southwest Research Institute, a private science outfit.) So to help him navigate the alien agency bureaucracy, he recruited to be his deputy and compass Todd May, a longtime NASA manager at Marshall Space Flight Center. An insider.

Todd, a classically trained materials engineer, had been with the agency for eighteen years by then, though working at NASA had never really been the plan. He was always going to be an engineer of some sort, and an Auburn University engineer if he could help it. His dad was an Alabama entrepreneur in the chemical and paper business, and Todd figured that one day he would do that, too. And his dad was a *real* worker, an inventive guy—held four patents in dual laminate plastic piping. Pre-May-père, the industry used pipes made of stainless steel because of its strength and resistance to corrosion. But resistance to corrosion was not the same as *doesn't corrode*. The secret to solving the corrosion problem, knew the elder May, was one word: plastics. The problem was that plastic lacked strength. So he invented a way of marrying a plastic lining to fiberglass outers so that a plastic pipe could be rhinoceros strong. That sort of bare-knuckle engineering suffused life at chez May. Once, when it snowed—which was almost never on the Gulf Coast, but it happened—his dad got this big piece of clear plexiglass, put it in the oven, and then bent it into the shape of a sled, and—well, it was the coolest sled in the neighborhood. When Todd was a little older, his dad made him a plexiglass skateboard, two layers, clear on the top and this marble pattern on the bottom. So Todd grew up around materials, had an intuitive understanding of their strengths, would even chew on things—just gnaw gently in an improvised effort to interpret intrinsic material properties, and it was normal, this homespun structural analysis, a thing you did as a kid.

When he enrolled at Auburn, however, he checked the box marked "electrical engineering." He knew even then that it was a mistake, and one day, while sitting in a circuits class, it hit him hard: you could spend a lifetime doing this, and you'd die of old age never having seen an electron actually traveling down a wire. He liked stuff he could touch, really feel. The very culture of electrical engineering felt wrong—right for some, but not for Todd. So, when reps from the materials engineering program came along and pitched the field of the physical, Todd called his dad and took the leap.

Until that switch, May's academic performance might have measured as mediocre to good. He was the class clown in high school and an infrequent presence in the classrooms of Auburn. He'd turn up to learn the exam dates, and, thanks to a photographic memory, would read the material the day before and appear for tests just long enough to bubble in the correct answers. His heretofore most comprehensive science at Auburn had involved the tasting of the salts and occasional ethyl alcohol experiments with his fraternity brothers. But when Todd switched—returned home, really—to materials engineering, he was all in. Devout and born again. Loved it. Straight-A student. Not only attended every class but *exhausted the course catalog*. He took every materials engineering class offered by the best engineering university on the planet, and when they finally made him don cap and gown, he'd completed his undergrad with work just shy of a doctorate. He was an ABD: all but dissertation. He'd aced every doctoral course, but minus that *D*, lacked the degree. Todd figured he could always go back later and pick it up, and anyway, he already had a great job lined up and was ready to get to work. Atlanta Gas Light, a natural gas wholesaler, had hired him to help work out the bugs in its system—literally. Actual bugs. The company's pipelines were made of a particular polymer that, it turned out, grubs found delectable and would eat right through. It was a public hazard, and they needed someone to reformulate the material such that it wasn't grub grub. It was a good materials job and they made him a good material offer. But before he started, a friend who worked at Marshall gave him some advice: forget fighting this boring battle against beetle larvae and come to NASA. The agency needed a guy like Todd, this fellow who looked like a linebacker, who had Einstein's eyes and von Braun's brains. And what kind of engineer says no to NASA? Especially an Auburn engineer. An Alabama engineer. It's where American space flight was born—Huntsville, of all places!—where von Braun's bevy built the Redstone rocket that sent Alan Shepard to space, and where was born the behemoth Saturn V that put Armstrong on the moon.

Todd joined the space program by way of Grumman, an aerospace outfit under NASA contract to build the ever-embattled Space Station *Freedom* (at the time little more than blueprints and moving speeches). Grumman paid a lot less than Atlanta Gas Light, but he had to contend also with fewer larvae. Once at Marshall, he worked in its Materials and Processes Laboratory, and he was in heaven. Materials: anything that has mass and occupies space. Processes: anything you do to it. From that lab, you could go wherever you wanted to go at NASA, work on whatever project you wanted to work on. It was right there in the name!

They didn't give Todd the keys to *Freedom* on his first day. His first project was to develop a new database organizing the extreme temperature properties of all the metals in NASA's materials archive. MAPTIS, it was called, an acronym for Materials and Processes Technical Information System, and, really, it was grunt work—the sort of thing you give the FNGs, but he loved it, too.

The idea was ... so everything has room-temperature properties. And different materials behave differently at high temperatures. Yield strength goes down (i.e., it might deform sooner), or it might get more ductile (i.e., it might stretch easier before breaking), or its actual tensile strength might lower (i.e., it might withstand lighter loads). Conversely, when some materials reach low temperatures, their crystalline structures shift, and they become brittle. Every metal has a ductile-brittle transition temperature. Drop below it, and trouble ensues. This means an aerospace engineer might design an aircraft in the steamy South and then fly it to Alaska and, oh, the landing gear snaps off—shatters like a rose dipped in liquid nitrogen. A materials engineer makes sure that doesn't happen.

May spent six months on the database, found every possible reference on every material Marshall ever used, added them to a file, and documented the source. It had to be thorough, and it had to be accurate. Launch, space operations, reentry—extreme temperatures were sort of NASA's *thing*. When he was hired into the agency full-time, he presented his *magno opere materiae* to the deputy director of the laboratory. Todd was proud, and his superior was impressed. It wasn't a huge thing in the context of moonshots and space stations, but it got their attention, would make a lot of jobs for a lot of engineers a lot easier going forward, and proved, if nothing else, that this man May could *work*.

IN 1992 SPACE station *Freedom* was reformulated. The program had been managed previously from facilities in Reston, Virginia, with development work divided among NASA centers various and sundry. The station design was expensive but credible—an expandable orbital platform for science, medicine, and manufacturing. In the near term, it would be a space laboratory, but it would eventually serve as a layover for astronauts on their way to the moon.[194]

You could assemble, service, and support lunar rovers and space-craft from there, and in the very long term, do the same for Mars-bound crews and cargo—a literal station, as for trains and busses. It would be pricey, yes, but not wildly out of line with what NASA did every day. When it was first formally proposed by Ronald Reagan, the agency estimated that the station would cost eight billion dollars.[195] That kind of cash request was not well met, but Congress appropriated sufficiently to stand up the program and see what might come out of it. Funding continued into the Bush administration. This wasn't driven entirely by a love for NASA or a whirlwind congressional crush on a flying bus stop. Between defense cuts and two landmark US-USSR nuclear accords—the 1987 Intermediate-Range Nuclear Forces Treaty and the 1991 Strategic Arms Reduction Treaty—we just weren't building nukes like we used to, which was fine from an everyone-not-dying-of-strontium-90-in-the-black-rain point of view, but, look, if the Commies somehow pulled themselves together, broke a beer bottle, and wanted to dance, we needed to be ready.[196] Defense contractors would have to stay warm. So a space station was win-win. It would employ tens of thousands from countless congressional districts across the country, and it would prevent the defense-industrial base from withering away. That NASA would get something it had been asking for since the days of von Braun was a bonus.

But by 1990, the Berlin Wall was sledgehammered and détente the new norm for foreign policy. The United States slouched into an economic recession, and deficits and the national debt were rising. As the plan for *Freedom* matured, meanwhile, the cost increased. Inflation didn't help, and NASA was now finally factoring in the cost of the space shuttle as well, which was, itself, pricier per pound to launch than anyone liked. In all, *Freedom* would cost thirty billion dollars more than previously planned,[197] and was now gnawing mightily into the planned Cassini mission to Saturn.[198] Cassini's twin, CRAF (the Comet Rendezvous and Asteroid Flyby mission), had already lost its race with the reaper. But Ivan was out there, waiting—what if? So station survived.

The Soviet Union did not, however, and suddenly, and for the opposite reason, the United States *really* needed a space station now, because the only thing worse than a Russian rocket scientist making missiles is a Russian rocket scientist out of work and weighing options.[199] *Why not freelance?* A man without a country could make a nice living abroad, with American antitheticals paying princely to learn long-range rocketry.[200] It had happened before. It's how *we* had a rocket program! After World War II, Germany's best rocket scientists absconded to America and the Soviet Union. Wernher von Braun was one of them and basically built our program from nothing. So the United States government decided to put those Russian rocket scientists to work helping build the space station. Bonus: the secrets of the Soviet aerospace sector were among the most coveted intelligence during the Cold War. That sort of information didn't come easily and it didn't come quickly and it certainly didn't come cheap. But now for a pittance—whether sixty million or six hundred million—pocket change either way—we could just . . . buy their scientists![201] And NASA could cut shuttle launches, too. The Russians could use their bargain-bin (but really reliable) rockets to help launch the space station *Freedom*. Well, there was no sense in being a sore winner. They'd help launch the *International* Space Station.

In 1993 NASA closed the Reston facility and relocated management of the station to Johnson Space Center in Houston. By then, eleven billion dollars had been spent, and the station existed mostly in a filing cabinet.[202] Todd joined its program office to take over as space station materials lead. He was the government pay grade equivalent of "just some guy," yet given a position of leadership in a high-stakes, high-profile project that absolutely needed to fly, and that kind of thing just *didn't happen at NASA*, but have you ever been to Houston in July? Have you ever sat in its traffic? None of the agency's senior executives wanted to move there, so they took a chance on Todd and other youngsters. Relative to the ever-graying agency, the program office at Johnson ended up teeming with indomitable twenty- and thirty-somethings. No one really had any experience with failure, and no one had limits—or at least, no one knew what his or her limits were.

The space station was modular. You built it one room at a time and launched each into space to connect like Legos, an ever-expanding orbital facility. Propulsion, airlock, docking port, laboratories, habitat, solar panel segments—it would be the most complex construction project since the pyramids, the modules divided between former mortal combatants with very different ways of doing things, and the first time the completed components could come together would be two hundred miles above the Earth and orbiting at eighteen thousand miles per hour. And it all had to work the first time.

That part was on Todd. To ensure that Russian modules worked with American ones, and that everyone remained on the same timetable, NASA established management elements devoted to each country's components. The idea was to stay au fait with what the Russians were doing so that there would be no surprises during development. In 1994 Todd transferred to the Russian integration office, and they later made him deputy program manager. He helped sustain synchronicity among engineers foreign and domestic, unifying structural models, thermal models, delivery dates, data exchange methods between modules, and hardware compatibility. They never stopped. When flying halfway around the world to Russia or back, the engineers would land on a Sunday and start on Monday morning. Smoking was still allowed on those international flights, and you would touch down with bloodshot eyes from hazy cabins. You'd work with the Russians for nine hours, drive back to the hotel, and muster for a ninety-minute daily debrief. Then you'd start thinking about dinner. Repeat. And the flight home, they called it the "baby flight" because American couples were ever keen to adopt Russian babies, and the whole way back, those babies would cry. You'd fly straight to Houston, not a REM among Todd and his crew, be up at six in the morning, have breakfast, and be at Johnson by seven for meetings at eight.

It was a hard haul. The Russians were way ahead of NASA in the space station business, having already built and flown *Mir*, their own modular station, for a full decade, beginning in 1986. Nobody spoke the same language, the cultures were alien, and everything had to go through interpreters. There was friction, fatigue, diplomatic formalities—but also great fun. These engineers, Russia's and America's, everyone was so young, and they just plowed through. And what engineers the Russians were! Classically trained, innovative, and you're bleeding together in the trenches, and part of this huge thing, and NASA, not necessarily fairly (but not necessarily inaccurately) described sometimes as ossified or entrenched, is somehow being changed by all this. You could just sense it. In Washington, the station was beset on all sides—too hard, too expensive, "welfare for engineers," a swindle, really, a boondoggle![203] *It would never launch!*—but in Houston, there was never any doubt that it would reach the launch pad, that it would work. No one was halfway. You were there and you were all in.

In 1998 the first module launched. CNN covered it live, all the operations, the countdown from Russia, everything. It was the first piece of the International Space Station, and others would follow. Todd returned to Marshall that year and took over as project manager of the Quest airlock module—the thing astronauts would use to exit the station for spacewalks. If he never did anything else in his career except work on that airlock until it flew, he would sit around and tell his grandchildren about it—that a part of *him* was up there.

In 2001 the airlock launched from U.S. soil on the space shuttle *Atlantis*. CNN made brief mention of it and went on with the day's news. For his part, Todd, riding high from his airlock achievement, was set to take over the habitat module, where astronauts and cosmonauts would dwell in their orbital manor, but before he could lay hands on it, budgets proved unfriendly, and NASA figured out as an agency that it just couldn't afford to build it. Right now the space station wasn't as important to the nation as Todd thought it was. He took it personally.

When a project flies, a project manager's work is done. May considered returning to Auburn to work on his dissertation, but was enticed instead to take a job working on a spacecraft to test Einstein's theory of relativity. It was called Gravity Probe B, and no mission in NASA's history had been so long in development—forty years, by that point.[204] The probe had gained a reputation as a star-crossed spacecraft sentenced never to launch and had been canceled in 1989, 1993, and 1995.[205] Todd would become its program integration manager. Clearly, he was the guy for the job, having served the role previously on a project where ordering lunch required an interpreter, and also having launched a module of a space station that most thought would absolutely never see space and was constantly under threat of congressional cancelation. For the next three years, May spent his time protecting and pushing along the development of the gravity probe, laying the rails on which it would glide to the launch pad. When it lifted off in 2004, Todd had another win under his belt—and, again, had to figure out what to do next.

Every morning when he opened his medicine cabinet, he saw a clipping he'd taped there of the Auburn Creed. It was written in 1943 by George Petrie, Auburn's first football coach. Petrie was a linguist and history professor—spoke five languages—and had seen his very first college football game in the 1880s while a student at the University of Virginia. He taught modern language and history at Auburn from 1887 to 1889, when the school was known as the Agricultural and Mechanical Institute of Alabama, then decided to go off to Maryland to earn a doctorate in history at Johns Hopkins University.

That is where Petrie really fell in love with football. He also became enamored of "scientific history," which emphasized fieldwork as much as bookwork. You want to tell a story, you dig deep, drill into the research, go to the places where events occurred and get local dirt under your fingernails. You sit across from the experts and break bread with the people who were there. There was nothing easy about that kind of work, that kind of history, but there was an honesty to it, an almost magic to it, a prestidigitation, a conjuring of the past as it was lived so that the future might know: this is what happened. Beyond secondary sources and thirdhand accounts. *This is what happened.*

Petrie returned to Auburn in 1891, determined to introduce both football and scientific history to the school, and he did. He coached for a single season. The Tigers won the first game they ever played, 10–0 over the University of Georgia, and finished the season 2-2. Not bad. He promptly handed over the whistle to focus on his professorial work for the next fifty years, not retiring until 1942 at the age of seventy-six. The following year, Petrie wrote the Auburn Creed. He wanted to craft something that reflected the qualities of the school he'd spent fifty years with, and something, perhaps, to serve as its lodestar.

The Auburn Creed begins: "I believe that this is a practical world and that I can count only on what I earn. Therefore, I believe in work, hard work." It continues, a quiet reflection on the values of education, honesty, truthfulness, God, and country—even the human touch, "which cultivates sympathy with my fellow men and mutual helpfulness and brings happiness for all." It ends: "And because Auburn men and women believe in these things, I believe in Auburn and love it."[206]

Todd—systematic, disciplined—still had every single test he had ever taken at the university, every sheet of green engineering paper, his notes, his derivations for engineering problems sometimes going on for three, four pages. It was Auburn that taught him the love of solving very difficult problems—problems that, at first glance, objectively *could not be solved*. The love of systematically breaking down a problem and managing one piece at a time until the job was completed. It was what he'd done, what he still did every day, with problems that were life or death for astronauts and cosmonauts, and for robotic spacecraft that did serious and pressing science. But he had unfinished business in academia. And just when he decided to go back to Auburn and complete his doctorate—he couldn't think of anything more interesting than being called Dr. May—his phone rang.

NASA wanted to move the newly combined Discovery and New Frontiers program office to Marshall, and Todd . . . we want *you* to be program manager.

This was—well, it was a decision to make. It was an honor. Planetary science missions funded by NASA were organized generally by expense and science return. A flagship-class mission such as Galileo or Cassini could be built and launched only once per decade—optimistically, because they plowed boldly beyond budget caps but did commensurate science. At the other end were the Discovery-class missions, their price capped around five hundred million per. They were the haiku to the Homeric epics that were flagships. Each solved precise problems, settled some specific celestial quandary, uncovered some unknown unknown.

The Discovery line allowed NASA to cast its net wide across the solar system with minimum investment. Each spacecraft opened doors. At the dwarf planet Ceres, for example, the Dawn spacecraft found hydrated minerals consistent with an ancient ocean sitting on its surface. Where did that ocean go? Was liquid still there today? It was anybody's guess, but because of Dawn, some future mission now knew what experiments to carry to figure things out. The Discovery program could launch a mission every two years, and maybe *two* missions every two years, to the moon, Mars, asteroids, comets, and who could say where else. Engineering pluck was the only limit.

And with that, Todd decided, the doctorate could wait a little longer. He was an Auburn man, and he was needed.

As far as NASA programs went, you couldn't do better than Discovery. Every three or four months, a mission was launching, or an existing one was encountering some strange new world, or another was being selected for further development or receiving science data from deep space. In exchange for his mission and management experience, May had been given the portfolio equivalent of a theme park: all roller coasters and cotton candy. Discovery had a superb success rate, and its missions just couldn't stop delivering these fantastic firsts.

The same could not be said for the fledgling New Frontiers, a new, middle mission class between Discovery and flagship, still finding its legs and with only a single project yet approved by 2004: a Pluto flyby called New Horizons being led by a scientist named Alan Stern.

Todd had heard the Pluto mission was having problems, and to get a grip on the state of things, he took time to visit the Southwest Research Institute in Boulder, which managed the mission's science. While in town, he went to Ball Aerospace, which was building one of the Pluto spacecraft's key science instruments. He went to Southwest Research's San Antonio office, where the New

Horizons payload was being managed. And he went to APL in Maryland, which handled the overall project and spacecraft operations. There Stern and the New Horizons team gave Todd a full project review.

Alan was one of the highest-profile planetary scientists in the field, tenacious and proven. He started his presentation with a slide of a postage stamp that had been issued by the U.S. Postal Service in 1991 as part of a series celebrating space science. The United States had been the first country to reach each planet in the solar system, and the only country to reach *every* planet—except one. PLUTO, the stamp read. NOT YET EXPLORED.[207] Oh, how that incensed Stern! This was more than a twenty-nine-cent collector's curio or postage payment for the water bill; it was a philatelic swipe at science itself! That stamp was unacceptable, and if Alan was driven before as an American explorer in the New Age of Discovery, he was implacable now and would hoist a Jolly Roger up the mast if that's what it took. An entire planet in *our* solar system unexplored? That would not do at all.

Todd sat quietly in the presentation, taking notes, taking in the New Horizons team, assessing Alan's boldness, the mission leader's brains and heedless devotion to this singular task. It infected Todd, too, right there, on the spot, because when you listened to Alan, you realized that this project was more than about flinging a probe three billion miles. This was a khaki-clad-tiger-tooth-necklace-binocular-at-the-ready-pistol-and-bullwhip-machete-jungle-slashing *adventure*. The wilderness must be explored![208]

By the end of the project review, Todd became convinced that Stern would launch this thing even if he had to fire it from a cannon, Jules Verne style. This wasn't business—it was a crusade. Stern had been planning one Pluto mission or another since the eighties, fighting fervidly and sometimes bitterly with NASA along the way. He and his team would dig in and draft a detailed proposal hundreds of pages long, and it would be declined, postponed, or offered only tepid support before cancelation. Again they would dig in, draft a detailed proposal hundreds of pages long, and it would be declined, postponed, or offered only tepid support before cancelation. And again they'd dust themselves off, and draft, again, yet another proposal, but there was always some better place to go than Pluto, some excuse to ferry funds elsewhere. It happened five separate times. New Horizons was number six, and though Stern could do this for the rest of his life—what is a bureaucratic behemoth next to a steadfast scientist?—celestial mechanics made this the last serious shot at getting a mission on the pad. If the New Horizons spacecraft didn't make its launch window of January 2006, Jupiter would be out of alignment for a gravity assist. The farther Pluto proceeded in its orbit, meanwhile, the greater the likelihood that its atmosphere would collapse—would freeze away as it eased outward from the sun—not to return for another two hundred years.[209] New Horizons simply had to launch.

After the review, Todd was convinced of something else, too: New Horizons was not going to launch on time . . . if it launched at all. Alan's team had a lot of problems, not all of which were within their control. The internal ones could be managed and corrected quickly, but as for the external ones, well, Stern and Co. were having trouble getting the attention they needed from NASA headquarters, who they sometimes tended to alienate. So Todd called Mike Griffin, who ran the space department at the Applied Physics Laboratory. Griffin was a smart guy, an important guy, held an undergraduate degree in physics; five master's degrees, in aerospace science, electrical engineering, applied physics, business administration, and civil engineering; and a doctorate in aerospace engineering. New Horizons fell under Griffin's aegis, and he needed to know what was going on.

The call did not go well.

TODD: Look, you've got some trouble, Mike. This thing is not on a path to success—

MIKE: I can't talk right now, Todd. I've got some congressionals in here. I'll call you back.

TODD: OK.

[Click] [210]

Todd placed the receiver in its cradle, sat back in his chair, and turned to Paul Gilbert, a colleague and program integration manager.

"He's not going to call back," said Todd.

Griffin didn't want to hear this news, Todd surmised, because it was bad and would be embarrassing for a lot of people, and he (i.e., Griffin) didn't want to be one of them. And he was big-dogging Todd! *Congressionals in there.* Right.

So during his lunch break, Todd decided to cross something else off his to-do list, and it wasn't quite as thrilling as exploring the cosmos, but he asked Paul to take him to NAPA Auto Parts. Todd needed to pick up a radiator hose for his 1998 Range Rover. The car was seven years old now and wasn't starting right, and Paul had figured out that the hose was leaking fluid onto a spark plug. Off they went, and while on the road, Todd's phone rang. It was a 240 area code. Maryland.

TODD: Hey, pull over so I can take this.

[Answers phone.]

TODD: Hel—

MIKE: You're saying to people we aren't on a path to success? Are you trying to get this mission canceled? Are you trying to get me fired?[211]

After Todd talked Griffin into putting down the gun, he spent forty-five minutes describing the things he had seen in and around New Horizons. Not that it *can't* be successful, Mike, but it's *not going to be* without an intervention. And Mike listened carefully and at the end said: OK, then. Intervene.

Click.

Todd ordered an independent ninety-day analysis of New Horizons, stem to stern, spacecraft subsystems to plutonium production. The review revealed six showstoppers, each of which made the others more menacing. It was like a kung fu flick, only this time the bad guys weren't taking turns. Results in hand, Todd went to Alan and gave it to him straight: You're not going to make it. And not long after that, Todd was back at Marshall, when Paul caught him in the hallway.

PAUL: Hey, our buddy has just been announced NASA administrator.

TODD: Who?

PAUL: Mike Griffin.

TODD [*confused*]: No, no, no, no. That's this other guy—

PAUL: No, no, Todd. *Our* Mike Griffin.[212]

They found a computer and googled it. It was their Mike Griffin. So the congressionals . . .

When you're the new administrator of NASA, you visit Marshall Space Flight Center, the heart of American rocketry. And the center director of Marshall had known that Todd—who certainly didn't merit the attention of the administrator on his own accord—had worked with Mike previously. Personal connections going a long way, he told Todd to put together a presentation on what they were doing at the center and how they were handling the Discovery and New Frontiers program.

When came time for the meeting, there were no empty conference room chairs on the ninth floor of the Marshall headquarters
building. With NASA administrator Michael Griffin sitting at the
head of the table, Todd began his brief presentation.

> TODD *(basically):* Good morning, sir! We've got this new
> Discovery and New Frontiers program office. We're using
> risk-based insight methodologies and lean management
> principles and—
>
> MIKE *(basically):* I don't want to hear any of that. I want to
> know what it's going to take to get New Horizons off the
> ground and launch in its window, and I need you to tell
> me anything I need to do. Who do I need to shoot to
> make it successful?[213]

This left Todd quaking, because Griffin's words were stronger
and more colorful than that. But it was also encouraging, because
Todd knew that Mike meant it.

So that weekend he wrote a letter to the *administrator of NASA*
explaining everything that needed to happen. And Griffin, as it
turned out, forwarded the letter to basically everybody. He tasked
the agency's chief engineer, Rex Geveden (a Marshall man), to help
Todd help Alan get this thing off the Earth.

They ate the elephant one bite at a time, broke the Pluto problem into its constituents and built teams to take on each of them.
Like any bureaucracy, NASA was slow to move, but when it did, it
had the mass and might of a freight train, and woe to anyone who
dared stand in the way. When contractors caused trouble, Todd had
the agency squeeze them into submission. Nuclear launch approval
was particularly thorny. The spacecraft New Horizons used a radioactive power source, solar panels being useless three and a half billion miles from the sun. It wasn't a reactor—that dream died with

Prometheus—but rather was driven by radioactive decay; the plutonium produced heat, which the generator converted into power. But any time you launched something radioactive, an independent panel had to be formed from members from the Nuclear Regulatory Commission, the Department of Energy, and interested parties, and they would write a safety assessment report. Somewhere along the way, the panel and the New Horizons team had locked horns and egos, and the panel affirmed its authority by insisting upon a malicious compliance with regulations—and when it came to nuclear, there were an awful lot of regulations with which to comply. Suddenly they were asking for things like obscure thirty-year-old Russian reports on plutonium reactions in certain environments. Well, two could play that mulish game, and the project team procrastinated. Which was fine, except the nuclear panel wasn't trying to meet a launch date. So Todd set the agency on finding someone to manage the process, and he focused on other things.

On January 19, 2006, New Horizons launched.

It made sense that Mike Griffin would hire Alan Stern away from Southwest Research Institute and have him run the agency's science missions. Like Griffin, Stern was a member of the diploma wallpaper club—held five degrees that covered every aspect of spaceflight: physics, astronomy, aerospace engineering, planetary atmospheres, and planetary science. And if Alan was breathing, Alan was working, which was good because at the Science Mission Directorate at headquarters, there was a lot of work to do. There were ninety-three flight missions in the science portfolio, about half in development. Alan would oversee all of it.

And he had an agenda. Mars, Alan believed, was overwhelming the program at the expense of the entire solar system.[214] There were the satellites: Mars Odyssey, Mars Express, Mars Reconnaissance Orbiter, Mars Atmosphere and Volatile Evolution; the rovers: Opportunity and Spirit, and the behemoth under construction, Mars

Science Laboratory; and a soon-set-to-launch lander, Phoenix. You were looking at more spacecraft at Mars than at every other planet beyond Earth combined.[215] But as wide as that footprint was, it was the endless thirst for dollars required by Mars Science Laboratory that was throttling the planetary science program. This was a zero-sum game: MSL's cost overruns meant that Discovery missions—prolific and profitable scientifically—would be delayed or canceled outright. Mars had killed at least one Pluto proposal previously and contributed to Europa's inability to get out of the gates.

Mars needed a reality check, and Alan had no qualms about adjusting its attitude. He didn't mind being viewed as the villain if that's what it took to bring balance to the planetary program and get costs under control. And viewed as the villain he was.

CHAPTER 6

Maestro

KARLA CLARK HAD SPENT THE LAST DECADE WARGAM-
ing Europa invasions—developing mission concepts—brooding over
notional spacecraft designs—science payloads—power sources—
trajectories—communications—mission operations—with each effort attacking the problem from some previously untried vector, and
when the Quad Studies started in 2007, she was the natural lead at
JPL.[216] When NASA wasn't writing checks for such studies, the
lab covered costs, and the mountain people of Saint Gabe had cultivated over the years all the engineering expertise and insights necessary to do this thing, were *ready to go*, wanting only a nod from on
high. But Curt Niebur had made it clear that this time in particular
Europa had better bring its A game because no nod was assured.
Yes, the Decadal Survey said DO EUROPA, but *NASA had already
tried;* success or not, that box was checked. Cassini's electrifying
Titan discoveries weren't to be dismissed lightly, and that moon's
team was hungry and motivated, wanted to know more. The Saturnian system had momentum, without even mentioning Enceladus
blasting its ocean into space. Europa has a subsurface ocean, sure, but

good luck getting through that thirty-kilometer ice shell. At Enceladus we could basically fly through the fountains, fill a bag, and ferry it home.

The point is, said Curt: *Europa, do not underestimate those teams.*[217] Karla got the message.

In the fifties and sixties, the United Kingdom experienced what the Royal Society in London called exquisitely and with uncharacteristic alarm a "brain drain."[218] Statistics seemed to suggest that science scholars were slipping across the Atlantic for more favorable funding and better research billets.[219] And why wouldn't they? If America's anticipated atomic-age Shangri-la of side-finned, fusion-fueled flying Cadillacs didn't materialize, then the post-Sputnik space age assured spots for scientists on every NASA moon base and Mars colony soon to see construction. The U.S. National Science Foundation, meanwhile, reported with zero modesty that the "American scientific community could continue to absorb foreign scientists at approximately their present rate of entry for some time."[220] The British government was spending more annually to subsidize chicken food than serious scientific research.[221] NASA, meanwhile, had a blank check and Wernher von Braun.

Those were the conditions that in 1960 brought Karla Benjamin's father, a Welsh research chemist, and her mother, who managed the household, to the city of Cincinnati, where Karla was born and raised. Dad did research for Procter & Gamble, hopped the globe on behalf of the conglomerate, and when foreign scientists came to town, standing dinner invitations kept the Benjamin household lively. Via these kitchen table cultural exchanges, these spontaneous scholarly symposia, Karla came into her own.

She attended Rice University for her undergrad, double-majoring in chemistry and chemical engineering, the first because she loved the science—and especially the quantum chemistry subfield—and the second because she loved the idea of gainful, meaningful employment, which the quantum chemistry subfield provided sparingly outside of academia's grind. After graduation, Hughes Aircraft Company in Glendale, California, made her a solid offer, and she accepted, working on flight batteries for communications satellites. Thus Karla Clark (née Benjamin) joined the American space program. And while working during the day on flight projects (i.e., things launched or launching to space)—no small task, spacecraft batteries, power being king in space exploration—at night she attended classes at University of Southern California, which had a great graduate program for working professionals. This time she studied and earned master's degrees in both engineering management and mechanical engineering with a focus on the thermal subfield.

In 1987 Karla jumped ship to Jet Propulsion Laboratory, just down the road from USC and Hughes. She was hired as a battery engineer for the lab's flight missions, including the prospective Cassini project, though it wasn't called Cassini at the time. The job overall entailed designing, procuring, and delivering spacecraft batteries. Since Hughes launched communications satellites all the time, she was one of a handful of such engineers in the entire lab with any previous flight experience. JPL projects were big but launched only rarely, and the battery group became closely knit, supporting the lab entire and representing it at other NASA centers across the country, as well as at agency headquarters and in various prestigious working groups that kept America at the forefront of space science. It was just an unbelievably good deal if you had ambition and knew what you were doing, and within three years, Clark had been to just about every NASA center and learned from peers working not only in batteries but also in complete power systems. In addition, she had joined the review panel for the Hubble Space Telescope and been thoroughly educated in how flight projects worked from a systems perspective. She was on her way in the world, and others.

Eventually the Cassini team did away with the batteries in its design, but a position for power subsystem engineer opened, and Karla asked the project management to take a chance on her, and they did. It was a pretty important job, and she held it for most of the spacecraft's development, eventually becoming assistant technical manager and then technical manager, responsible for a team of twenty.[222] By the time Cassini launched, she was responsible for its power subsystem, having seen it through assembly, test, and launch operations (or ATLO, and pronounced that way), the final, critical phase of the project.

In 1997 lab leadership asked her to become project system engineer for the nascent Ice and Fire program to develop low-cost missions to three difficult destinations: Pluto, which required a spacecraft to fly twenty-three times faster than a bullet for a full decade, culminating in a precision flyby lasting three minutes at most; the sun—specifically, the twenty-five-hundred-degree deep interior of its atmosphere, which randomly reaches outward explosively in every direction; and Europa, which was considered *the hard one*.[223]

Clark was responsible for everything from the look of the spacecraft to their essential needs—computer systems, power systems, structural design, deep space communications, launch vehicle, science instruments—and she drew on trade studies of each to pull together basic mission concepts. It was Karla's first detailed introduction to Europa, and from the start, the radioactive badlands surrounding the Jovian world—there was more radiation there than would be found immediately after total thermonuclear war—vexed and confounded the Ice and Fire team.[224] Any spacecraft's computer and delicate scientific instruments would need heavy shielding, which was doable but for a thorny mandate from NASA headquarters: the Europa orbiter was to fly directly to Jupiter; there could be no gravity assists along the way.

Few impediments could have been more severe. For a spacecraft to reach the Jovian system with enough speed to eventually achieve orbit around Europa, it had to either launch from a powerful rocket (which NASA lacked, limiting spacecraft to a space shuttle deployment) or be absurdly light (which the required radiation armor rendered impossible). JPL engineers dashed out hastily written equations in chalk before driving fists against blackboards in fits of despair.

Nothing for NASA was *ever* free . . . except for gravity assists. Ordinarily, the agency could compensate for the meager speeds of heavy spacecraft by taking indirect flight paths and using planets encountered along the way to yank and shove the robotic pilgrim outward, inward, or onward.[225] The laws of physics being immutable, and the salient numbers known, NASA's orbital dynamicists could do this all day, running the numbers to sling spacecraft precisely, one planet to the next: free propulsion from Isaac Newton.[226] It was incomparably the best bargain in space exploration.

But then television tabloid journalism got involved, and everything became complicated.

In 1997, while waiting at Cape Canaveral for liftoff, the Cassini mission was beset suddenly by political protest. Cassini carried three radioisotope thermoelectric generators, which were powered by the decay of plutonium 238. The plutonium wasn't of the *Back to the Future* variety—a disquieting drop of Scary Substance Indeed into a homemade flux capacitor—but rather was stored in a ceramic form, wrapped in iridium, and caked in graphite. It could not corrode, or be obliterated by heat, or vaporize, or disintegrate as an aerosol, or dissolve in water. It was made to withstand not only the explosion of the rocket carrying it, but even a catastrophic reentry into Earth's atmosphere. Because it couldn't vaporize, in a disaster situation, no one would inadvertently breathe it in and develop superpowers or extra appendages. In fact, it was designed so that you could even eat the stuff.[227] The human body could not absorb it.

But ten days before three and a half million pounds of rocket thrust put inches between Cassini and Earth, a much smaller number—sixty, as in *60 Minutes*—nearly nailed NASA to the ground. The CBS TV newsmagazine aired a feature on the soon-set-for-Saturn spacecraft, Steve Kroft starring in the segment. The correspondent's opening line: "On October thirteenth, a Titan IV rocket is scheduled to lift off from Cape Canaveral carrying seventy-two pounds of deadly plutonium; enough plutonium, in theory anyway, to administer a fatal dose to every man, woman and child on the face of the Earth several times over."[228]

And it got only worse from there. Cassini was an afterthought in the story, and interviews from experts were interspersed with comments from . . . nonexperts, to be kind, but very well-spoken nonexperts, whose contributions—the generous ones!—included lines such as: "What gives anybody, including the federal government, the right to risk the population's death or—or injury just for space exploration?"

The segment featured a plutonium expert from the Department of Energy stating flatly that even if the rocket, spacecraft, and graphite-sealed, iridium-wrapped, ceramic plutonium blew up on the launch pad, it was literally impossible for the debris to do what protesters said it would. But just to be balanced, Kroft's menagerie of doomsayers described in lurid detail what plutonium—not in the form used by NASA, which you could safely sprinkle on your breakfast cereal, because, again, *you could eat it*—could do to the human body. Among the highlights: "it can produce pulmonary cancer" and "you could have numbers like one hundred thousand or more people who develop lung cancer" and "if there is such an explosion, you can kiss Florida good-bye."

Kroft even found a former NASA employee ("He's neither a

scientist nor an engineer," admitted Kroft, "but . . .") to lament publicly his role in endangering lives for such frivolities as space exploration. "I feel guilty, quite frankly," bewailed the penitent insider.

To seal the deal, Kroft intercut the story with snippets of an interview with Wes Huntress, head of NASA's planetary program, who had presided over the successful landing of Mars Pathfinder only months earlier.

"This is from your own environmental impact statement," said Kroft to Huntress—the tone of the host solid but affable, his countenance hard but eyes somehow benevolent. "I want to read you a couple of things from it."

Huntress was a pioneer in the study of interstellar clouds and one of the world's foremost experts in planetary exploration, but he was not exactly tabloid-TV material, and after the cavalcade of activists arguing compellingly and without interruption, he seemed less than confident in his responses.

Quoted Kroft: "If there's an accident it talks about, quote, 'removing and disposing of all vegetation in contaminated areas, demolishing some or all structures and relocating the affected population permanently.'"

"If there should be any such accident," said Huntress, accurately but unhelpfully.

Replied Kroft, "I mean, that sounds fairly drastic . . ." and Kroft waited patiently for Huntress, in possession of rope necessary to hang himself, to fill the silence, which *60 Minutes* interview subjects always did, and he did, and did.

"Well, the—what they're probably talking about mostly is—is the damage on site, near the—near—near the launch pad because there's clearly, when one of these things goes, a lot of damage near the launch pad."

And after Huntress tap-danced and staggered—*this guy didn't even know what his own official Armageddon report said!*—and at last swung gracefully from the gallows, well-honed doomsayers followed up, explaining precisely how Life as We Know It was drawing to a close, and kiss your babies tonight because our foolhardy quest to conquer the cosmos—Saturn! This pointless mission to a gas giant, whatever that meant—will leave mutated survivors fighting for the last canned goods on ransacked store shelves.

Worse yet, Cassini would take a second swing at the peaceful people of planet Earth! If it didn't blow up on launch, it was set to follow a VVEJGA trajectory to boost its way toward Saturn: that is, two swings by Venus (V, V), and then it would *play chicken with the Earth,* and if something went wrong . . . (but if all went well, from Earth [E] to Jupiter [J] for a gravity assist [GA]).

The Clinton administration really did not have time for this but dutifully absorbed the panicked letters and optics of protesters grasping concertina-topped chain-link fences on Cape Canaveral's perimeter, while on the inside, police lined up in body armor and carrying riot shields stared silently, just waiting to—what? Open fire? Brandish batons?[229]

Nevertheless, NASA went forward with its reckless rocket launch likely to leave only cockroaches crawling the Earth (or whatever some future species would call this planet), and things were fine, as they had been for previous launches dozens of times over. But the message from headquarters to those filing future space missions: if you must launch radioactive material, *do not* plan trajectories taking the spacecraft back to Earth for a gravity assist. Nobody needs the headache.[230]

Which meant, for Karla and company, years-long discussions on potential trade-offs for the Europa Orbiter mission, as it came to be called. They analyzed other trajectories, other launch vehicles—*anything* to get more mass for a suitable science return. What hardware do you make "rad-hard"—impervious to radiation (but expensive)—versus simply wrap in "dumb mass" (i.e., big blocks of

cheap protective shielding)? What was the absolute smallest science payload possible? Ultimately, they found a relatively happy
medium: a spacecraft that could launch direct and achieve the minimum science required to make a Europa expedition worthwhile,
and NASA loved it, and then the cost doubled, and in 1999 Ed
Weiler shot it dead. Just like that.

WHEN PROJECT PROMETHEUS came along, Karla was already
at work on a conventional Europa mission, building from the previous Europa Orbiter effort. She had, by now, more engineering
experience on that moon than most, having led or been part of
four separate studies. When she learned what the Prometheus
people planned to propose for Europa, she saw problems. She told
John Casani point-blank that his engineers were underestimating
the amount of mass necessary to pull off this sort of mission. The
shielding, the instrumentation: Europa wasn't the place you para-
chuted into with only a bowie knife and moxie—you went there,
you packed for war. She told him this again and again until finally
he pulled her aside and said gently but firmly to quit complaining
about it, and either come over and help him on JIMO or leave him
alone.

It was not a difficult choice. If there was something to learn
about doing big things in space, John Casani could teach you. What
hadn't he done? The first American spacecraft to land softly on the
moon? The one that would figure out if the descending *Eagle* would
actually land or just . . . keep landing? That was a John Casani
probe. The Ranger missions to map the moon? Casani. The Mariners, Surveyors, Pioneers, Voyagers, and Vikings? Also Casani.
Galileo and Cassini? Casani and his three fellow pillars of JPL had
touched everything, had seen it all, success and failure. The chance
to work with such an engineer and learn from him? Karla didn't

know if the project would last three months or fifteen years, but she agreed on the spot.

And she was not disappointed. Watching John Casani work was a master class in teaming and project management. The way he built bridges between scientists and engineers. His interaction with aerospace contractors. To see how he managed the political part of Prometheus, unifying major NASA centers (erstwhile and elsewhere mortal combatants for slices of the agency's pie). Integrating the U.S. Navy reactor cadre. The way he treated people as a team of interlocking professionals in service of the impossible, and how he forged bonds between them—Karla absorbed every lesson she could.

What she learned remained relevant after the later deaths of Prometheus and JIMO. With the battlestar concept filed away indefinitely, and eight years now removed from the Great Gravity Assist Panic of 1997, NASA was willing again to stipulate on mass and radioactive power sources. This gave engineers a freer hand to incorporate heavy shielding, saving the mission millions otherwise spent hardening individual spacecraft electronics on a microscale, and they applied their new liberties to a small, internal study that resurrected the orbiter concept of old. Even as the engineering made headway, however, the Europa project was missing something other than NASA's money—some spark, a wild card, a part that could make for an exponentially bigger sum.

Enter Robert Pappalardo.

It was a real coup in May 2006 when Gregg Vane, a manager in JPL's Solar System Exploration Directorate, hired him. The lab needed a scientist to put Europa on a war footing, and Bob's arrival was a sign from above that management meant for this mission to happen. It was better still that he and Karla hit it off from the start. The lab had made it clear that if a mission flew, Bob would be the project scientist. (They made no such promises to Karla, though she aspired to be the project manager by launch.) The two roles were complementary. The project scientist—always a scientist—oversaw all decisions affecting the project science. The project manager— usually an engineer—was in charge of delivering the spacecraft on time and on budget ("time" and "budget" both being defined by the Powers That Be). In Bob's view, both were simultaneously in charge, by necessity—ensuring the scientific integrity of a science mission required coequal footing with the one delivering said spacecraft.[231] In keeping with this logic, a compatible project manager and project scientist could achieve anything. If they hated each other, however . . .

Bob worked in the science building and Karla worked nearby in Building 301, mission formulation. Her knowledge of How Things Work was as good as anyone's at the lab, and she knew everything necessary to get the agency to bite on a mission concept. When it came to icy satellites, meanwhile, he had revised the story of the Europan ice shell, hypothesizing that it operated by way of solid-state convection: i.e., Europa's ice shell is like a lava lamp, with relatively warm, slushy spots in the lower shell rising upward, and cold, hard regions at the top sinking lower. And while top-tier scientists could sometimes be abrasive, Bob didn't talk down to anyone. The pairing was a sign from above that these endless explorations of Europa mission concepts were not exercises in wheel spinning. You didn't entice someone like Bob Pappalardo away from the soft life of academic tenure if you didn't plan to use him.

Europa science when Bob met Karla was much improved from when Bob met Carl. What started as an image of a stunning, scratchy, blurred ball taken twenty-five years earlier by the camera of Voyager 2 was now a real world in space starting to make sense. There were maps now, and you could slap them on a desk, point with a flourish at features, plan your attack. *What is this and why?* There were hypotheses for Europa's inscrutable lineaments—an appreciable achievement for a world that once made no sense at all, lacking anything comparable in the known universe. And things had names now! On Earth, Africa and Everest and Loch Ness and the Seine and the Amazon and Egypt and the Pacific—they just *always were*. But Europa was tabula rasa. So there was Cynthia Phillips, a second-year graduate student in the nineties at the University of Arizona—an affiliate member of the Galileo imaging team, lowest-ranking person in the room, and doer of grunt work— and there was no real role for her here at this stage in her career, and so she made one for herself, taking data beamed back to Earth from the spacecraft Galileo and uploading them to the file server for scientists nationally to begin to study, and while she was there, she pieced together the pictures, and she named things. Just like that! Craters, it was decided, would be named from Celtic mythology. That area is called Deirdre, said Cynthia.[232] There is Maeve. That is Gráinne. Millennia from now, when wondered by all why we called this Europa mining town Maeveton—we call it that because Cynthia decided that that would be its name. She drew heavily on heroic women in particular. She pulled in data from the Voyager archives and pieced together the best map of Europa in existence and the baseline for the mission going forward.

But a map of what, exactly? Europa's giant slabs of ice were definitely pulling apart. You need never have audited a geology class to see where they separated, and that something rose from below to fill the resultant cracks. But here was a problem on Europa: while there were all sorts of places where the sheets were plainly pulling apart (called extension), there was no evidence of places where they pushed together (called contraction). The whole moon was made of these giant sheets—there were no blank spots on the map—so, by definition, they had to push together somewhere, but for twenty years, nobody could figure out where.

Louise Prockter and Bob Pappalardo figured it out.

That their breakthrough involved extension and contraction seemed appropriate given that their relationship oftentimes worked the same way. The Europa community was small, and the upper echelon of icy satellites scholars smaller still. You worked closely with each other, saw each other at conference after conference around the world. You saw colleagues more often than you did some family members, and sometimes more often even than family members with whom you lived. Clashes could be familial and thus titanic. Louise and Bob both wanted the best science possible, and when they disagreed, *they disagreed*. When Louise asked Bob to meet her at the 1999 American Geophysical Union conference in San Francisco about something she had discovered in the Galileo data, he was hesitant; they hadn't really spoken in months.

The conversation went something like this:

—Bob, I think I found something.
—Whatever.[233]

But he agreed to meet with her. It was the cusp of Y2K, and maybe all the world's computers would stop working and the apertures capping nuclear silos would open and end the dreams of our ancestors. Or maybe not—it was in the hands of COBOL developers now—and meanwhile, the annual geology conference went on as normal. (Geology played the long game, after all, would survive doomsday in any event.) At AGU, dozens of talks were given simultaneously in fifteen-minute bites, one after the other eight hours a day for a full week—thousands upon thousands of scientists revealing results to colleagues, with ancillary meetings held throughout for various concerned parties, and poster presentations were spread across one million square feet of the Moscone Center. It could be crowded.

The two met at the top of a quiet, remote set of escalators, taking seats on an adjacent staircase. Louise pulled out her laptop and brought up an image of Europa's surface. *Right there, Bob: that's a fold in Astypalaea Linea.* (Latin for "line," and referring to long line-like geology.) *It's compression.* Louise was a geomorphologist: her job was to look at a surface and tease out its history. Bob, meanwhile, was of the Carl Sagan school of extraordinary claims requiring extraordinary evidence, but he saw it, too: there was something there.

And together now, they pressed forward. You've seen something, but *what does that mean?* They wrote the paper quickly—collaborated brilliantly—understood each other and the implications of the work, this key part of the Europa story, and they found more evidence yet at Libya Linea. The paper was published in the journal *Science*, where neither of them had been published previously or so prestigiously, and suddenly they were doing interviews with beat journalists in the national press.[234] It was Dr. Prockter's first-authored-paper debut.

Scores more would be written post-Galileo, as scientists squeezed the sum of Europan scholarship dry. Bob had published on solid-state convection in Europa's ice shell, and he developed the science of Europa's subsurface ocean. Geoff Collins, their former colleague at Brown, and by then an assistant professor of geology at Wheaton College in Massachusetts, was looking at chaos regions on Europa and how hydrothermal vents on Europa's seafloor would and would not affect the ice shell above.[235] Greg Hoppa, a principal system engineer at Raytheon and a former student of famed orbital dynamicist Richard Greenberg at the University of Arizona, was working out the origin of the moon's cycloidal ridges—bizarre, connected scallops, like the open sea as depicted in a child's drawing, hundreds of miles across—helping explain the way daily tidal stresses made tectonic features on Europa.[236] Little by little. This explains that. That tells us this. How much heat does this motion create? How would that affect the ice shell?

Meanwhile, back at the lab, Bob toiled tirelessly on the most ambitious activity of his life: he had decided to write the book on Europa. It began as a series of papers he drafted while standing up a Europa lab for JPL, summarizing where the planetary science community was in its thinking about the icy moon. Alan Stern's 1997 textbook, titled *Pluto and Charon: Ice Worlds on the Ragged Edge of the Solar System,* which similarly assessed the state of thinking for those two worlds, had helped organize and galvanize scientists interested in the Plutonian system. Indeed, Stern's book, as part of a wider push by guerrilla outer planets scientists parachuting into space conferences to explain their work and what more needed to be known, led ultimately to a decisive endorsement by the Decadal for a medium-sized mission there, which led to New Horizons. It was clear to Bob that Europa scholarship had reached a similar level of maturity, and he reached out to the University of Arizona Press, publishers of Stern's book, pitching one similar but for Europa exclusively. The negotiations were quick ("You'll write it for free." / "OK."), and Bob approached two coeditors: Bill McKinnon, a professor of earth and planetary sciences at Washington University at St. Louis (who also coedited Fran Bagenal's Jupiter book), and Krishan Khurana, still of UCLA. Three weeks later, they had written a proposal and submitted it to the publisher. It was accepted shortly thereafter. At no other time in human history could these books—the Europa book, the Pluto-Charon book, the Jupiter book, and a growing library of others—have been written. It was only with the launch of spacecraft and space observatories that the knowledge within their pages was attainable.

When the Quad Studies had started, NASA headquarters formed a science definition team for what Bob Pappalardo and Ron Greeley, who would lead it, would call Europa Explorer. A science definition team comes up with the science goals for a prospective mission. Team members are not doing new research but rather organizing and taking full measure of the collective knowledge of the planetary target in question. Each scientist brings his or her expertise to the table (e.g., geology, magnetic fields, composition) and says: Here is what we know. What are the biggest things we don't? What are the most important questions that this mission should answer? How do we answer them? They use these questions to build a "science traceability matrix"—a straightforward but densely detailed grid listing the science goals of a mission and how to achieve them—and move forward from that.

Bob was essentially the assistant to Ron, who was conductor of this symphony of scientists, each cell and section adding a rich and complementary layer to the composition. You wouldn't call Ron a "strong leader," because it implied something he was not. Rather, he was a leader with gentle fortitude, and he prodded you in the right direction without your realizing it. This was even more impressive since Ron could have crushed you, whether with the weight of his experience or the force of his intellect or his blunt power in what was, after all, a political program. He could have dictated terms for the mission. But he did not. He'd been there from the start, in human and robotic exploration, in the field of planetary science, understood the whys of how things worked, had defined some of those whys in the first place, and perhaps that shaped him. When Ron first set foot on a NASA facility, Apollo science was practically a blank piece of paper. By the time he left, he had helped choose landing sites for moon missions and had pioneered the science of lunar caves. And it was just the beginning: of his career, of the space program, of true exploration, of a new realm of scientific inquiry.

The real start, though, was in Mississippi in 1957. Before that, there was Ohio. California. Texas. Alabama. Colorado. Greeley was an air force brat; he came from lots of different places. Ron had wanted to be a geologist, had known it since he was eight, and it gave him clarity of purpose, focus. The Greeleys would drive cross-country, back and forth, one base to the next, and when they'd roll through the mountains, where rock was cut away for road, you could see the stone stratigraphy, and it fascinated him. But Mississippi was where he met Cynthia—Cindy—and it all proceeded from there. They were in high school then, and he was the lifeguard at the Great Southern Country Club pool in Gulfport. She was there to swim. They struck up a conversation. He would be a junior when summer ended; she a sophomore, barely sixteen. He had lived all over the country. She was from here. She had a date that weekend. So did he. But after that? Yes.

So Ron and Cindy planned it out while they spent that overlapping year together. Her parents, wise and genteel, wouldn't allow them to get engaged while she was still in high school. It was just . . . too much. He graduated, enrolled at Mississippi State University, a five-hour drive away, and during her senior year, he drove home to see her every weekend. The day she graduated from high school, Ron pulled out an engagement ring and proposed. She said yes. He was studying geology at Mississippi State, and she studied chemistry at Mississippi State College for Women, just twenty-five minutes away. The distance was more bearable, and the plan was to graduate and marry before he pursued his doctorate. She would have finished school by then, too, or have been on the cusp, and they could really get started on life.

During Cindy's freshman year, his sophomore, he was headed her way to pick her up for one of their regular dates. It was always hot in Mississippi, but in March 1960, on that short stretch of Highway 82 between Starkville and Columbus, laced along the foothills of the Appalachian Mountains, at night it could drop into the midthirties easy, and there's not much between the two towns (or *in* the towns, really), but that night, there was another car, and tires crossed the yellow line, and the vehicles met head-on. Ron was in a sports car, and this was when automotive safety features consisted entirely of whose car was heaviest, and Ron's was not. His head absorbed the worst of it, the concussive blow crunching it in places, slicing it in others. The emergency response team rushed him to Columbus Air Force Base (his father still an air force man), and they took one look at Ron, the surgeons on call, and they knew they could not save this one. They flew him to Maxwell Air Force Base in Montgomery, Alabama, where maybe there was a chance.

Maxwell was a training base, its lineage beginning with a special school in 1910 set up by a man named Wilbur Wright, who felt certain that his heavier-than-air flying machine would be of benefit to civilian and soldier alike, and that those with surnames other than Wright should know how to fly them. The Wright school didn't last long, but its legacy made it a natural choice, much later, for an air force training center. By the sixties, airmen and officers of every specialty converged there for advanced training, including physicians, and its hospital was first rate. When Ron Greeley arrived, this son of an air force officer, the prognosis was unfavorable, and the attending physicians knew it, and the race was on to stabilize the patient and find specialists who might be able to piece Ron back together.

He spent the next three months in that hospital room. Cindy was there, too. And when he could talk, they talked, and having had this close a shave with the reaper's blade, they made some decisions. Life is too short. Yes. Anything can happen. Anything. Do we want to wait? No. When the finest doctors in the U.S. military had finally made Ron whole again, he was released, and he and Cindy married immediately.

Cindy would make sacrifices for this change in plans. She quit school, found a job at the university press, first as a typist and then as an editor. She worked on things like Mississippi agricultural brochures, of which there were many, and she did it all: design, type-setting, and print prep. She continued taking classes on her lunch hour but belonged more to the Student Wives Club. Ron, mean-while, back on his feet and in need of money to feed his new family, joined the Reserve Officers' Training Corps. Everybody in those days did a stint in the service, it seemed, and certainly Greeley the Elder was an example to be followed. Anyway, that monthly thirty dollars from ROTC really helped the young student marriage. The downside was that eventually he'd have to serve two years on active duty, but they would cross that bridge soon enough. Global sanity seemed to be holding, and America's presence in Vietnam consisted barely of a battalion of advisors. If President Kennedy upped that number even *tenfold*, that was still less than a division's strength.

Soon enough, Ron graduated with a bachelor of science in ge-ology, and the Student Wives Club threw Cindy a small ceremony of her own. Ron had to choose whether to knock out that two-year commitment to the army or postpone things to pursue a master's degree. He went with the master's.

They were a serious couple, always had been. Before they'd even married, Ron and Cindy made their first major purchase to-gether: a canoe for sixty-five dollars—sixty-five!—exorbitant!—but they were outdoorsy, she a Mississippi woman, he a geologist. She taught him the water, and he loved it, learned to love boats. They ran rapids, paddled down rivers and tributaries, and found camp-ing spots. Her folks had a house out on Cat Island in the Gulf of Mexico that they used for vacations, and they swam and skied and paddled away the summers.

When it came time to choose a doctoral program, it boiled down to the University of Nevada or the University of Missouri at Rolla. Nevada made him an offer, but he liked the micropaleontol-ogy program at Rolla and its attendant fellowship, and geography (Missouri was closer to home) put Rolla over the edge. Ron started teaching as part of the program, first as an assistant, and he was good at it, and soon stood on his own. Micropaleontology entailed studying tiny fossils under a microscope. If you were a geologist in the early sixties, that was about the best you could do financially, because it meant money from the oil industry. When Greeley graduated in 1966 (his dissertation was on lunulithiform bryozoans), the military deferrals ended, and he had a nine-month window before having to report for training at Fort Holabird in Baltimore. Ron took a job at a Standard Oil office in Lafayette, Louisiana, where he put his expertise in micropaleontology to good use, and where, when his eventual army hitch was up, he'd have a good job waiting. Standard Oil subsidiaries and partners would bring in these giant cores drilled by oil rigs and extracted from deep within the Earth, and his job was to take slices from the core and study them under the microscope. He was looking for foraminifera (undersea invertebrates, though you just called them "bugs"), and if particular bugs presented in the slice, that would tell the drillers that they were boring the right holes—that oil was likely nearby. No bugs, and the rigs would relocate to begin the coring process anew.

Ron did this for nine months and absolutely hated every minute of it.

It was almost a relief in 1967 when he reported to Army Intelligence School. (Almost.) They put him through an intensive course on aerial photography—"imagery intelligence," it was called—in which reconnaissance aircraft flew over an area and took standard and infrared images for later analysis. The training was geared to prepare him for Vietnam. Global sanity, it seemed, was never the best bet. By now, troop numbers were sky high—nearly a half million U.S. military personnel were there—a hundred thousand more than the year before, and thousands would spend the rest of their lives in Southeast Asia.[237] It was going to happen, Ron's tour, and he would experience that army marching cadence that begins:

Got a letter in the mail / Go to war or go to jail . . .

When Ron finished intelligence school, he got his letter in the mail.[238] He opened it. He scanned it.

. . . Greeley is assigned . . .

. . . Presidio . . . San Francisco . . .

. . . attached to the Space Sciences Division . . .

. . . National Aeronautics and Space Administration . . .

. . . Ames Research Center, Moffett Field . . .

There weren't many geologists in the U.S. Army, but being a geologist didn't prevent you from going to Vietnam. There weren't many imagery intelligence officers in the U.S. Army, either, and being one pretty much *guaranteed* you would go to Vietnam.

But there also weren't many people in the army who could *read* the words *lunulithiform bryozoans,* which sounded like something moon related and could possibly be of benefit to perhaps the only institution in America that had a higher priority than the ongoing conflict overseas.

Got a letter in the mail, read Ron's letter, essentially. *Go to Ames or go to jail . . .*

JFK had called for an astronaut on the moon by decade's end, and his successor, LBJ—a real space hawk from the start—was going to see it through. But despite the urgency, and like every federal office, NASA had a maximum head count for civil servants. To meet the impossibly accelerated goal of a crewed lunar landing, the agency exploited a loophole that said that military service members tasked to NASA didn't work against the agency's head count limitation. And a guy like Lieutenant Ron Greeley was perfect for the moon mission. NASA now had miles of surface terrain photography from the Ranger, Surveyor, and Lunar Orbiter reconnaissance spacecraft. Geologically speaking, the moon was one big, baffling mystery. Where was it cement-solid and where was it limestone-porous and crumbly? Was it covered in places, meters deep, with ash-like cosmic dust? And what would astronauts do when they got there, anyway? What valuable geology could you do at the various proposed landing sites? Furthermore, the astronauts were going to take pristine samples back to Earth. What defined a good sample, and how could you teach a bunch of test pilots to tell the difference?

Ron wasn't a moon expert, but this was the sort of gee-whiz geology that not even his eight-year-old self could have imagined because of its fantasticality and preposterous ambition, so, of course, he was beside himself. And even without considering the Southeast Asian alternative, as far as army hitches went, an assignment to Ames was about as good a posting as any soldier was ever likely to get. A bonus: where the doctors had reassembled Ron following the car crash as a young man, scars remained, and he had grown a beard to help conceal them. But Mother Green sanctioned no facial hair—this isn't the navy at sea, Lieutenant. NASA, however, didn't care. Grow it as long as you'd like, Dr. Greeley! And that was his rank at Ames: doctor. It turned out that his only actual connection to the Presidio was the PX (for tax-free grocery shopping) and to pick up his paycheck. He didn't report to anyone. He didn't file reports with anyone. He didn't wear a uniform. He just . . . showed up at NASA like a standard-issue civil servant. If you didn't see his personnel file, you would never have known.

They had driven from Missouri to California, Ron and Cindy and Randall, their firstborn, packed in a blue Chevy station wagon, and rolled across rolling prairies, waved at the passing amber waves, and drove through mountains carved away, stratigraphy showing, Ron's life, full circle. Cindy found work near Ron at the naval air station in Sunnyvale. She was an administrative assistant; they called her the secretary of the navy. The air force was building a test center for satellites, and the navy was in charge of its construction. Cindy mostly handled paperwork related to that project.

It was interesting stuff. Artificial satellites were still a pretty new invention and the Defense Department didn't want any single contractor knowing everything there was to know about one of them being built, so each floor in the new building was sealed off, keeping things physically compartmentalized. Cindy also took night classes that year at De Anza College in Cupertino, California. Deciding finally that it was her turn to go back to school full-time, she enrolled at San Jose State College, but changed majors from chemistry to history. She soon, at last, earned her degree.

And there they were: the historian and the moon scientist, making lunar history. At night, the two of them would sometimes go outside and look up at the moon and just think . . . *wow*. Not only about the Apollo program, but where their lives had taken them. They'd look up, and they knew it would happen, the moon landings—had no doubt. What they didn't know was that with Neil Armstrong's first step on the moon and his second sentence on its powdery surface—an instant geologic analysis—this would not be the capstone of some brief, bracing phase of Ron's career in geology but rather would mark the beginning of something new entirely. After his two-year army hitch ended, Ron was supposed to go back to Lafayette and look for bugs, but you put men on the moon, be among the first to apply scientific rigor to something so spectacular and unsullied, unpack the geology of a pristine world unspoiled by humankind, and you look around yourself at Ames, see not only what they're doing—they—*us*—what *we're* doing—and not temporarily: Armstrong was the starting point—and he knew Armstrong! There were astronauts to train and more celestial geology to do, and not only here—von Braun was talking about Mars! Knowing now what Ron and Cindy knew, having done what they did, you don't pack the station wagon and head back south to pick petroleum-portending bugs from oil rig cores, you just don't.

Job openings at Ames were scarce, so he applied annually for independent research grants, and he was successful. Their new life unfolding before them, Ron and Cindy bought a little gray Eichler house on Somerset Drive in Cupertino—red door, glass from floor to ceiling—for thirty-seven thousand five hundred dollars.[239] By now, Ron was doing deep dives on proposed Apollo landing sites and training the astronauts of Apollos 15 and 17 to conduct geology experiments. He was one of a small group of such trainers, including a guy named Jim Head, who later ended up teaching geological science at Brown University.

In 1977, a decade after his arrival at Ames, Ron joined Arizona State University as a full professor and brought with him a Regional Planetary Image Facility: a NASA library of maps, texts, and tomes.[240] Among his Tempe colleagues, there was resistance to this Greeley guy setting up shop and shaking things up. He'd unbalance the department, they warned—turn us into a school of space exploration! But you met Ron and felt . . . better about things. He was a gentleman, soft-spoken. By then, he had been part of multiple missions to the moon; had run the Apollo Data Analysis program, the Mars Data Analysis program, the Mars Geological Mapping program, NASA's Planetary Geology and Geophysics program; and on the Viking mission to Mars, had been in charge of geological mapping and helped certify the landing site for Viking 2. Now, as a lead on the Galileo imaging team, he was standing up the Jupiter Data Analysis program for when that spacecraft finally got off the ground. So at least he knew what he was doing.

When visiting the Phoenix area, Carl Sagan was a regular dinner guest. There he was, in an animated discussion with Ron and Cindy's ten-year-old daughter, just so good at *talking* to people. Of course, when Sagan was off to visit Hollywood and talking to Johnny Carson, somebody had to do Carl's work for him on Viking. *Carl can't work today. He's doing* The Tonight Show *again.* Ron was among those who had to fill in. You weren't bitter about it, exactly. It was more of an annoyed-eye-roll sort of thing. Carl was popularizing the work, and that brought great benefits. But on the Viking team, the science was relentless and sleep sometimes in short supply, and those doing Carl's work weren't above grumbling about picking up their celebrity's slack.

Ron and Cindy were friends not only with Carl but also with the entire mission team. Later, on Galileo, Ron and Jim Head had split the planning of the Galilean moon imaging campaigns, with Ron receiving Europa—what he called the "gem of the solar system."[241] Of course, the mission was postponed for so long, and then lasted for so long, that it turned into this careerlong thing for most of the team. Ron, Cindy, and the scores of Jovian scientists from around the world would attend meetings in Germany or France or California or wherever, gather for dinner parties, and they would see one another's children growing up. You really want to bond with someone? Watch their kids grow from pigtails to parenthood.

BY THE TIME Curt Niebur at NASA headquarters asked Greeley to co-lead the science side of the Europa flagship competition study, Ron had not only been around the block; he'd mapped it, paved it, built its public transit, and taught half of the block's fellow travelers how to drive. He had advised everyone from the French to the Kuwaitis, been part of almost forty major committees and science definition teams (chairing the majority of them), been part of flight projects to Venus, Mars, Jupiter—even Earth, sending an experiment to space on the shuttle. But by now, his beard was white and he knew as well as anyone that were this Europa mission to prevail—were NASA, indeed, to spark welding torches and begin cutting metal tomorrow on a spacecraft—the years of development ahead, coupled with the nontrivial travel time necessary to achieve orbit around Europa, meant he'd never see it happen. He was pushing seventy. At best, he would be retired, traveling the world with Cindy. You become a scientist to answer questions, to ever elevate the threshold of collected human knowledge. You publish papers, get grants. But do it long enough, and you realize that raising the threshold is only half of it; your real job is to make sure that science doesn't stop with you, that the threshold is ever rising. The Dark Ages are always one day away.

He didn't make some maudlin show of any of this, of course, not Ron. He was too stoic for that. So as scholars since Socrates had done, he made it his mission to impart all he had learned so that others might be able to answer the questions he would never know were even asked.

Robert Pappalardo was the sorcerer's apprentice, *der konzert-meister*, the first chair of the violin section, and in ways worthwhile and subtle, Ron was handing over the baton. Bob, would you help organize this group? Perhaps you could iterate on the meeting agenda? I'm thinking of this researcher and that as co-leads of this part of the project—what do you think? Relax, Bob—what happened at that meeting was normal; every committee has to go through it. Would you prioritize this list? Give this talk. Lead this section. What do you think of these notes? And those?

Up front, the two men huddled and worked through Europa questions big and small. Some were sophisticated, but some involved "first principles," e.g., Why are we using an orbiter? What other mission scenarios can achieve the desired science? Do we keep the option for a lander alive? What about an impactor? What instruments would we need for each? It was brainstorming, mind mapping, standing before blackboards, arms crossed and eyebrows furrowed, and approaching the study with a beginner's mind.

The basic objectives of Europa Explorer weren't that different from those enumerated by small, previous internal lab studies; or the JIMO science definition team years earlier; or the Decadal Survey;

or the Europa Orbiter proposal from 1998. The goals were refined further during regular meetings of the Outer Planets Assessment Group. Meanwhile, the left and right parameters of the mission were set in stone by Curt Niebur at headquarters: among them, a three-billion-dollar cap, standard radioactive power sources, and no "miracle" technologies.

To develop Europa Explorer, there would be four meetings of the eleven-person science definition team, the first convening at JPL on February 19, 2007.[242] The average meeting might last two and a half days, with science discussions, instrument presentations, lectures led by team members, and guest talks by experts unaffiliated with the team but with insights that could help the team better develop the mission. Louise might give a forty-five-minute master class on Europa's geology. Don Blankenship, a geophysicist and Antarctica scholar from the University of Texas at Austin, would explain how an ice-penetrating radar would work and what it could do. Krishan Khurana would talk magnetic fields. Professor Christopher Chyba of Stanford University would tell you what you needed to know about astrobiology. Later meetings would look at such issues as planetary protection: If Europa has life, how do we protect it from Earth microbes that have stowed away on the spacecraft? And as study lead, Karla sought constantly to find the sublime intersection of what scientists wanted and what engineers could actually do.

The core of the Europa mission, they determined together: an ice-penetrating radar, a camera, and a composition instrument (the latter to figure out what the moon was made of—what those stripes were, and why the different hues). Those data—Europa's makeup, its ice shell in three dimensions, and surface imagery to understand its geology—you didn't even need to prioritize them. In fact, Ron specifically insisted that they *not* be prioritized, as they had already agreed that the mission wasn't worth flying without every single one of them. Greeley was adamant that if there was no need to begin a difficult discussion, don't.

Karla was ever astonished by the quiet intensity of the scientists, and the way Ron let them work. It was a dream team, for sure, which meant that you put a stick of chalk in anyone's hand and had him or her debate the relative merits of some abstruse subsect of an abstruse subsect, suddenly you're dealing with a weaponized mind, and if there was disagreement, how could you possibly bring them into accord? And there was Ron, this quiet force, never raising his voice no matter the room's temperature, this wise grandpa saying, *OK, let's talk. What I'm hearing is . . .* and suddenly—magically—maddeningly—when you really thought about how easily he did it—he pulled the whole thing back together, the tangents collapsed recursively, and Ron had somehow absorbed every argument, placed each one in context, finding the areas of overlap, finding the places of mutual disagreement and dispatching them, simplifying, simplifying, simplifying—he'd even write on the board *K-I-S*—the KIS principle: Keep It Simple (this was Ron Greeley; he'd never add an obscenity like "stupid")—and by the end of the meeting, everyone was content and optimistic about the work they'd done and the work yet to do. A dozen scientists in total agreement? That's hard—but he could do it. Karla had been doing this for twenty years, but to see Ron Greeley work? It was spellbinding.

A SCIENTIST ANSWERS to humanity. What she does, she does for the benefit of all humankind. It is the loftiest goal. But she's also trying to make a living and not lose her job. She wants to get her mission approved and her spacecraft designed and built, or later, if she's on an instrument team, she wants to keep her instrument on that spacecraft so that all of the engineers at her institution don't lose their jobs. NASA wants to obtain the best science, but it also needs to keep civil servants employed at its many centers. NASA must also keep the White House Office of Management and Budget happy. The agency might choose a spacecraft or instrument that isn't capable of carrying out the highest priority science but is low-risk and unlikely to go over budget and cause problems down the line. NASA must keep the public happy; it must engage the taxpayers and excite them about what is happening in space. It must also keep Congress happy. So the agency is thinking about a lot of things beyond whether a science goal is responsive to the Decadal Survey.

Three weeks before the final flagship competition study reports were due to NASA headquarters, Curt Niebur got a panicked call from an engineer at the Applied Physics Laboratory. A manager there had had lunch with Alan Stern and mentioned the flagship studies. Somehow the cost came up—three billion—and Stern stopped him cold. Why are you coming in that high? I only have two billion for this thing. Curt, the caller explained, management here is presenting the two-billion-dollar cost cap as a *directive from headquarters.*

It was like a bomb had gone off in Curt's head. He was at the Lombardi Comprehensive Cancer Center in Georgetown with his wife when the call came. Seven weeks earlier, doctors discovered a two-centimeter something in Susan's right breast. She had been having difficulty breast-feeding their newborn.[243] The boy, their second, was five months old when the problems started. She thought maybe he wasn't latching properly—left worked, right didn't. Her lactation consultant suggested finally that she give up on the right and let it go back to normal. But it didn't go back to normal. It hardened and hurt, had the texture of an orange peel. She wasn't worried, really—there was no lump—but she went to see her obstetrician for reassurance. She didn't receive it. He had never seen anything like it, he said, and ordered her to a specialist.

The mammogram revealed something, they did a biopsy, and the next day, she was told that she had inflammatory breast cancer, invasive, fast spreading. There was no time, so her options—the list inclusive—were: chemotherapy. Two weeks later, they started. The nausea arrived promptly, the splitting headaches. Another two weeks, and her hair came out in clumps.[244] She shaved her head.

Susan had left NASA the year before, wanting to work from home and spend more time with her first child, who was one at the time. An agency manager didn't want her to telecommute, so she chose her family, and started a home-based consulting firm to help companies put together good proposals for missions in the Discovery and New Frontiers classes. She had run the Discovery program from headquarters for four years, so she certainly knew what the agency wanted. She was her own boss now, worked her own hours. Meanwhile, she blogged every step of her cancer treatment and built an online community of survivors. Susan didn't do half measures. She kept in her office a talisman: a Lego minifig that she called Princess-Who-Can-Defend-Herself. The Lego princess wore eyeglasses and carried a sword. But now in addition to blogging, doing her job, and raising her sons, Susan was also doing things like explaining to her toddler why Mommy's hair was falling out, and she's at the store or the traffic light, and people look-don't-look when they see the hair or lack thereof, and it hurts sometimes, a lot, and she's carrying around stage three cancer and there is no stage five, and she knows math.[245]

The day Curt got the call from the APL manager, they were at the cancer center for a blood draw. Susan's counts were down. She could feel it. She had taken her two-year-old on their weekly playdate and had to leave early, light-headed, and when they got home, she nearly passed out on the couch. It was all so frustrating for Susan, to slow down. She'd never not worked, never not, well, reached for new heights to reveal the unknown for the benefit of all humankind (as was the NASA vision statement). Curt accompanied her to the treatments, tried to lighten the mood. He would have quit rather than missed one. His boss, Jim Green, was a saint, though—take whatever time you need, stay home. The grandparents helped with the little ones. There was no balance—it was all happening so fast, each minute focused on whatever needed attention for the next sixty seconds.

And in this set of sixty seconds, while waiting for the latest counts, suddenly the flagship studies—two years of planning by Curt and eight months of intensive work by the study teams—had been cut by one billion dollars. It was a catastrophe.

When he got back to the office, Curt sat down with Jim, and they tried to figure out a way forward. There was no time for the study teams to descope, or "strategically abandon," key objectives of their missions. Even slashing instrument payloads to the core wouldn't get them down that far. It was an especially severe blow to a Europa spacecraft, given the heavy (and expensive) shielding required for it to survive marination in the Jovian radiation belt.

The next Monday, Curt flew to Boulder for a workshop called Ices, Oceans, and Fire, which brought together researchers studying the outer planets of the solar system, and their moons in particular. Ron, Bob, Louise, and Dave Senske (who co-led the Ganymede study with Louise) were there, and they weren't thrilled. The big problems were: 1. we just lost one billion dollars, 2. our respective laboratories still want these missions to fly and *will* carve one billion dollars from our spacecraft without our consent or advice, 3. but those sorts of cuts would remove core science from the spacecraft, 4. and we are not going to sign off on the science value of a two-billion-dollar mission without full face-to-face consultations with our science definition teams, and 5. that will take months to organize and will push the study delivery dates well into 2008.

The discussion went on until ten thirty that evening, and, ulti-mately, all Niebur could propose was that they descope their mis-sions as best they could. Don't be drastic, he added. You cannot validate the appropriateness of billion-dollar descopes by the study deadline. Just do your best and make mention in the study reports that you can get lower given more time—and that the plan, indeed, is to go lower—but, again, look, we need more time.

THE FINAL REPORT of the Europa Explorer mission study came together on November 1, 2007, assembled physically from a hun-dred stacks of pages arranged in a grid on a conference room floor, Bob orchestrating it at this level, piecing the book together one section at a time. Karla's job was to give NASA everything neces-sary for it to say yes to Europa. To win this thing for Jet Propulsion Laboratory.

And she was proud of the Europa Explorer report. The thing was massive: hundreds and hundreds of pages, filling a three-and-a-half-inch-thick binder.[246] You could use *it* as radiation shielding. Objec-tively speaking, all they had done was written a book, but, really . . . it was more than that. It was the first time in all the years she had been doing this that Karla knew without a doubt that this mission could go forward. That they could build it tomorrow and deliver on time. That they could travel to Europa, get back the science within the promised cost of less than three billion. They were ready to execute. It was a political decision now.

A couple of weeks after Karla handed in the study report, Curt pulled her aside. He had read it, of course, but needed to ask her, honestly, in that soft Breese accent of his, Do you really think you can do this mission for this cost? And she told him, honestly: If we can control the instrument costs, yes.[247]

It was always Europa's to lose, and Karla's report was everything

Curt had hoped for. In fact, each study submitted was better than it should have been. Ganymede, Enceladus, and Titan . . . well, that last one was truly extraordinary, like something out of Victorian-era science fiction. As planned, Curt convened independent review panels of scientists and agency personnel, one for technical, one for science, and they studied the studies, then issued replies at the close of 2007. Only one should have emerged, but two came out on top: Europa and Titan.

At a three-hour meeting, Curt briefed Alan, as well as the chief scientist of the Science Mission Directorate and the division directors of planetary science, heliophysics, astrophysics, and earth science. Alan was especially taken by the Ganymede study led by Prockter and Senske. But since there wasn't enough money to get *any* of the missions going at their current price points, they wouldn't decide just yet.

So for the teams, each of which had whole worlds at stake, things got ugly.

CHAPTER 7

The Death Star

A DREAM MELTS INTO A RING, A RINGING. TODD MAY'S eyes open. He's awake, he's asleep—he's—the phone is ringing. Dark. It's night. Where's—nightstand—there, green numbers, 3:30. In the morning? His eyes close. It's three thirty. The phone is ringing. He reaches for it.

Hello.[248]

*thumpthumpthumpthumpthump*Did you read that report?*thump thumpthump*.

Alan. It's Alan—*thumpthumpthump*—he's *awake*? Three thirty. Read the report. Report? The Smithsonian thing? It arrived yesterday—yesterday?—it is three thirty a.m.—yesterday afternoon on his desk—*thumpthumpthump*—what is that sound?

No, Todd says, I haven't had time to read it yet.

*thumpthumpthumpthumpthump*Well, our comments are due in five hours*thumpthumpthumpthumpthump*You really need to review the report*thumpthumpthumpthumpthump*.

Todd is awake now. I—the thumping—is—ah, Alan is on a treadmill.

I'll get on it, May says.

Todd dressed, drove to headquarters, and by four thirty in the morning was in his office, reading the report. The Smithsonian Institution had created, in conjunction with NASA, the National Science Foundation, and the National Oceanic and Atmospheric Administration, an exhibit on the Arctic, and a whistle-blower accused the museum's director of acquiescing to Republican coercion to alter the exhibit and minimize the human element of global warming.[249, 250, 251] Alan was all in on earth science, had made enhancing its budget a priority, and wanted to make sure the agency went to war in a subsequent report on the subject, making NASA's affirmative stance on anthropomorphic contributions to climate change explicit. Todd's first meeting of the day was in three hours. He wrote his comments and left the binder on Alan's desk.

At nine he was back in his office, and Alan walked in.

—Did you actually read this?
—Sir, said Todd, I did the best I could in the time I had.
—Look at this! said Alan, flipping midway through the thick report. Read that! That's not acceptable! We can't allow—
—*I did the best I could with the time I had!*

That is what it was like to work with Alan Stern. He wasn't cruel or unkind. In fact, he was a really great guy, honorable. A genius. But that really great guy, honorable man, genius, really liked to work, seemed never to rest, read everything, knew everything, and seemed to think that everyone else ran at the same pace, which, I mean, Todd had been working since he was a fourteen-year-old boy, *loved* working and revered the work ethic, and it had taken him far, from chicken processing plant to launching a space station—so it's not like he wasted weeks on a recliner in front of the television.

But compared with Alan it sometimes seemed that he barely rated as animate because he did waste time occasionally on tasks like . . . sleep.

No new Discovery mission had been chosen since 2001—six years and counting.[252] The cadence of new missions to Venus, the moon, asteroids, and beyond thus ground to a halt. When the Mars Climate Orbiter and Mars Polar Lander were lost in 1998, Dan Goldin and Ed Weiler scraped together the funds to get Mars back in the air, and it had to come from somewhere.[253]

It particularly annoyed Alan that, in his view, Jet Propulsion Laboratory had misrepresented the price of Mars Science Laboratory. Every dollar over budget came from some other mission that didn't cause the problem. NASA's planetary science program had once managed to explore half the solar system simultaneously. Now? Only two major non-Mars missions were in active development: the medium-class spacecraft Juno to study the Jovian interior and poles, and the Lunar Reconnaissance Orbiter to map the moon—and the science division wasn't even paying for the moon mapper![254] Exploration Systems, essentially the human spaceflight support directorate, was footing the bill. The rest of planetary's dollars were going—of course—to Mars.

Alan had a plan to change that, though, and it involved adjusting Curt Niebur's Quad Studies, which were baselined for three billion dollars. Alan had been looking closely at the books and realized that if he brought the Mars program in line and simultaneously imposed some serious fiscal discipline on the wider planetary science division, he could find two-point-one billion dollars—not to study more mission concepts, or to get things kickstarted and hope for funding down the line. No, for two-point-one billion, he could *launch* an outer planets flagship. Better still, the European Space Agency had offered to help; it wanted to get in on the science, too.

So, although the Quad Studies were intended to take four potential missions and settle on a single destination, Stern directed a second round, reducing the competitors to two: a Europa mission (doing as much of the Ganymede mission science as possible) and a Titan mission (doing as much Enceladus science as it could). It would be a shootout—one with an unequivocal winner. The best study to hit the two-billion-dollar bull's-eye would move immediately into development.

But first he had to get the Science Directorate on sound fiscal footing. Based on his Pluto experience, he knew a balance was possible and was stunned by how resistant the people at headquarters were to change. It was as though they *liked* things broken and wanted to keep them broken. Alan had Todd and his team run the numbers. In the previous four years, missions across the agency had gone *five billion dollars* beyond their allotments.[255] Get budgets under control, and you could do more missions of every size to targets across the solar system. And he wanted desperately to do more missions. So he would fix this, and if that meant being seen as the bad guy, then he'd point his death ray at Metropolis with pride.

Stern's first success in imposing fiscal sanity was the Kepler space observatory, which had been selected in 2001 with a budget cap of about two hundred ninety-nine million dollars and a mandate to launch in 2006. But when Alan took over science in 2007, Kepler was still on the ground and two hundred million over cost.[256] And they needed more.[257]

Alan offered to the Kepler team all the time it needed. Take years if you want, decades, generations, but if you need another cent, I will cancel you. And two months later, a miracle! A coincidence, surely! The Kepler project found a way to solve the problem, no new money required. The mission launched in 2009 and went on to discover thousands of exoplanets circling hundreds of stars.

So Alan's philosophy worked. Applied more broadly, when

projects needed more money—*really* needed it—he would pay for overruns by taking from other projects in the same portfolio (e.g., if a Mars mission went over budget, he would pay for it with funds allocated across all Mars missions). Overruns would then proceed outward, concentrically. If a portfolio needed more, they had to tap funds already allocated to other projects in the same division (e.g., planetary or astrophysics), and if that wasn't enough, go even larger: the entire Science Mission Directorate. It was a way to spread the pain equitably.

That's how he would handle the Mars Science Laboratory overruns.

Headquarters had received word that one of the rovers on Mars might not survive the winter because there was dust on its solar panels, which, along with the dim seasonal skies, would effectively put the craft into a hibernation from which it might not wake. The way Alan saw it, Spirit was built to last, but if it *happened* to die . . . you could reduce the project workforce and thus the Mars budget. Even if the rover survived, the Mars program could sustain a budget cut because a sleeping Spirit wouldn't exactly need a full mission team working round the clock. So Alan proposed a twenty percent cut in the rover budget and sent a draft letter stating such to the program manager at Jet Propulsion Laboratory. *Somehow* that letter leaked to the community, which, over the weekend, was radicalized en masse into an army of berserkers.[258] This wasn't some penny-ante project like Kepler that Alan was poking with a stick. This was Mars, and nobody was going to wait around for meetings or negotiations.

Enter the administrator of NASA. Mike Griffin told Stern to just give them the money. Alan suggested instead that Mars Science Laboratory be put on ice for a while—find another launch window, find a team who could do it on cost. The Mars program had an annual budget—why not just slow the project's development to keep the other rovers running? Make Mars live within its means rather than allowing it to eat off everyone else's plate. Or they could descope the mission.

Sure, it was a tough call, but Alan had made it and expected Mike to back him up.

Griffin had made some tough calls himself—had made, in fact, the toughest call Alan had ever seen. It happened before the launch of New Horizons. Because of Pluto's distance from the sun, you needed nuclear power—the same sort Cassini had carried a decade earlier (nearly grounded by alarmists). Despite Cassini's success, nuclear launches had continued to be a giant headache for NASA, and this was no exception. Myriad federal agencies including the Department of Defense, the Environmental Protection Agency, NASA, and the Office of Science and Technology Policy had to be satisfied before the spacecraft left Earth. Then the rocket itself needed a certification of flight readiness—another arduous process involving the agency, key contractors, Kennedy Space Center, and the Applied Physics Laboratory.

It was during this process, at the Lockheed Martin plant where the rocket was built, that engineers happened to be testing the same sort of liquid oxygen tank that New Horizons would use in its first stage. It was a perfectly normal test—aerospace contractors conducted hundreds, if not thousands, of them in order to certify rockets. On this test, though, the tank ruptured, and the accompanying BOOM impeached every future Atlas launch, including—especially—ones with nuclear material onboard.

Lockheed and NASA mounted an investigation so detailed that it reached the molecular structure of the tank, searching intently for the cause of the breach. Of greater, immediate importance: NASA had to determine whether the spacecraft should be allowed to launch.[259] This would have been fine, except that the rupture occurred in mid-2005, and New Horizons was required to launch in January 2006. If it missed its two-week launch window to Pluto, the mission would be pointless, leaving New Horizons a seven-hundred-million-dollar monument to dashed dreams.

One week before the launch window opened, after dozens of meetings at NASA centers, there was a final gathering at agency headquarters. A hundred or so people met in a single sweltering room, SRO. Mike Griffin was there, as were experienced engineers, high-level managers, center directors, and other agency brass. Alan Stern was there as principal investigator of the New Horizons mission, and Todd May was there as head of the Discovery and New Frontiers program office at Marshall Space Flight Center.

James Wood, chief engineer of the Kennedy Space Center, ran down everything that was known and unknown. He gave the launch a GO.

A question-and-answer followed, and a discussion among the high-level NASA executives followed that. And then it was time to poll the room.

Rex Geveden, the chief engineer of NASA, was also GO.

Mary Cleave, a former astronaut and the head of NASA science, said it wasn't worth the risk. If the launch went bad, she didn't want her fingerprints on it. NO GO.

Andrew Dantzler, the head of the planetary science division, voted GO.

Bryan O'Connor, another former astronaut and head of NASA safety and mission assurance, voted NO GO.

This continued for some time, but when you got right down to it, when the head of science and the head of NASA safety join forces and say, "No, do not launch this nuclear-powered science mission because it is *too dangerous to do so*," you are not in a good place.

Finally, it was Mike Griffin's turn. He was of course the NASA administrator, but before that, he was head of the Applied Physics Laboratory's space division. Like Dan Goldin, two administrators earlier, he was also a veteran of the Strategic Defense Initiative. He was, in short, as smart as anyone in the room—and he was usually in rooms with very smart people indeed. Griffin argued the case before them, repeated the critical points made and added a few of his own. Finally, he said to the NO-GOs in the room: Your advice is just that: advisory. We will fly New Horizons.

GO, he said.[260]

If you are Alan Stern (who was, of course, a GO) your knuckles would have been white if you had a table to clutch on to, which you didn't because you lacked the rank to warrant a seat at the table. Look, had that launch gone south—and not only because of the tank, but if it had gone bad *for any reason*—Mike would have, at a minimum, lost his job. It would have ended everything the administrator hoped to do at NASA, shattering his dreams of putting astronauts back on the moon, of planting an American flag on Mars. A launch incident would have made the front page of every newspaper in the world: NASA NUCLEAR DISASTER LAUNCHED AGAINST ADVICE OF HEAD OF SAFETY. Mike Griffin's neck would have been unburdened of its head.

But he had been right, and New Horizons was flying even now on its way to Pluto. It was the culmination of Alan's life's work, and it happened because Mike Griffin made the tough call. And now it was Alan's turn, and he needed Mike to validate his wisdom and authority.

But Mike shot down Stern's idea. Keep the budget overrun out of the press, he said. Let's figure out how to pay for it. And that nauseated Alan. NASA's culture could be shaped, but it would take a long time and required an adherence to principles. Alan Stern's greatest strength was that he was uncompromising—that's what got the Pluto mission launched. But Alan Stern's greatest weakness was that *he was uncompromising*. His tragic flaw was almost Shakespearean. So there it was: They weren't going to turn off Spirit or Opportunity or any of the orbiters. They weren't going to delay, disrupt, or discipline Mars

Science Laboratory. Indeed, Mars would feel no pain at all, which meant the only place to get the cash was outside the program. Other worlds. Other missions. Good missions.

It was 2008. Alan's oldest child was twenty years old, and his youngest, fourteen. When they were growing up, he told them that integrity comes when you look in the mirror. Regardless of whether somebody *else* knows if you did the right thing or the wrong thing, *you* know, and you have to look at that face. And that's a very high standard. He couldn't, he decided, have looked himself in the mirror if he punished the innocent by rewarding the guilty. There was a way to thread the needle, but Mike didn't want to do it, and Alan understood why. Mars was Mars. Still, Alan didn't want to be the firing squad. After a sleepless night, he got up, wrote Mike an email and said: I don't think I'm your guy anymore. I don't think I can do what you want me to do, and I think you ought to find somebody who will.

He knew Mike would fire him. He knew it would cost his career deeply. But he was still proud that he had done it.

He got his response.

Your resignation is accepted.[261]

TODD WAS AT Jet Propulsion Laboratory when he heard what had happened, and he heard it from Tom Gavin, pillar, who delivered the news with characteristic directness.

Your boss just quit.[262]

It was a Monday morning. May knew it had been a long weekend for Alan because of the Mars memo. Things with Alan always tended to come to a head around budgets and finances. And though Alan had succeeded with projects such as Kepler, Todd knew that Mars was different. The Mars community was large, and people succeeded in groups. They would work together and protect their interests.

Still, in disbelief, he called Alan, who confirmed the news. And when May returned to headquarters, he had a new boss, just like that.

Four years earlier, when Todd had been hired to help finish Gravity Probe B, the spacecraft had been in development for forty years. The running joke was that everyone at the agency had worked on it at some point. It was a rite of passage—a space explorer's *annaprashana*. The mission had been canceled and then revived three times. By the time Todd came on board, there was this guy running the science division, Ed Weiler, and it seemed like it was the man's life's work to kill the probe.[263] Ed threatened to terminate it seven separate times during Todd's tenure alone, and Todd just thought, *Wow, what a mean guy.*

Four years later, Alan was out and Todd's next meeting was with Alan's replacement . . . Ed Weiler, who was being brought back to lead the science directorate. Todd didn't know what to expect, but Ed asked him immediately to stay on as deputy.

You have a job as long as I'm here, said Ed.[264]

Todd thanked him, but he never wanted to make a career in DC, and said as much. He had four small children at home in Alabama that he missed. Weiler understood. May agreed to stay on for the Mars landing of the spacecraft Phoenix, and that would be it.

He was glad he'd decided not to leave immediately. For one, he had never personally experienced the "seven minutes of terror"— the agonizing time it took for a spacecraft's signal to travel from Mars to Earth with news that it had landed safely. And the morning after Phoenix's successful touchdown, Todd flew to the University of Arizona to be with the science team for the first images to arrive from the Martian arctic. Some of the men and women had been working, *dreaming,* their entire careers to see something like this. Nobody had ever glimpsed the north pole of Mars, and there it was.[265] The terrain was broken up in this soccer ball pattern—just as the scientists had predicted. People saw the images and started crying. Phoenix would finish what Mars Polar Lander had started ten years earlier. It was the perfect way for Todd the engineer to end his run at agency science.

While concluding his tour of duty at headquarters, Todd asked Ed about Gravity Probe B. Ed admitted that the cancelations and resurrections were all a tactic. Kabuki theater. He knew Grav B was hard and that there would be overruns, but the cancelations forced those working on it to find better solutions to otherwise expensive problems and sent a message to the division that missions had to be run with accountability. Perhaps more importantly, the cancelations showed the science community that Ed would do whatever was necessary to keep mission overruns from harming other projects. There was a leadership lesson there.

Long ago, when Todd applied for his first senior executive job at the agency, the position announcement read: "NASA seeks executives with a sense of daring."[266] Todd thought, How awesome. How cool. I mean, sure you'd see that on army recruiting posters, but what kind of organization specifically sought executives with a sense of *daring*? It just resonated with Todd. When he was a kid, he had a bicycle that didn't have handlebars. There was this big hill in his neighborhood on this road, Spanish Main. It was like a quarter mile long, or seemed that way at the time, and he would ride down it on that bike because he knew the centrifugal force of the wheels would keep everything lined up like a gyroscope, and it wouldn't get wobbly unless you were going slow. You had to go fast. You went fast, you didn't hit the brakes, you just had to go. No hands—hands in the air! And if you went over the front, it was going to be a bad day. But that—Todd connected with that when he saw the announcement. That's NASA. That's what's cool about it. Everything they did had a sense of daring to it. And now Todd would have to find his next thing.

CHAPTER 8

Auto-da-fé

BY SEVEN IN THE MORNING, KARLA CLARK HAD USUALLY switched on her office lights in Building 301 and pulled up her email. There were about a hundred buildings, S, M, and L, at JPL, each distinguished by some role in the spacecraft and mission development pipeline.[267] Across the street from 301 (Mission Formulation) were 170 (Spacecraft Fabrication) and 179 (Spacecraft Assembly). Just down DSN Road was Building 230, the Space Flight Operations Facility—the Center of the Universe—where at any given time, messages might be sent to or from spacecraft scattered across the solar system. On the northern edge of the campus stood Building 150 (Environmental Testing), where a piece of hardware might be subjected to simulated space in a thermal vacuum chamber. Arrive early enough in the morning, and an errant mule deer might still be grazing the courtyard alongside ground squirrels, the morning mountain mist on Saint Gabe making everything ethereal, enchanted. Those kinds of mornings, before the daily grind of meetings and intractable problems and So Crazies It Might Work, it was a place where anything could happen. You might even build a spaceship to find space whales in alien oceans.

Karla started each day returning letters to scientists and engineers who punched in at different hours. The scientists would roll up a little later in the morning and work, sometimes, until eight or nine in the evening. Her overnight inbox could thus be considerable. Next was a team meeting with key engineers, and they'd go over the plan for the day and then separate, because much of the team worked only part-time on the study; you came on to something like Europa because it was a passion, but your *travaux alimentaires* might be Mars or management or whatever. For Karla, Europa was both. Then there were lots of meetings with lab higher-ups to discuss workforce issues or technical findings, and one-on-ones to walk through problems and find solutions, or figure out who had enough experience at the lab to know said solution. (Here, Mars Science Laboratory could especially impede progress; the singular expert on some abstruse subsystem might be tied up for weeks working through Red Planet problems, and Karla had to get in line.) Later in the day, she would interact with Bob Pappalardo to keep science and engineering happily married. It was all about keeping the Europa story moving forward, turning PowerPoint into hardware.

Ed Weiler's reinstallation as head of science was bad news for Europa compared to Alan Stern, who'd wanted to fly an outer planets flagship. Ed, Karla knew firsthand, had once shot Europa Orbiter dead and would have killed JIMO had he been around post–Sean O'Keefe. Overall, he had a frosty relationship with Jet Propulsion Laboratory—didn't trust it, didn't trust its cost estimates (which, to be fair, were routinely, absurdly low). And with the mess he now had to clean up from Mars Science Laboratory's latest overruns—look: Ed was in no mood for more fictional flagship funding figures.

But while Ed's return brought storm clouds for the outer planets with his rumblings for *another* Mars mission beyond MSL, his return also brought, almost perversely, some very good news for Europa, because he immediately eliminated Alan's mission cost cap. Changed the rules, just like that! He didn't want to hear about two-point-one-billion-dollar flagships because he didn't have two-point-one billion dollars. We weren't switching off a Martian rover. We weren't deferring, killing, or carving into Mars Science Laboratory. Regardless of who won the shootout, Europa or Titan, the mission would require a New Start. And if we were going to go through that—getting the administrator, the White House Office of Science and Technology Policy, the Office of Management and Budget, the science community, and Congress on board at the same time for the same mission—then we may as well do more than present a bare-bones spacecraft with a marginal science return. No, he said, he wanted each team to find its respective mission's sweet spot. We weren't going cheap. We weren't building a battlestar. We were going for some elegant in-between.

"Sweet spot" didn't exist in the literature—certainly not in the volumes of spacecraft guidelines filling libraries at Jet Propulsion Laboratory, the Applied Physics Laboratory, or NASA headquarters—but what Ed meant was this: on a graph charting the increase of a mission's cost relative to its science return, the more science investigations you added to a spacecraft, the more money you needed for the mission. Even the smallest change to the spacecraft could require more mass or more power or more shielding or a bigger rocket or more ground support, and thus more money. Such a graph should reveal a serenely straight slope, increasing gradually as instruments were added. But at some point, there would be a tipping point, and the cost curve would kick upward exponentially: the science return would inch while the price point suddenly yarded. Just before the moment of a massive price spike, *that* was the sweet spot. The entire phase two Europa study would thus be devoted to finding the sweet spot and building a mission around it, from spacecraft to orbital trajectories. All of this was true for the Titan study as well.

There was a twist, however. Before he had left, Alan put in place a contingency plan that would keep a Europa or Titan mission alive, he hoped, *no matter what*. After completion of the Quad Studies, Alan moved both missions to JPL and asked the European Space Agency to partner with the lab for the shootout. Two European teams would each develop a spacecraft contribution to the Titan and Europa missions.

As a matter of science, the Europeans had a keen interest in the Jovian system and were developing a mission there called Laplace.[268] They were similarly developing a Titan and Enceladus mission called TandEM.[269] If NASA and the European Space Agency could each have a spacecraft in the Jovian or Saturnian system at the same time, the orbiters would work together, talk to each other, correlate data. Rather than double the science return, you'd get maybe double and a half—for free!

As a matter of politics, Karla considered Alan's plot a masterstroke. The Europeans didn't do things like the Americans. Their deep space missions were meticulously orchestrated by necessity because, unlike NASA, they didn't have to get the White House on board; they had to get more than *twenty* White House equivalents on board from each of the European Space Agency member states. And they had to parcel out the work just so, making sure France got a piece, and Italy got a piece, and Denmark got a piece, and Germany, and . . . which took a lot of work, so when they committed to a project, they committed firmly and for the extreme long term.[270] NASA, in comparison, seemed sometimes to the Europeans like a junkie genius lab partner: when the Americans were good, they were very good, but when they were bad . . .

As a matter of engineering, a lot of Europa team members at

JPL wondered what, exactly, did anyone think the *Europeans* could bring to the project? But Karla had seen the greater power of a European partnership while working on Cassini. That spacecraft was bedeviled from the start, so very, *very* (and eventually) VERY expensive, with Congress, the White House, or both ever ready to zero it out. But the Europeans had invested a half billion dollars in Huygens, which required Cassini to, first, carry it to the Saturnian system and then, once separated, act as a communications relay during its (i.e., Huygens's) descent to the Titanian surface.[271] Cancel Cassini, and you would effectively cancel Huygens. Every time the U.S. government gripped firmly its red pen to strike Cassini out of existence, the prospective fallout from Europe stayed its hand.[272] With respect to Cassini, the European Space Agency was Amalthea to NASA's Zeus.

As a matter of international diplomacy, a broken commitment by NASA to ESA wouldn't offend one obscure office in Paris; it would offend half the European continent. It would be an international incident. The United States could scarcely afford to do so now, in 2008, with free trade treaties under negotiation and active military alliances overt and covert in Iraq, Afghanistan, Africa, and Southeast Asia. Ed Weiler might be quick to kill a Titan or Europa proposal he hated. Congress or the White House as well might decide we had more pressing priorities (Mars), but with national prestige on the line and the urgency of averting a transatlantic dustup, hostile hands would be stayed.

As an added insurance policy, Jet Propulsion Laboratory agreed to a strategic partnership with the Applied Physics Laboratory to develop both missions in concert.[273] APL would get about a quarter of the workload in exchange for Pasadena taking the lead. (APL didn't have the resources to tackle an entire flagship, though its U.S. Defense Department contracts gave it greater experience in dealing with radiation.) Given the history of competitiveness and bad blood between the two laboratories, NASA headquarters was shocked to receive their signed memorandum of agreement. The Capulets and Montagues had friendlier relations. Neither laboratory had ever partnered with another in this way for such a project, so to prevent any bar fights, APL hired veteran engineer-slash-manager Tom Magner to be Karla's APL counterpart.[274] He had worked for Weiler at Goddard and had a good working relationship with JPL. Magner could fill a lot of cracks in the foundation.

The new joint Jupiter expedition would be called the Europa Jupiter System Mission, or EJSM. The joint exploration of Saturn would be the Titan Saturn System Mission, or TSSM. Privately, the engineers at Jet Propulsion Laboratory lamented the very idea of working with their East Coast colleagues. It's not that they considered the Applied Physics Laboratory to be junior varsity; it's that they considered APL to be even less than that. APL was the little brother who kept hanging around, begging for attention. The only outer planets mission launched by the Laurel lab was New Horizons to Pluto, and that was practically a tinker toy next to the mighty starcraft unleashed on the cosmos by JPL. It was cute that the Applied Physics Laboratory wanted to play, but this was fourth-and-goal. It was no time for handholding and mollycoddling, I mean, first Europe, and now this? *It was all politics!* lamented the engineers, not entirely inaccurately. In fact, the whole of the space program limped along on such arrangements; they were why single rockets were built across several states. More stakeholders meant more protection. Leaders of both labs knew that Maryland senator Barbara Mikulski, the powerful ranking member of the Senate Appropriations Committee, represented APL. She shielded the projects in her state once they were green-lit by NASA. Of course, the scientists and engineers at the Applied Physics Laboratory had their own long list of criticisms of the profligate JPL, which was a vortex of waste relative to the tight ship they ran; which had an impressive record of success—if, that is, you ignored all the failures; to say nothing of the Pasadena lab's perennial shortage of key personnel and its perpetual surplus of delayed missions.

And so, with a new "sweet spot" set of rules and a new pair of partners, Europa and Titan stepped into the O.K. Corral.

ON FEBRUARY 6, 2006, the new NASA budget was announced. For planetary scientists, it seemed as though the End of Days was again upon them. Agency science was set to lose in 2007 over three billion dollars from its long-term road map—a severe blow to a directorate whose entire annual budget was just over five.[275, 276] Solar system exploration alone absorbed two-point-nine-nine billion dollars of that cut.[277]

Bush administration officials, driven by desperation, were funneling funds instead to pay for a vehicle—*any* vehicle—that could carry humans into space as part of the Constellation program, its station-moon-Mars mission sequence. After the heartbreaking disintegration of the space shuttle *Columbia* on the first of February 2003, the fleet spent thirty months on the ground for inspection, and it was fast clicking with everyone that age and employment had rendered every shuttle a potential tragedy with wings. It wasn't reckless (yet) to keep them operational, but America's space trucks were definitely flying on borrowed time. Human spaceflight being NASA's raison d'être, this was a shaky foundation for ongoing operations. Unless astronauts acquired a new set of wings and boosters and soon, Americans would be forced to ride in Russian capsules on Russian rockets, the ignominy inconceivable. We won the space race for *this*? Can it be said even that we won if we were *trapped on Earth* while the losing team could just come and go as they pleased?

From his office in Houston's Seventh Congressional District, just off the newly expanded Katy Freeway—smooth, black-topped, and twenty-six lanes of Lone Star freedom—there was the honorable John Culberson of the U.S. House of Representatives, member of the appropriations subcommittee with jurisdiction over NASA, who knew where he wanted the agency to be one hundred years from now and how he wanted to get it there. It was just self-evident to him: all roads ran through Europa, and he personally, repeatedly wrote missions there into the reports accompanying the federal budget, including the most recent.[278] If there were things swimming in Europa's ocean and NASA found them, the agency would also find the sort of support not seen since Apollo, and, well, cue the gospel choir. But as it had done with JIMO, headquarters said thanks for the funds, and yeah, great advice, Mr. Culberson, but we'll take it from here. And once again, Europa saw nary a nickel.

THIS DID NOT GO OVER WELL WITH THE HONORABLE JOHN CULBERSON OF THE SEVENTH CONGRESSIONAL DISTRICT OF TEXAS, MEMBER OF THE U.S. HOUSE OF REPRESENTATIVES APPROPRIATIONS SUBCOMMITTEE ON COMMERCE, JUSTICE, SCIENCE, AND RELATED AGENCIES. That tremendous freeway near his office? Well, it was he who built it, fulfilling a campaign pledge to do something about the traffic. John Culberson—he was a guy who got things done. No money for Europa? Well, that aggression would not stand.

Lacking the chairman's gavel, however, there was only so much he could do. The latest budget brutalized the agency, and he sent an open letter to scientists and engineers urging them to call their members of Congress.[279] He threw rocks at NASA headquarters for ignoring his *specific congressional directive* to go to Europa. He had cultivated a real contempt for NASA's astigmatic leadership, whose fecklessness and apprehension knew no bounds, and also for the White House Office of Management and Budget, which somehow made NASA leadership look audacious in comparison.

Why were we allowing accountants to run an agency of science and exploration? Why were we wasting the greatest space program in the world?

Amid these cuts to planetary science, the thing that Europa had going for it was unity. Sure, the outer planets community argued internally about science and exploration prospects and priorities, but it kept those things in the family now. The Outer Planets Assessment Group was working as intended, and public pronouncements were issued routinely now with one voice. OPAG listed Europa as the "consensus priority target" for the community, and that was that.[280] Moreover, the planetary Decadal had declared Europa the top target for space exploration, and we were still in that decade. It was an unambiguous, unimpeachable affirmation from the entire community—even those who would not benefit personally from a Europa mission. There could be no higher recommendations than these. Planetary scientists the field over were in lockstep. Europa: DO IT! And NASA, while not legally obliged to DO ANYTHING AT ALL, was at least morally compelled to follow the consensus of the field, and, short of that, give some defensible reason it was rejecting the informed, unified, and harmonious opinion of the community it served.

Enter Jonathan Lunine of the University of Arizona.

Each year, the crushing majority of the planetary science community gathered in Texas for a weeklong, field-strong confabulation called the Lunar and Planetary Science Conference. In March 2006—a year before the Quad Studies started—fifteen hundred scholars from across the country and around the world gathered to present papers, network, give talks, and organize working groups.[281] Even NASA's science leadership at headquarters attended annually to face the community in a town hall session. On the first day of the conference, as was tradition, an influential scientist would give a keynote address called the Masursky Lecture—a sort of master class on some science, celestial event, or pressing planetary issue. Topics tended to trend along the lines of "Kuiper Belt Binaries: A New Window on Runaway Accretion" or "Mercury as an Object Lesson on Approaches to Planetary Exploration."[282] In 2006 the Lunar and Planetary Institute, which sponsored the conference, asked Professor Lunine to deliver the keynote. Jonathan's talk was titled "Beyond the Asteroid Belt: Where to Go Next in the Outer Solar System, and Why."[283]

He did not intend for his talk to be controversial. Provocative? Well, sure, but only in the way that stating an unspoken fact can sometimes provoke discussion. After a kind introduction and warm applause, Lunine opened by running through the spacecraft to explore the outer planets of the solar system, Pioneer 10 in 1972 through Cassini today, which was studying Saturn even as he spoke. He listed the fundamental questions that could be answered only by reaching beyond the asteroid belt: How did the solar system get this way? Next slide. How did the giant planets and their moons form? Next slide. Are any of them habitable? Next slide. And yet, he noted, there had been a literal failure to launch since 1997.

Jonathan was an instrument lead on the first real Europa study likely to leave Earth (it didn't), so it was nothing personal, what he was about to say, but maybe because of that Europa association, what he said had a particular sting. While we've been wrestling vainly with Europa and its impenetrable ice shell and its ferocious radiation environment, there have been new and exciting developments elsewhere in the solar system, and specifically at Saturn, he said. Ladies and gentlemen, since we haven't launched a spacecraft anyway—since we have not settled, even, on a spacecraft concept— maybe we need to rethink where we are going.

"All plans for exploring the outer solar system are in complete ruins at the moment," he said.[284] Next slide. If microbes, merpeople, or sea monsters swam in the ocean of Europa, how, exactly, were we going to get through an ice shell fifteen—or maybe sixty!—miles thick to study it? We didn't know where to land because we lacked adequate surface resolution. We didn't know how to drill through the ice, and if we figured that out, we didn't have space submarines to swim the subsurface seas. And anyway, how long, exactly, would all this take? "Maybe it's the right target to go to first in the outer solar system," he said, "but we've tried." Next slide.

HOW ABOUT ENCELADUS, read the slide, and across the audience, jaws either dropped or clenched and eyes on pensive faces darted left and right—what the—what was this guy *doing* up there? We'd already endorsed—*as a community*—Europa! Somebody shut off his microphone! Jonathan explained that unlike Europa, Saturn's moon Enceladus was expelling its ocean into space. You didn't need a spacecraft to penetrate an ice shell to get to the ocean. The ocean was penetrating the ice shell to get to the spacecraft. Next slide.

But there was an even better choice, he said. The best choice. The most obvious choice. Why not Titan? The Saturnian moon had dunes, lava domes, a liquid-carved crust, not placid weather but veritable *tempests* that carved the soil! There were standing lakes of liquid methane on its surface. Organics, he explained, were everywhere on Titan. Next slide.

Titan wants us! he said, and put it right up there on the screen, projected in PowerPoint in large letters sans serif. TITAN WANTS US! It's easy to get there. We know the surface conditions. And it promised "the archetypical motivator of space travel: to explore a strange new world." Next slide.

Lunine listed the benefits and negatives of exploring Europa, Enceladus, and Titan. Europa's ice shell and computer-melting, thermonuclear-war-level radiation were, in his view, pretty big minuses. Titan's singular negative? Well, we hadn't yet planned a mission to get there—but we could. "We have no baggage or anything that's weighing us down in terms of thinking about what we might do."[285]

Jonathan wasn't oblivious to the crowd's reaction. He was a professor—lectured for a living—had been doing so for a very long time. And as he spoke, he noticed the talk was . . . not going well. Standing room only gradually gave way to standing room and a few chairs in the back . . . and along the aisles. People were leaving, just walking right out. He finished his talk to silence. During the question-and-answer period, he noticed entire rows in the rear opening up, and then the front: a crowd, dissolving before his eyes.

After the talk, he emerged unmolested from the ballroom—he didn't need a police escort or anything—and crossing the hallway threshold, he smacked face-to-face with lapel after lapel of these Europa buttons or stickers worn now by his colleagues. They (i.e., the paraphernalia) had proliferated, apparently, during his talk. And while he was happy to continue the discussion, he had a flight to make to Rome, where he was working with Italian colleagues on a project, and he slipped to his car and hit Beltway North for IAH.

Lunine was right about one thing, later conceded those who listened. Because NASA had not maintained a careful cadence of exploration, the outer planets, lighted for three decades by one spacecraft after another, would almost certainly go dark after Cassini. The worst-case scenario was upon us, and it was unlikely that a spacecraft could even begin construction before Cassini plunged permanently into Saturn. During the decade of darkness that would follow, planetary scientists would slip away from the field and into the arms of industry. The longer NASA took in getting behind an outer planets flagship, the fewer graduate students would focus their studies beyond the asteroid belt. What would be the point? Better to be an employed Mars researcher than a starving Europa scholar.

But the Titan community, grumbled scientists, had no real mis-

sion concept.[286] Its science wasn't mature—it wasn't even complete!
It wasn't like they were wanting for new material: Cassini was still
there and would be for years to come, transmitting tantalizing Titan data practically daily. It took years to formulate the correct *questions* to ask of a world you wish to explore, let alone develop a way to
answer them. By the time a Titan study reached the project phase,
we could have a spacecraft at Europa, or at least flying there.

ON CHRISTMAS DAY 2004, two years before the Masursky
Lecture and Jonathan's act of sedition, Cassini, its orbital position
just so around Saturn, released the lander Huygens.[287] It would
take twenty days for Huygens to arrive at Titan, and in the runup,
its science team gathered at the European Space Operations Centre in Darmstadt, Germany—Europe's equivalent to the Center
of the Universe at JPL—to prepare for the first spacecraft landing on an outer planetary body. As Huygens descended to Titan's
surface, it would beam back to Cassini—which was within range
and recording relentlessly—everything it saw, sniffed, scanned, and
struck. Once the probe alighted on Titan's topsoil or ocean (the
probe floated . . . just in case!), it would live for maybe three or four
minutes before passing through nature to eternity. The data upload
complete, Cassini would swing around and point its dish toward
Earth and transmit the complete collection of Huygens's Greatest
Hits.[288]

Lunine was an interdisciplinary scientist—basically a free
agent—on Huygens and, like everyone else, really wanted to be
there for the probe's initial images as it drifted down to titanus
firma. Nobody had ever *seen* the surface of Titan before—not with
telescopes, space observatories—not even from orbit, Cassini's
camera stymied by Titan's opaque atmosphere shrouded in hydrocarbons. And if you were a Titan scientist, vexed by its flaxen

veil—oh, you developed models, analyzed the data, saw a surface with your mind's eye as clearly as deaf Beethoven heard Opus 125 on the page—but you just didn't know for sure—and you wanted to be the *first* to see it.

It was freezing that day in Darmstadt, a city south of Frankfurt, renowned as the beating heart of German science, with research centers, universities, technical institutes, and, on a hill to the south, Castle Frankenstein—that one, yes—which, if nothing else, loomed as a menacing scientific totem to ward away hubris. The Huygens imaging team members were holed up in a tiny trailer adjacent to the operations building. They needed to work undisturbed, analyze incoming data in an isolated setting, didn't need managers, dignitaries, and press peering over shoulders, smudging monitors with greasy fingers and asking questions as though speaking to tour guides. Jonathan wasn't on the camera team, but he scored an invite by its principal investigator to join them in the trailer and bear witness to the horizon to which he had dedicated his life.

Though the Huygens antenna was oriented and powerful enough only to reach Cassini, radio astronomers back on Earth did the math and realized they could track its carrier signal—that is, they could tell *when* Huygens was talking to Cassini, though not what it was saying. If nothing else, they would know if it was working and for how long. That was exciting in itself. So on the morning of zero hour, as Huygens plunged from the tip-top of Titan's atmosphere to touch down on solid surface, the Titan team gathered to watch the signal of its able emissary. Descent took just over two and a half hours, and slowdown involved a heat shield and three parachutes— the last chute smaller than the second, designed to speed things along lest the lander ran out of battery before actually touching the surface. In all, Huygens slowed from fourteen thousand miles per hour to ten miles per hour and, on gentle touchdown, zero.[289]

Engineers expected Huygens to survive three or four minutes

after touching Titan's surface, and there were cheers all around when zero altitude was reached and the signal survived. What a plucky little lander they had fashioned! And three minutes elapsed, and it was just great—marvelous!—and then four, and that was even better. Every second on the ground would yield extraordinary data. And then five minutes elapsed, and could you believe how well we built this thing? And then six and seven and then *thirty* minutes elapsed, and still, there's Huygens chatting happily with Cassini, and this went on for an hour, and then *two* hours, and everyone's looking now for Rod Serling, cigarette in hand, submitting for our approval one Huygens probe, weight: seven hundred pounds, with a delicate payload of six scientific tools designed to decipher a murky, mysterious world.[290] Engineers expected its battery life to last no longer than three hours, almost all of that dedicated to the descent. It was now approaching hour five.

After three hours and fourteen minutes sitting on the surface, Huygens at last transmitted its final packet of data.[291] Cassini had long stopped listening; it had crossed Saturn's horizon hours earlier and, as planned, swung round and phoned home. It takes more than an hour for a signal sailing from Saturn at lightspeed to arrive at our planet. At midafternoon, it reached the Deep Space Network on Earth. The "recording" played. And there was nothing. Silence. Dead air. And at that moment, the Titan team weathered the woes of the crews of the Mars Polar Lander, Mars Climate Orbiter, and Mars Observer. The moment of betrayal by a promised signal.

Or maybe not. After five minutes of terror, data began to arrive from the *second* channel of the probe's communications device.[292] Huygens, time limited (Rod Serling notwithstanding), spoke to Cassini on twin channels. For whatever reason, one of Cassini's two channels did not play properly, and the science team would thus receive data from only one of them. Which was OK—the channels were redundant! That was the *whole point* of redundancy.

But members of the imaging team, congregated together, arms folded, gazing downward, would look up, lock eyes, exchange glum, knowing glances, their insides now melted, turned to goo. The camera system was *not* redundant. It used both channels. The idea was to get twice the data during descent. Now they would have half.

The cold, convivial dawn in Darmstadt became a raw, dolorous evening. It took time for the Huygens data to be split off and sent to the various science teams: the atmospheric structure group, the surface science team, the mass spectrometer section, &c. It came from space as a single, eight-hundred-million-mile beam, the data, and the imaging team trudged to its trailer, a brief walk through a piercing German January chill, to wait. When it was time, their parcel downloaded, they huddled around the computer, and one-by-one, one image per second, they became the first human beings to gaze beneath the Titanian haze. They were blurry at first, the images, fuzzy from the murky sky, but slowly a world resolved below. The images weren't in order—you'd get ground and sky and haze and hillside at incongruous intervals—but suddenly a shot filled the screen that caused Jonathan to gasp and scream, and the entire imaging team recognized what they were seeing, and hooted, hollered, and hurrahed. They were staring at a set of braided river channels. *Liquid flowed on Titan.* Maybe now, maybe long ago, but it *flowed.* There were hills. A stream. It was—these were features you could recognize. And there were features that you could not—in the plains, there was this thing that looked like the tail of a dragon. And as deeply and distressingly as it had set in, evening gloom dissipated, gave way to the unfettered joy, the frontier excitement of science, discovery, and exploration. No human eyes had ever seen what they saw. No human eyes might ever again see it in such detail. The wider Huygens team, the press, and the public all needed to know now what the camera group knew, what they had

witnessed, so at ten o'clock at night, the imaging team began building a mosaic to go out to the world. They pieced together Titan from
different heights, eyes unblinking, the team running on reserves, the
coffeemakers mainlining caffeine into their systems. This world they
were building was so recognizable and yet so alien, exotic, unnatural.
Mountains and lake beds, floodplains and channels where rivers
flowed—it looked both like Earth and not. Titan was a world not
of water but of hydrocarbons, and its haze, its sepia skies, the low
light levels, the enigmatic chemistry—it was all the same, like here,
but made of different stuff. Where Huygens settled on the surface
were these rounded stones made of . . . ice? . . . perhaps, from Titan's interior. The scientists were doing analysis on the fly here. It
all reminded Jonathan of the lowest level of Dante's hell: a place
not of eternal, inextinguishable flame but of blistering cold, and
the pit above shaped by landslides, earthquakes, and erosion.[293, 294]
Titan was so familiar—Jonathan couldn't look away—and yet so
bizarre: Earth through a mirror darkly. Titan felt like a place that
was somehow *wrong* to see.

AFTER A DECADE of leading strategic science committees, sitting on prominent panels, and just being an agency go-to guy, Jonathan Lunine, post–TITAN WANTS US! was suddenly persona non
grata'd, a troublemaker, and for more than a year, his phone simply
stopped ringing. But his reputation at Jet Propulsion Laboratory
remained quite good. He knew how to do mission science, which
is why, when JPL inherited leadership of the shootout studies in
2008, Curt Niebur asked him to lead the American half of the
Titan science definition team.

Ralph Lorenz of the Applied Physics Laboratory had led the
initial Titan proposal for the Quad Studies. An engineer and planetary scientist, Lorenz had worked on the Huygens lander and, by

the time of the Titan study, had written a half dozen books on the Saturnian moon, physics, spaceflight, and planetary missions of exploration. Never dry tomes, either, but imaginative works with literary flair. Accordingly, what he and his co-lead came up with was extraordinary: the Apollo program if Jules Verne had been in charge.[295] An orbiter that would map Titan from space, scan its surface composition and atmosphere (by doing so returning four times the Titan data that Cassini could hope to achieve); a lander that would drift to the surface by parachute, touch down, and collect close-up data and take seismic measurements; and a Montgolfière—a hot-air balloon—that would drift across Titanian skies, passively studying its atmosphere, imaging its surface at a meter scale, and going whither the wind would take it (and thus returning meteorological measurements as well). It could float across the lakes, capture waves and tides, and amble aloft and across Titan's gentle mountaintops and valley bases.

Even the arrival was ambitious. You could use Titan's atmosphere to enter orbit by way of something called aerocapture, which involved arriving at a target body, dipping into its atmosphere, and letting the resultant aerodynamic drag slow the spacecraft to a suitable speed.[296] Once decelerated, the vessel could again exit the atmosphere and enter a suitable orbit around the moon itself. This was critical for the Titan mission because of the astounding speeds at which spacecraft traveled through space: in this case, four miles per second. Historically, to enter a body's orbit, a spacecraft would fire thrusters to slow down. The problem was that such propulsive braking required you to bring boatloads of fuel because those thrusters were going to be blasting like crazy to slow this thing to some reasonable velocity. But by using aerocapture, the Titan mission wouldn't need all that fuel and could use the thus-available mass for other things. It wasn't some harebrained scheme: aerocapture was proven, rated even for human spaceflight in a little program called . . . Apollo.[297] Alas, when Alan Stern narrowed the studies from *quad* to *duo*, an inexplicable decision filtered from headquarters forbidding Titan from using aerocapture, which really annoyed Ralph because it meant the spacecraft wouldn't be able to carry as many instruments, and would therefore do less science. Moreover, Jet Propulsion Laboratory would build the orbiter, but the European Space Agency would build the Titan lander, as it had done for Huygens. The Montgolfière would go to the French for reasons perhaps poetic: the brothers Joseph-Michel and Jacques-Étienne Montgolfier invented the hot-air balloon in the late eighteenth century.

Well, right away, Ralph figured, that was that. All the sexy stuff on his mission was going to the Europeans! Ralph himself was born, raised, and educated in the United Kingdom and was stridently Scottish, so it was nothing personal. But in this arrangement, he worried, Jet Propulsion Laboratory would be far less invested in making the Titan mission a success versus that of the Jupiter Europa Orbiter, which would be built almost entirely at the lab. So he expressed pointed disinterest in leading the next step. He had done a fine job the first time, thank you—why reassess the mission with excessive constraints that could only frustrate everyone involved? He would still take part but would not take point. Thus the appointment of the apostate Jonathan Lunine.

In a sense, Lunine was born on Titan. He had a tough childhood. His father was an alcoholic, and the disease would prove fatal, though it didn't happen all at once. There was a long decline first into dysfunction and then to divorce, and in the meantime, to support her family, Jonathan's mother went back to work as a dancer—a Rockette—at Radio City Music Hall in Manhattan, a job as grueling as it was glamorous.[298] At the rehearsal hall, you stepped on a scale sometimes daily and had to weigh just so. Had to smile just so. Had to kick, tap, pirouette, jeté just so. You rehearsed this rigid, Swiss-watch choreography all the time, and when you didn't get it right, or—and this was too horrible to fathom, but it happened—if someone made a mistake during a performance, you repeated rehearsals between shows. And you're doing this while still wearing the luminous Rockettes smile. (Not too broad, not too narrow. Just so.) The money wasn't great, and the work was seasonal, but you were a celebrity. She loved it, and because she was now a single mom with two children, each of those thousand daily breadwinning kicks counted, kept her family fed, under a roof, in clean clothes, and warm on winter nights.

Then came the toe problem (she was a dancer; it was just part of the job), and she was on painkillers for that, and it sure didn't help her overall health, and then one day she was walking the family dog and she slipped and broke her shoulder, and a gloomy situation got grim, and fast. A dancer with a broken shoulder cannot dance, is not even a dancer, is—what is she? A Rockette with a broken shoulder cannot be a Rockette. There were ten thousand women waiting in the wings to weigh just so and smile just so and kick and tap and pirouette just so, and Jonathan's mom was soon on disability and then unemployment.

Jonathan, fourteen years old, escaped a dreary life in the pages of *Sky & Telescope* magazine. Every month, he waded through issues of galactic import—the goings-on of spacecraft launched and in development—major moments in astronomic history—the transits and occultations of planets circling the sun—observatories now open and what they've seen—notes from such recent meetings as the Lunar Science Conference in Houston—serious science: the early evolution of stars, the activities of comets at perihelion—astrophotography—a celestial calendar for the month ahead. He saw mention of a book by this astronomer at Cornell named Carl

Sagan, titled *The Cosmic Connection: An Extraterrestrial Perspective,* and he rushed to the bookstore to buy it, money crumpled in his hand. The store didn't have it, so he had to wait for it to be specially ordered, shipped, and delivered, and when he received it, he'd bent the spine in seconds.

This book was everything. The words were so eloquent that, unprompted, he would read passages aloud to his mother. Jonathan loved science fiction, but this was just so much more than a moment's entertainment. He saw a glimpse of his life that might be, of studying and exploring the cosmos.

His mother encouraged Jonathan to write this Carl Sagan fellow a letter, but Jonathan felt that he couldn't do that—I mean, you didn't just write to guys like *Carl Sagan,* people who could write books like *this,* but she really pressed the point, and because deep down he knew he wanted to write it, to reach out, to somehow connect to this person (she knew it, too), he relented. Jonathan poured himself into the letter, explained his love of astronomy, of the book, and he asked, simply: How do I do this? How do I become an astronomer? I'm not rich. How do you even *pay* for something like college? He found the right address, folded the letter, stuffed the envelope, stamped it, and sent it.

In April a letter addressed to Jonathan—well, it was so much more than that: a manila envelope, nine by twelve, thick with something more than a single sheet of correspondence inside— arrived in the mail. He traced his finger on the return address. Center for Radiophysics and Space Research. Cornell University. C. Sagan. He opened it, and there was a letter in there, two pages, dated April 18, 1974, and it was typewritten, single spaced: "Dear Mr. Lunine . . ."[299] It opened with a thank-you for your letter, and explained that all good colleges and universities, including Cornell, offered scholarships to students in financial need, and it went on to describe what he should study in high school if he wished to be an astronomer, and what happens in college, and enclosed were reprints of recent scientific papers from the Mariner 9 mission to Mars—Mariner 9! The first spacecraft to orbit another planet! And there was Sagan's signature, scrawled carefully, just so, and Jonathan was holding this, reading it, and it was just—I—what— he'd never seen a real science paper before—the odd two-column format, the peculiar typeface, the diagrams, and those paragraphs: dense as neutron stars. And Jonathan was holding the letter, the science papers, and feeling the texture of the thing—the thickness of the paper—and he smelled it, took it all in, and he just looked at it and the words, and he read the letter and the papers again and again. He was now a part of something. He was connected to this thing: astronomy, space exploration. He lived on a spinning ball of rock ninety-three million miles from the sun, and he had this. This thing. These pages. This was his. He was part of it.

Thus began a continuing correspondence with Sagan, whose advice was solid, and Jonathan followed it for a couple of years, but the seventies were on the downslope, and the space program and science more broadly were not doing well, and there was a stereotype out there, and Jonathan heard it, and he couldn't shake it: of the person with a physics Ph.D. driving a taxicab. And Jonathan had to be realistic here. He was smart and he needed to make something of himself, needed to make sure he didn't bring the financial circumstances of his adolescence with him into adulthood, and he agonized over it—it felt wrong, but he was set on it—and he decided to become a medical doctor. He held doctors in high regard—they saved lives!—and he appreciated the gadgetry, the high technology. And—very important—it was a guaranteed job. So he read books on the heroes of medicine and volunteered at Manhattan's renowned Mount Sinai Hospital, where he worked for an oncologist, making graphs of data for her and sending out reprint requests. He wrote to Sagan about this decision, and Carl was gracious, responding in his customarily wry tone that medical research carries with it *some* of the frontier excitement of astronomy.

In his senior year of high school, Jonathan discovered that his pediatrician had gone to Cornell—Jonathan had dug up an article the doctor had written in 1930—and when visiting his office, found out that he had *up-to-date Cornell course catalogs.* As Jonathan flipped through the pages, his pediatrician said in passing that, hey, I have these tickets for a talk by Frank Drake, a Cornell astronomer, and do you want to go instead of me? Jonathan nearly said no—his destiny was in medicine. The offer even offended him on some level. *I'm going to be a doctor—what don't you understand about that?* But for whatever reason he said OK.

Drake, one of Sagan's colleagues at the university, had worked on the search for extraterrestrial intelligence in the sixties and seventies, and in that evening's lecture, he outlined the latest science from the radio telescope at the Arecibo Observatory in Puerto Rico. And once again, there was Jonathan, on this ball of rock swinging round the sun, and there was this guy Drake speaking, revealing Arecibo findings that were helping to unlock the secrets of the universe, and with every word and across each passing minute, Jonathan could feel the physician's life slipping away like a lab-coated ghost leaving his body. He was too shy to approach Drake after the lecture, though he wanted to, but as he stood in the elevator, going down, he made his decision—or rather, unmade his decision and reverted to the first. He didn't want to be a medical doctor. He'd *never* wanted to be a medical doctor, not really. He was going to be an astronomer, follow in the footsteps of Carl Sagan and Frank Drake. And that is what he did.

Despite his continual, coincidental connections to Cornell, Lunine didn't go there, just couldn't afford it. He went instead to the University of Rochester, in western New York. During his last two summers, he worked at the Lunar and Planetary Institute in Houston, which organized the Lunar Science Conference, the proceedings of which he had read in the *Sky & Telescope* magazines of his youth. (The conference, later renamed the Lunar and Planetary Science Conference, would host his heresy and undoing.) After earning a bachelor of science in physics and astronomy, he cast his eyes across the map to figure out where to go for graduate school. Jonathan applied to Cornell, MIT, Caltech, and the University of Arizona, among others, and was accepted by each of them. It came down to Caltech and Arizona, but Caltech put on a better show, and so Caltech it was. He packed his '72 Datsun 1200—a prototypical college student car, standard transmission, aftermarket air conditioner that, when you turned it on, caused the car to lurch as though you had just applied the brakes. There were twelve-inch wheels on the thing, and he learned that them there were pretty small wheels, son, because he had a blowout on the Joad-like drive to the West Coast, and the guy at the service station put a standard size tire on it. It was August and somewhere past Houston, the nonfactory air conditioner gave out, and west he went down Interstate 10, through the Chihuahuan Desert and the Sonoran Desert and the Mojave Desert, and it was a rolling lesson in how a sweat lodge can alter one's frame of mind. From Amarillo to Flagstaff, then through Phoenix and on to Los Angeles, he survived the West, his horse a Datsun with a bum wheel and broken AC.

His dissertation at Caltech was titled "Volatiles in the Solar System" (volatiles being things that like to vaporize and become gases). The finished product was a behemoth at three hundred twenty pages, and came in three parts: 1. a theoretical study on the thermodynamics of clathrate hydrates (water ice formed in the presence of high-pressure methane or other gases); 2. the evolution of Titan, Saturn's cryogenic moon; and 3. a model that he put together with his thesis advisor and another professor proposing a global ocean of ethane and methane on Titan. There was some poetry in all this: Carl Sagan had once proposed an ocean of methane on Titan, but data from the spacecraft Voyager 2 didn't jibe with pure methane. Jonathan's ethane-methane ocean solved the problem. That ocean part of his thesis was completed early—in 1983—and Jonathan submitted his model for publication. Cornell that year was, coincidentally, having a conference on satellites of the solar system. Jonathan put in an abstract, though not on that topic—he focused more on the mundane. He flew out there with his thesis advisor, they having together hatched a plan in which he (i.e., Jonathan) would surprise everyone with his ethane ocean talk.

It was a great conference. They scheduled his talk for the evening, lowly Lunine, and his advisor (in collaboration with a Cornell physics professor) bought Jonathan a couple of beers to ease things along. After downing them, he took the stage, stood before a nice-sized crowd, and gave his talk on the ethane ocean model. The Q&A period led off with a question about whether organic solids would sink or float in an ethane-methane ocean. Jonathan was about to reply, when he was interrupted by a voice from the back of the room—this lilting, mesmerizing baritone—and up sprung this man declaring boldly that he knew the answer.

It was Carl Sagan.

He climbed onto the stage and placed on the overhead projector a transparency listing the densities of various organic solids, and comparisons to the density of an ethane-methane ocean.

Jonathan's emotions slid across an internal spectrum from resentment to acceptance: if he was going to be upstaged by anyone, it may as well have been Sagan. There was also an element of curiosity and suspicion: nobody knew that Jonathan would be discussing the ocean or the details of his model, or would have had a transparency at the ready . . . unless Sagan had been one of the referees on the paper Jonathan had submitted for publication. The boy who'd

written the famous astronomer a fan letter a decade earlier was now
officially his colleague.

THE ROAD TO the final shootout selection gave both teams a nice
stretch of time to really work up some acrimony and indignation.[300]
The shootout ran for about eleven months, culminating in each
team handing in a NASA study report, a European Space Agency
study report, and a joint summary report explaining how the two
missions would work together, should they fly together, which they
weren't required to do. The NASA reports alone for each mission would be door stoppers—well over four hundred pages—and
collectively, you had the spacecraft equivalent of *Infinite Jest*. The
shootout would culminate in full presentations by each team at Jet
Propulsion Laboratory, during which the respective mission's science, technical design, management, and cost would be discussed,
and panels sick of lugging around these foot-high stacks of reading
material would keep drilling down into details, divining fact from
science fiction. The Titan team would talk on December 9, 2008,
and the Europa team two days later.

The stakes were just so high: the future of outer planets exploration for ten, maybe twenty years—the proposed launch year
for the winning mission being 2020, and then the travel time, and
then the prime missions. Jonathan, especially, felt drawn into what
he came to see as an arena. Titan had defined his academic and
professional career, so he was aggressive in criticizing the Europa
Jupiter System Mission . . . even when he wasn't strictly *invited* to
make those criticisms. The Titan tribe developed a slick marketing
campaign of videos and brochures that really sold its vision: the
Montgolfière floating above, waves lapping on the shoreline below.
Titan science since the heady Huygens descent had exploded. Not
only had scientists discovered elaborate lakes of liquid methane,

but so too had they found a vast network of capacious seas.[301, 302] The Quad Studies submission for Titan exploration called for a lander. The shootout study would build a boat! And a balloon. And an orbiter.[303] And while Titan science swelled and ripened, so too did that of Enceladus, tiny Enceladus, which, like the sculpture of its mythological namesake at the Palace of Versailles, jetted water vapor (and salt and silica grains) into space from vast stripes of boiling slush. From there, it snowed upward. Saturn was where the action was, and the Titan team castigated the Europa mission in NASA hallways, grumbling to anyone who might have some say in the matter, "Well, I'm just glad we're not building a Christmas tree like the *Europa mission*," and just letting it hang. And that's how they felt, the Titan people. Galileo barely launched. Cassini barely launched. You're telling me this Europa mission is going to launch, no sweat, by the numbers? You think Congress won't step in and start scrutinizing this battlestar sure to be annihilated in mere months, at best, by the radiation that engulfed their little ice moon?

For their part, the members of the Europa team just could not believe what they were hearing. *Christmas tree?* Cassini was still getting data from Titan! And the Titan people, meanwhile, were developing the most ostentatious, convoluted mission in the history of space exploration. Boats and balloons and orbiting satellites? Good luck keeping even one of those under cost. The Europa team had been developing its spacecraft, science, and a broader mission for a full decade. There were no surprises here, and Europa's science was mature enough to bottle and store in a cellar. The Europa team knew how to study its target, understood the thrill of its target, the unanswered questions, the implications of an ocean beneath the ice. But the community and NASA leadership had been hearing about it for so long that it seemed, maybe, like old news? And with the Titan team promising golden-age sci-fi versus some boring old Europa orbiter, it was just discouraging. If you were on the Europa team, you just knew it: the Titan team was going to pull this thing off.

BOB PAPPALARDO FELT most acutely the suspicion—if not animosity—radiating from the Titan team, and his troubles began with a parting directive from the outgoing Alan Stern. Missions in development didn't always blow budgets because of engineering obstacles. Sometimes, said Stern, it was because of basic inexperience on the part of project leaders. So, he directed, in order for a principal investigator to propose a large mission, she would be required to have served either a four-year stint in some mission leadership role or two two-year tours as such.[304] Managers needed to know how to make difficult decisions and twist arms and sometimes fire respected colleagues when situations warranted. However, in a field as young as planetary science, there was a problem that Alan had perhaps not considered, and Susan Niebur, formerly of NASA and now running a consultancy, spotted right away . . .

It had been a tough six months for her, Curt, and the kids: first, chemotherapy to rid her body of cancer cells and reduce the tumor burden in preparation for a double mastectomy, which she had, to remove any undetected cancer cells that survived the slow poison drip. Then there were the tattoo sessions to mark permanent aiming points for another seven weeks of radiation therapy.[305] She felt great physically, relative to the previous couple of years, but she carried mortality now on her shoulders, and there were things she still wanted to do before dying.[306] Not that she had given up! Not that she had accepted it—dying—or that she *was* dying. But she was in an in-between state. She didn't know—*couldn't* know—if her body was growing cancer elsewhere. And she was a scientist. Inflammatory breast cancer had a ninety percent recurrence rate in the first five years. So much of her body was being blasted with radiation that lymphoma, a cancer of the immune system, was now a likely side effect.[307]

Susan's family was foremost in her life, and the preponderance of her time went to her children. But in the twenty-fifth and twenty-sixth hours of her day, she founded a group called Women in Planetary Science in hopes of correcting a punishing gender imbalance in the field. In part, she was frustrated that so many of her colleagues with children were unable to attend the Lunar and Planetary Science Conference each year because it offered no provisions for day care or nursing mothers. The annual gathering wasn't only about the science presentations you were missing; you skip a major conference, and you lose out also on the networking opportunities such assemblies facilitate, and your career consequently suffers. To get things going, Susan organized a breakfast, and twelve people responded, and the night before, she expected maybe eight to show up—you know how it goes—and the next morning, one hundred and eight women were there.[308] So the project was off to a promising start. She launched a website for the group to highlight news relevant to women in the field, encourage discussions, and share the stories of women who had pioneered planetary science.[309]

And right away, she identified that years-long mission leadership requirement as a big problem in need of a solution.[310] You have a field that is predominantly male, with mostly men as mission leaders, and *from the start*, practically no women were eligible to propose missions—ensuring that the problem would perpetuate. Moreover, the exchange of business cards being an important element when teams first organize, no women at conferences (see: no nursing rooms, no child care) meant no way to network your way onto teams and work your way to leadership positions to overcome the . . . and so on.

Back in the San Gabriel Mountains, the leadership requirement wasn't doing Bob Pappalardo any favors, either. He would never put his professional problem in the same galaxy as the one Susan was working to solve, but given the paucity of outer planets missions relative, especially, to Mars, outer planets scientists also hurt for opportunities to gain the requisite skill set to propose missions. If Bob, Europa's study scientist, was going to be its project scientist (should Europa win the shootout), he would have to find a way to solve his lack of project leadership experience posthaste.

The lab was one step ahead of him on this one.

Bob was told he needed to meet with Charles Elachi, the lab director, which was a little unnerving. The meeting would be about Cassini. Were they kicking the camera lead off the team? Bob wondered. That was Assistant Professor Pappalardo's goal all those years ago, after all, before being brought into the lab, what he had written down in that *what-do-you-want-out-of-life* workbook from the self-help section of the bookstore: lead scientist on a camera instrument. He wanted it to be to Jupiter, though Saturn would work. But maybe that wasn't the plan. Maybe they wanted him to be a Cassini deputy? No, that was too much.

So he met with Elachi, and the director gave it to him straight: Cassini is moving into a different mission phase, we're replacing its project scientist, and you're our guy.

Pappalardo was stunned.

By 2008, the planetary science community numbered in the thousands and growing, and the number of flagship planetary missions that launched in a single lifetime *might* reach five, and that number was shrinking. Project scientist of a flagship mission simply was not a job one aspired to. I mean, maybe for Europa, sure, but he had been building that mission his whole professional life. That was one thing, and even then, the lab wouldn't put it in his contract. But Cassini!

Well, of course he said yes, but right away, Bob voiced concerns about his obligations to the Jupiter Europa Orbiter and the joint Europa Jupiter System Mission study, and he was also still working on his book on Europa. But Elachi blew off his concerns: You have two deputies on each job, Bob, and two coeditors on the book. Figure it out. In the case of the Europa study, there was Louise, who also supervised the Planetary Exploration Group at APL, and Dave, who worked down the hall in Pasadena. And of course he had Ron Greeley, who was the American cochair of the joint science definition team, and Karla Clark, the study lead. On his book, he still had Krishan Khurana, codiscoverer of Europa's ocean, and Bill McKinnon, who was revered in the field.

But coverage was one thing. The Cassini project team seemed . . . disorderly to Bob. No question: the spacecraft was getting things done, but in its own way. Cassini's culture was a bit insular, and Bob, everyone knew, was an outsider. He hadn't fought the battles, wore none of the scars of the mission's tortured development. What happened was this. From the Saturn orbiter's inception, Congress and the White House and sometimes NASA headquarters had repeatedly carved into Cassini's budget, making pitiless demands for fewer features, greater simplification. And with each request, Jet Propulsion Laboratory had to streamline the spacecraft, more, more, keep going—where else can you cut costs?—and eventually, the lab solved a big part of its fiscal tribulation by bolting the scientific instruments directly onto the orbiter, thus removing any moving parts. Articulated arms, gimbals, servos: all were eliminated, which cost cuts and complexity, but at a different sort of price. With this final, fixed design, if you wanted to point an instrument at something, you had to point the *entire spacecraft* at it, along axes x, y, and z.[311] Consequently, during flybys, you couldn't use every instrument at once. For Titan's flybys, for example, if you wanted images, you couldn't use the radar. If you wanted to use the radar, you couldn't use the mass spectrometer. You wanted composition, you couldn't use the camera. And so on. During the six-year-plus cruise to Saturn, the rancor manifested. Recognizing that they wouldn't be able to accomplish all the science they had proposed, instrument teams fought for flybys among themselves, and not in the staid, genteel manner you imagine of scientists arguing for more mass spectrometry, but rather, with pointed fingers and that breaking-a-stick gesture. It was fear driven. But even after Cassini's prime mission was completed, everyone's chief science goals achieved, that tension never evaporated.

Bob took over as project scientist just as the Cassini mission entered its "extended" phase, when stretch goals could begin to be addressed. His first order of business was to choose a new orbital path around Saturn that would do the most and highest-quality science. What should we encounter and when—which instruments should be used—what would do the best measurements for the most disciplines—and so on. With limited time, instrument constraints, and a finite number of orbits, you couldn't do everything; Saturn had at least fifty moons, plus its rings, plus the planet itself, which defied explanation. In any mission plan, someone was going to lose. So the various working groups—icy satellites, Titan, rings, magnetospheres, &c.—submitted their preferred science, and project management plotted each on a chart to find that harmonious middle, the path that did the most science for the most people.

Bob felt sick when he saw the results.

Although there were many good ways forward, for the broadest spectrum of science to win, Titan had to lose. Other aspects of the Saturnian system, the results revealed, had higher scientific priority.

The shootout made a bad situation so much worse. The Titan and Europa missions were both good, doable, could fly tomorrow—and one was definitely going to fly—and you wanted it to be yours. But the stressful studies spawned mistrust, and already, the Cassini old guard were experienced knife fighters, and here was Bob the Europa guy put in charge of Cassini science having to tell Titan that, oh, by the way, I'm about to kill some of your hopes and dreams. He could practically hear the cries of conflict of interest.

Bob called Jonathan Lunine and explained the situation. Jonathan understood, acknowledging that the settled-upon plan was not as good for Titan but was the right answer overall. Bob asked whether he would be willing to *get on the speakerphone* and say that during the discussion. You're a Titan person. The Titan people will trust you.

Despite the open hostilities that now defined the outer planets community, Jonathan never thought of Bob as anti-Titan. And he came quickly to see that Bob's demeanor was right for the project at a pivotal time. So Jonathan said yes, and when Bob broke the news to the Cassini discipline working groups, knowing full well that his word would not be enough, Jonathan jumped in and explained the situation on behalf of the project scientist.

The shootout had placed Bob in an unbelievably bad spot as a matter of management and leadership, and despite Jonathan's interceding on his behalf, some Titan scientists complained formally, and others just to themselves while stabbing forks into Tupperware lunches. The well hadn't been poisoned intentionally, but you still needed to boil the water before drinking it, now.

THE TWO TEAMS, Titan and Europa, descended upon Jet Propulsion Laboratory and presented their respective cases to the evaluation teams in December 2008. They gave the context for their studies, mission overviews from the American side and the European side, and presented a concept for what a joint effort would look like. They discussed technicals, mission designs, operations scenarios, flight systems, programmatics—everything you could ask for—and held long question-and-answer sessions where panelists could ask for more details yet. Everyone involved had given so many talks, so many briefings, so many Q&As and PowerPoints, that, if they were being honest, the presentations felt almost anticlimactic. And then they waited.

Two months later, NASA and the European Space Agency announced their decision.

NASA chose . . . both missions![312]

That's how it was worded in the press release, that "officials decided to continue pursuing studies of a mission to Jupiter and its four largest moons, and to plan for another potential mission to visit Saturn's largest moon Titan and Enceladus." But below, the details clarified. The first to fly would be the Europa Jupiter System Mission. Meanwhile, technology development for the Titan Saturn System Mission would continue apace, and that flagship would launch a little later.

There was a sun-sized asterisk: the next Decadal Survey process had begun. A steering committee was working even then to assemble panels to make recommendations, and the whole thing would take about two years through to its announcement. Nothing would move into mission phase until the Decadal dropped. Which meant that for a Europa mission to go forward, shootout victor or no, it would again have to come out on top. But Europa came out on top in the last Decadal, and nothing had changed in ten years, so there was no reason to think the next one would be any different. Europa was well on its way.

When Jonathan read the decision, he felt for a moment elated—both missions were selected!—but after pacing around his living room for about an hour and thinking this through, it hit him: Europa had won and Titan had lost. Headquarters, when he called, was so noncommittal as to what came next for Titan development that they might as well have not even answered the phone, and he soon realized that he and his team had been given a form of the "Hollywood no," which was worse in a way because there would be no closure, no putting this behind them. They had been so close, come so far, left prints on the brass ring where their fingertips had grazed it, and now it was over. He would never likely see the surface of Titan again, would never see the cryogenic seas he had hypothesized. His fears were later confirmed. Though Cassini continued, the Titan Saturn System Mission had essentially been indefinitely deferred, JIMO style, the team handed a cardboard box and told to clean out its desks before leaving for the day, but nobody even bothered with the pink slip.

Jonathan was embarrassed. There was no other word. He had stuck out his neck during the Masursky Lecture and courted animosity in the community with his perceived insolence and aggressive evangelism of Titan, his love—so thrilling a target . . . how could we lose? They had this exciting and compelling mission. And he just felt like a fool. Maybe, he rationalized to himself, it was for the best. If a Titan flagship had gone forward, headquarters would almost certainly have killed Cassini to foot the bill. And Cassini was a great mission. But, wow, so was the Titan flagship. And now it was over. We were going to Europa.

CHAPTER 9

Grand Theft Orbiter

AND JUST LIKE THAT, KARLA CLARK WAS OUT.

The decade she had spent on the Europa problem—her management of a dozen Europa mission studies—more than anyone else on Earth—her fathomless insights on the Jovian radiation belt and survival therein—none of that mattered in the end.

Jet Propulsion Laboratory wanted a mission that NASA would buy, and Karla had delivered. Most of the Europa team thought she would be made project manager once the thing went forward. Karla knew better. The reality at JPL was that she did not yet have the experience on paper necessary to be made manager of a project that size. For a multibillion-dollar mission, lab leadership would typically put in place someone who had delivered a spacecraft in at least the Discovery or New Frontiers class, or their Mars equivalents. But it was the same problem planetary scientists faced when the four-year-leadership mandate came down from headquarters. The dearth of missions overall and the mostly male makeup of the lab meant that Karla never had hope—not really—of ever managing such missions.[313] So it was unclear even to her what would happen—what her role would be—once the study became a "pre-project," and later, once it reached the developmental milestone called Key Decision Point A, a formal project.

It was the outright hostility by management that took her aback, however.

Managers in the Solar System Exploration Directorate, where she worked, came and went as if by custom, and once the shootout ended, Jupiter Europa Orbiter the last spacecraft standing, the management merry-go-round happened to stop with an aerospace engineer named Keyur Patel, who was previously project manager on the Discovery-class Dawn mission to the dwarf planet Ceres. He had been with the lab for twenty-three years by then—about as long as Karla—and the two had history. Karla didn't respect him as a manager or engineer, and she was pretty sure he didn't respect her, either. It was obvious right away to Karla—and just about everyone else, if the rumor mill was to be believed—that Keyur wanted to be the project manager for this mission. And who didn't? The first spacecraft ever to orbit a moon other than that of Earth? The first spacecraft to orbit so deeply and for so long and lavishly in the Jovian radiation belt, second only to *the interior of the sun* as the most dangerous place a spacecraft could travel? The spacecraft likely to determine the habitability of an ocean world?

Keyur didn't just take the job, though; it was so much more insulting than that, Karla felt. As the study pivoted to a pre-project posture, he started staffing the team without consulting Karla, who, as study lead, was still in charge. Worse yet, if such a thing were possible, she didn't even know many members now under her aegis, and after reviewing their credentials was unconvinced that they were a good fit for the team. Karla had learned project management from John Casani himself. She knew how to build a team, and *this was not it*. Some appointees, she felt, were outright unqualified for the jobs they were being given. They were, however, part of the lab's thriving good ol' boys network, which, she knew, she was not.

Her project—her life's work—suddenly under threat by those she saw as unfit, she confronted Keyur about this, and his response, as she took it, was that this was not her project at all and that he was not interested in her opinion on the matter.[314]

Karla almost unclipped her official Jet Propulsion Laboratory access badge and dropped it on the table right there, but she didn't. She would see what happened next. She started, however, looking for a new job that very day.

There was a sign of hope when the lab assigned Tom Gavin—one of the lab's four pillars—to the Europa team as an advisor. He and Clark, too, had a history, but a good one. He was Karla's mentor during the development of Cassini when she led its power subsystems. Gavin was seventy years old and had come out of retirement to take this job. He was not a long-term project manager, and he never pretended to be. The Europa Orbiter still had a good ten years of development work ahead of it before leaving the launch pad. She hoped in that time to be made the deputy project manager and prove herself to laboratory management and NASA headquarters. Sometime around the final design and fabrication phase of the spacecraft, she imagined, she could take the wheel. Tom would be in his eighties by then, and though he was indefatigable, the man had to stop eventually—it gave Karla hope.

But, of course, hope is not a plan.

BEFORE BECOMING A JPL legend, Tom Gavin lived in a little house in a little town called Upper Darby that was near Philadelphia and known for absolutely nothing at all. He was born in December 1939, the last month of the last year of the Great Depression, and it was Tom, his mom and dad, three brothers, and a sister, with one sibling lost in childbirth. His home was full, his father drove a subway train, and they were poor.

In those days, the archdiocese of Philadelphia provided free education for grades kindergarten through twelve, including free books, and the family took advantage of that opportunity. When his older brother went to nearby Villanova University, Tom decided that he, too, wanted to go to Villanova, and when he was old enough, he applied and was accepted. He studied chemistry and math. He couldn't live at school because the Gavins just didn't have that kind of money, so he was what they called a "day hopper," and missed a lot of the college experience. During semesters, Tom worked twenty-five-hour weeks as a supermarket checker. During summers, he worked for a Titleist distributor, unloading eighteen-thousand-pound shipments of golf balls from eighteen-wheelers. It was hard, all of it—the relentless pace of sweat and study that only the working class ever experienced—and it didn't help his grades, but that's just the way it was at a private university when your collar was blue. Four years later, he had earned his degree and was ready to achieve his life's ambition.

When he was ten, Gavin saw a movie called *Destination Moon*. The film's plot involved a convocation of aerospace executives, engineers, and investors planning a private expedition to outer space. America had the rocket talent, but the American government lacked the vim, and rather than waste time slinging satellites into Earth's orbit, the expedition decided to just get on with it and plan a piloted moon mission, *in flagrante* Buck Rogers. One of the first serious works of speculative fiction to be committed to film, *Destination Moon* was based on a book by Robert Heinlein, who also cowrote the screenplay, which probably didn't hurt. And there it was, right there in theaters, in towns big and small, for the young and old, and especially for a boy from Upper Darby, how it was going to happen. What it would take to get us there—my God, *the moon!*—and what we would find once we landed. It depicted with realism the stresses of launch, the equivalent of astronauts (the word didn't yet exist) floating in microgravity, a spacewalk gone awry, and a lunar landing. The first words by the first man to press prints into lunar soil: "By the grace of God in the name of the United States of America, I take possession of this planet on behalf of and for the benefit of all mankind."[315] The film's foresight bordered on clairvoyant. This was 1950—three years before Wernher von Braun's *Das Marsprojekt* was published in the English language in the United States and seven years before Sputnik. Though the film was an exercise in educated guesses, in Heinleinian fashion, the script nailed it, anticipating artificial satellites, reusable rockets, the reentry methods of the space shuttle—even the motivations of the space race itself and that it would be considered a *race*. ("We are not the only ones who know the moon can be reached. We are not the only ones who are planning to go there. The race is on, and we had better win it . . . there is absolutely no way to stop an attack from outer space.")[316] It foresaw the woes that the spacecraft Cassini would encounter forty-seven years later (the same Cassini on which the ten-year-old from Upper Darby would one day serve as spacecraft manager): the launch of the (atomic-powered) *Destination Moon* spacecraft, *Luna*, is initially scrubbed because of alleged safety and environmental concerns. ("While it is admitted that no real danger of atomic explosion exists, a belief in such danger does exist in the public mind . . .")[317] Stretches of the film involve contractors and administrators sitting in rooms and having meeting after meeting, while engineers work their slide rules and bend metal on a spacecraft (another word that did not exist in this context). The U.S. government wouldn't foot the bill for the moon expedition; yes, even in science fiction, the space program couldn't get the money it needed! ("It's peacetime, Jim," says an ex-army rocket man to an aerospace contractor, "the government isn't making that kind of appropriations!")[318] And while Neil Armstrong's first words after taking one small step were loftier, globally inclusive, and laced with greater poetry, a phrase from our cinematic star voyagers—"benefit of all mankind"—would be the wording in the law establishing NASA eight years later, and would even appear in the agency's mission statement, updated to reflect the evolving English language: "benefit all humankind."

After seeing that film and glimpsing the wonderland of humanity's space-faring future, Tom Gavin read every work of science fiction he could get his hands on, every book about space in the library stacks, wanted desperately to be a part of the human enterprise that would one day take us there. And as far as timing went, his could not have been better. The year he graduated from Villanova, NASA's Mercury program had just put Alan Shepard in space, and President Kennedy would throw his hat over the wall and declare that "we choose to go to the moon in this decade and do the other things, not because they are easy, but because they are hard."[319] And with that, it was policy, destination: moon, and the American space program needed engineers—now. To staff quickly, battalions of aerospace companies from California canvassed the country, putting on career expositions in major cities. Tom, whose first postcollege job had him working as a chemist for Electric Storage Battery Company in Philadelphia, was rescued from the terrestrial trade by an advertisement in the *Philadelphia Inquirer* for one of these job fairs.[320] He attended, copies of his résumé in hand, and the Lockheed Corporation expressed immediate interest. Before he made any sort of commitment, however, Tom met a man at the fair named Bill Shipley, who worked at a place called Jet Propulsion Laboratory in a California town called Pasadena. Shipley was no simple starched shirt sent to collect résumés for the real engineers to later examine. He was, it turned out, a respected research engineer rising quickly in the ranks of lab leadership—having just been named chair of the NASA subcommittee defining how robotic space exploration could support the nascent Apollo program.

Why do you want to do this? asked Shipley, and Tom explained everything he knew about astronomy, the movie, everything he had read, and I mean, the future is in space travel![321]

Shipley offered Tom a job on the spot.

A few months later, Tom, his pregnant wife, Betty, and their five-month-old son packed everything they owned into a 1956 Ford convertible, drove west, and replanted their life. Thus it was that in 1962 Mr. Tom Gavin premiered at Jet Propulsion Laboratory.

They didn't give him the keys to the spaceships right away. His first few jobs involved doing quality assurance work on electronic parts. The lab was gearing up for the first mission to Mars—a pair of spacecraft called Mariner 3 and 4—and early on, Gavin began developing an expertise in hardware reliability in general and semiconductors in particular—what they could handle and what they couldn't. He was soon screening *every* electronic part used in the Mariner instruments, striving for total reliability under the inhospitable conditions of space. The Mariners would speed by Mars at eleven thousand miles per hour, scan it, and continue on into the inky void. It was a one-shot deal. So much was at stake—no one knew *what* to expect on its russet surface. Life? Maybe! The Mariner camera couldn't capture animals scurrying about, of course, but it might give us the broad strokes of what the die-hard creatures of cruel, cold, crimson Mars were up against.[322] So while the theoretical mattered, a lot of Gavin's research required reading about the practical, and he zeroed in on the Minuteman 1 intercontinental ballistic missile program. Now there was a program where reliability mattered! And not only to aerospace engineers, who had a financial interest in flying solid hardware, but also to the survival of the human race, which had gone all in on the concept of Mutually Assured Destruction. The Minutemen had to launch on command the first time, every time; they were, after all, the only thing standing between peace and the mine-shaft gap. The only difference between the Minuteman and Mariner, from Tom's perspective, was total hardware lifetime: a Minuteman had a very short-duration mission, launch to flash. Mariners 3 and 4, on the other hand, had to survive ten months, from launch to the complete transmission of flyby data.[323] So he studied everything that Minuteman overseers did programmatically. How did they achieve true reliability? How did they know their systems were totally dependable?

Mariner, in the end, was a success, and the lessons Tom learned became part of the lab's institutional knowledgebase.[324]

Three years after the 1965 Martian flyby, the mountain people of Saint Gabe wanted desperately to build spacecraft for what was being called the outer planets Grand Tour. Astronomers did the math and realized that in the late seventies the planets of the solar system would align just so, and that pairs of spacecraft launched at just the right speed at just the right time could hit every world, Jupiter to Pluto, in a single go. And it was now or never, or at least, now or one hundred seventy-five years from now—the next time the solar system would so align.[325] Virtually nothing was known about the moons of Jupiter or Saturn, or the planets themselves, for that matter, and zero was known about Uranus, Neptune, and Pluto. To accomplish this feat of exploration, the lab decided to develop a new class of spacecraft: true flagships of the National Aeronautics and Space Administration to be called, internally, at least, the Thermoelectric Outer Planets Spacecraft, or TOPS.[326] (They would think of something loftier later.) It would take four ships to do the job, spacecraft redundancy already being the standard practice by NASA and the lab; only Mariner 4 reached Mars, for example. Mariner 3 died hours after liftoff, when half of its protective payload fairing—the rocket's nose cone covering the spacecraft during launch—failed to detach. (This was outside of

Tom's reliability purview.) Unable to deploy solar panels, and with power being king in exploration, the spacecraft never recovered.

Bill Shipley, the TOPS study lead, brought Tom on board. When TOPS was conceived, everyone at the lab was busy with Mariners 8 and 9, which would be the first spacecraft to orbit Mars, and Vikings 1 and 2—a set of landers and orbiters that would touch down to study Martian protozoa, plants, and prairie dogs. Tom was placed in charge of the TOPS team responsible for the long-term reliability of electronic parts. It was a significant challenge because, unlike the Mariner spacecraft, which had to last ten months (itself a challenge), the TOPS spacecraft would have to survive seventeen years. (The farthest planets of the outer solar system were very far away indeed.) No one had ever designed a spacecraft to fly functionally for so long or so far from Earth. In every respect, it was unlike anything ever attempted. And after the TOPS team did a lot of worthwhile work on a then untackled problem, when its anticipated cost was calculated, the total was . . . unwelcome by Congress and the Nixon White House: one billion dollars, and at the worst possible time. The Apollo program was effectively over, and a thing they were calling a "space shuttle" was the shiny new object, looking likely to consume much of the agency's dwindling dollars. Moreover, what wasn't eaten by the shuttle would be swallowed by a proposed "orbiting space telescope."[327]

The lab didn't give up, however. TOPS had already paid dividends: in the course of studying the spacecraft concept, Tom's team carried out the lab's first large-scale work on integrated circuits, and that in itself was exciting. Out of TOPS came Mariner Jupiter-Saturn—twin spacecraft now, instead of four, and designed to visit exactly two planets: Jupiter and Saturn. The spacecraft used Mariner designs and Viking orbiter subsystems whenever possible, reducing risk and freeing up cash. And though only Jupiter and Saturn were the goal, the lab quietly asked the U.S. Atomic Energy

Commission to develop batteries with life spans longer than ten years. And if, say, one of these spacecraft *happened* to keep going past Saturn—flew for years and years and years—and, look, the whole point of the mission was to launch because of an exceedingly rare planetary alignment—if one of these spacecraft *happened* to encounter Uranus and fly a little father and then encounter Neptune, well, why, what a happy coincidence! (Only Pluto would be missed.)

So Project Voyager, as it was branded, emerged like a phoenix from the ashes of TOPS. It was less expensive and less ambitious, which spoke to the hearts of White House officials and congressional overseers. Though funding was ever imperiled in emaciating budgets—in the aftermath of Apollo, NASA's budget was halved—the officially trim Voyager effort rolled right along.

While most senior engineers at the lab weren't about to leave guaranteed work—Mars missions were sure things—for this flighty outer planets adventurism, Tom asked to stay on and became the Voyager spacecraft mission assurance manager, which is just what it sounded like: his job was to make sure this particular So Crazy It Might Work actually worked. (A *twenty-year lifespan*? Yeah, good luck with that, Tom.) And right away, and though Gavin was no dilettante, he realized he was in way over his head. The Jovian radiation environment alone was enough to overwhelm even the most experienced engineer, and years of Tom's life would soon be devoted to tackling the problem. He played out all the everythings that could possibly go wrong and then brainstormed how to guard against them. The Voyager team made quiet progress on their mission—the group bonded as it worked on the impossible, really became a family—and soon study became spacecraft, blueprints to hardware, with a course heading that would take them to every outer planet in the solar system, sans Pluto, and then, driven only by gravity, onward for millions of years. These two spacecraft

would be the emissaries of humankind. To that end, an astronomer of some celebrity named Carl Sagan, as well as Frank Drake,
author Ann Druyan, and a handful of others, assembled what they
called a "Golden Record" of humanity: a recording of laughter, the
ambient sounds of Earth, greetings in multiple languages, the music of Mozart, Beethoven, and Chuck Berry, and an astonishing
assortment of other data related to our primitive little species. Each
of the twin spacecraft carried one of these gold-plated records . . .
but it wasn't the only message they carried.

A few days before the engineers buttoned up the Voyagers for
stacking onto their respective rockets, Ray Heacock, the spacecraft system manager, called the flight team together. He presented
small titanium plates and an electric pencil, and invited each member to inscribe a message of his or her very own—separate from the
Golden Record—to send to the stars. And everyone did. When
Tom took the pencil, he engraved his name, his wife's name, each
of his children's names, and "Godspeed Voyager." It wasn't profound, he knew. All that mattered were the names of his family.
They would soon be flying evermore through star systems and the
breathtaking beyond. *Out there*. When all of them had finished
their engravings, the plates were tucked into the spacecraft thermal
protection material and sewn safely inside. Baggage checked, the
two spacecraft Voyager launched in August and September 1977.

Tom's grandfather came to America on the first of June 1903.
He settled in Philadelphia, was penniless, and worked as a laborer.
His father drove a subway train. And Tom had now been part of
something that would likely outlive the human race.

SEEING THE SPACECRAFT from stacking through launch kept
Tom at Cape Canaveral for four months, and when he got back to
the lab, he strode to the Voyager mission support area, swiped his

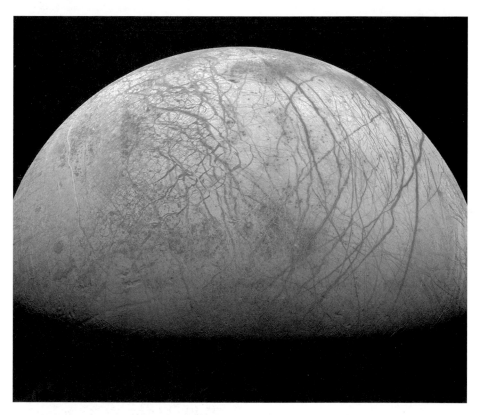

Europa, the icy moon of Jupiter.

Robert Pappalardo, Louise Prockter, Geoff Collins, Aileen Yingst, and Jiganesh Patel at a Galileo press conference at Brown University.

Carl Pilcher, Ronald Greeley, Clark Chapman, Gerhard Neukum, Richard Greenberg, and Richard French study planetary images while planning for the Galileo mission, 1979.

The spacecraft Galileo, seen from the window of the space shuttle *Atlantis* as it begins to deploy for Jupiter.

Rep. John Culberson at JPL for a Europa mission brainstorming session.

Susan Niebur seated at her desk at NASA headquarters.

Launch of the spacecraft Cassini on October 15, 2007.

NASA / JPL / KENNEDY SPACE CENTER

Curt Niebur discusses Europa at a NASA press conference.

Karla Clark talks with NASA administrator Charles Bolden about the Europa mission. Astrobiologist Kevin Hand and study lead Robert Pappalardo look on.

Jim Green, director
of NASA's planetary
science division,
at a Cassini press
conference at Jet
Propulsion Laboratory.

NASA / JOEL KOWSKY

Alan Stern, associate
administrator of the
Science Mission
Directorate, discussing
the NASA budget
in 2007.

NASA / AMES
RESEARCH CENTER

Todd May, manager of the Space Launch System rocket, and later the director of Marshall Space Flight Center, takes questions from the audience during a NASA center meeting.

NASA / EMMETT GIVEN

Sunlight glinting off of Titan's north polar seas.

NASA / JPL-CALTECH / UNIVERSITY OF ARIZONA / UNIVERSITY OF IDAHO

Edward Weiler, associate administrator of the Science Mission Directorate, discussing images from the Hubble Space Telescope at NASA headquarters in 2009.

NASA / BILL INGALLS

NASA deputy administrator Lori Garver shakes hands with President Barack Obama at NASA Kennedy Space Center.

NASA / BILL INGALLS

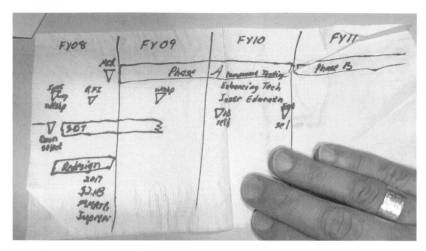

An early Europa mission schedule, sketched by Curt Niebur on a cocktail napkin during the planning for the Quad Studies.

CURT NIEBUR

An artist's rendering of Cassini observing a sunset through Titan's atmosphere.

NASA / JPL-CALTECH

Steven Squyres presents the 2013 planetary science decadal survey
at the 42nd Lunar and Planetary Science Conference, dooming
NASA to another thirty years of Mars flagships.

Dave Senske and Robert Pappalardo lay out the Jupiter Europa Orbiter science team report in 2010.

L. M. PROCKTER

Tom Gavin and Robert Pappalardo seated in the audience of the Ganymede orbiter presentation to European scientists in Paris, 2011.

L. M. PROCKTER

Rep. John Culberson speaks at a NASA budget hearing in the U.S.
House of Representatives.

NASA / AUBREY GEMIGNANI

Princess-Who-Can-Defend-
Herself.

CURT NIEBUR

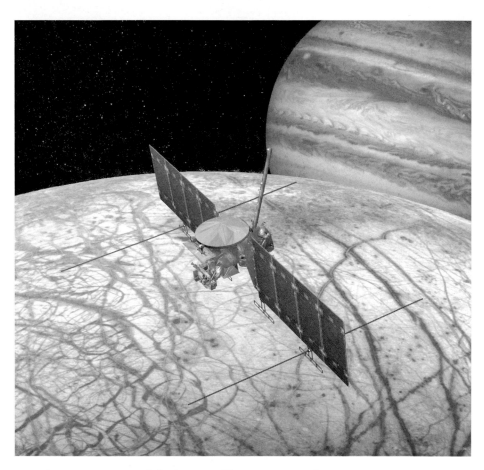

A concept image of the Europa Clipper spacecraft.

Robert Pappalardo poses in front of the famed photo of William H.
Pickering, James A. Van Allen and Wernher von Braun lifting
Explorer 1, the first American spacecraft to orbit the Earth.

NASA / DUTCH SLAGER

Louise Prockter receiving the Masursky Award for meritorious service
to planetary science, presented by the American Astronomical Society's
Division for Planetary Sciences.

HENRY THROOP

A photo of the Europa Project Science Group at their first meeting at Jet Propulsion Laboratory on August 10, 2015.

NASA / JPL-CALTECH / THOM WYNNE

access card to get in the building, and . . . nothing happened. He swiped again. Nothing. So he pressed the intercom buzzer to have someone let him in, and a voice on the other end answered, and Tom said: This is Tom Gavin. I'm trying to get in the building, and the incredulous voice responded: Can I help you? Who are you and why are you here?

And now Tom understood the nature of the job. Voyager wasn't his career. It came and went. And it was a hard lesson to learn, inducing a surprisingly strong sense of loss. The hundreds of people who had come together to turn a dream into a space-borne explorer—the story ended, just like that! And you moved on to the next thing.

His next thing was called Galileo, which he worked on from 1978 until 1989. Tom was in charge of reliability and especially radiation, and it was like being in purgatory, because the spacecraft was tied to the beleaguered space shuttle. After a brief sojourn running the hardware assurance division for the lab, he was conscripted into the problematic Cassini program as spacecraft manager. This was also his introduction to the money side of spacecraft development, the scalpel an unwelcome bit of instrumentation, where previously Tom had been concerned chiefly with engineering requirements.

He was next made deputy director of space and earth science programs, and came on just as the ill-fated Mars Climate Orbiter and Mars Polar Lander (collectively: Mars Surveyor 98) missions were wrapping up work. When the first either missed the planet completely or burned up on entry (neither outcome considered nominal), and then Mars Polar Lander became the Mars Polar Crasher, Gavin found himself suddenly in a leadership position during the darkest time, it certainly seemed, in the laboratory's history.

Over the decades, Tom had seen many vessels lost in space, and it was just part of the job, and everybody understood that. "Beware, Diomedês! Forbear, Diomedês!" shouted Apollo in the *Iliad*. "Do not try to put yourself on a level with the gods; that is too high for a man's ambition."[328] But that was exactly what Jet Propulsion Laboratory *existed to do*. The lab's motto: Dare Mighty Things, and with audacity came catastrophes along the way. It was the business they were in. Everyone accepted that fact, at home, on the Hill, at headquarters. But you lose two spacecraft consecutively while trying to do things faster and cheaper at the apparent expense of better, and the letters *J, P,* and *L* are stamped on the sides of each of those lost spacecraft, and it's the lab that absorbs the blows. Tom was a company man, and the life's works of so many great engineers were being tarnished in the name of saving a few bucks.

To restore morale and forestall future disasters, he began an initiative to craft a comprehensive, unified set of principles by which the lab would do design, verification and validation, and operations for all flight systems. He partnered with Tony Spear, who had led and launched the successful Mars Pathfinder in 1996, and the pair sharpened their pencils and put to paper what they thought were the lab's design principles. Then Tom commissioned every engineer in technical areas to get together and write down what *they* thought the lab's design principles were. What came out of the review boards was a document called, unsurprisingly, "Design Principles."[329] (Tom called them principles rather than rules because it is harder for engineers to complain about a principle.) Because this study of best practices had emerged from the bottom up, as opposed to being imposed from the top down, it was adopted quickly. It wasn't management saying, *Here is what thou shalt now do*. Rather, it was engineers themselves saying, *Here is how we ought to do things*. Tom's team also developed a set of flight project practices to which all teams would adhere, a report that said: This is how we run a mission, and why we must. This is how we build reliable hardware, how we review progress, how we avert disaster and safeguard taxpayer funds. These are the key decision points that every project encounters along the way, and the reviews that will determine whether said project progresses to the next. They titled it "Flight Project Practices."[330] Collectively, it took three years to draft the documents, which would quickly become sacred texts at Jet Propulsion Laboratory.

In June 2009 Tom Gavin retired. It was star-crossed Mars Science Laboratory that finally did him in. The director of the lab assigned Tom to devote himself fully to the Mars mission. All day, every day and into the night, he would work and then go home exhausted, fall into bed, and then wake up in the middle of the night thinking, *What about this? What about* this? *Is there some solution for that?* He was a seventy-year-old man! And it affected his home life. Betty had been putting up with his hours at the lab and his constant travel for nearly fifty years. She was a saint. It was . . . he had done enough.

So when Tom felt he had implemented the changes necessary to place Mars Science Laboratory on a path to success, he went to Elachi and said, "Charles, I'm tired."[331] He had done all he could.

Gavin asked officially to step down in April 2009, with full retirement in June. He agreed to take a long vacation and maybe work, if they needed, a couple of days a week—to act as an advisor to the Mars team, or wherever else his expertise might be needed. But he would do it on his terms. And when he returned, the Powers That Be asked him to mentor the Jupiter Europa Orbiter study team, and he agreed. Most of his career had been spent in the outer planets. He knew the radiation environment, knew how to build a spacecraft, and this kind of role, advising a study team, was ideally suited to his vision of a part-time, postretirement life at the lab.

KARLA CLARK WORKED with Tom on Cassini, had learned her philosophy of decision making from him, just as she had learned teaming from John Casani. In those Cassini days, Tom had been very, very good at making decisions. He wasn't afraid to shoot from the hip, and if you didn't know any differently, you would have thought his calls were arbitrary, that he was dictatorial, leaping headlong to conclusions with insufficient information. Part of it was his gruff exterior, his conspicuous working-class Philly accent, the clip and cadence of his speech. The warmth was there—here was a good man, a happy man, a jovial man—but if you didn't know that, a simple "Good morning!" from him might seem like a direct order: *Have a good morning.* But Karla knew better. She knew that Tom's choices were made with extraordinary, almost extrasensory insight. The man was very smart and very experienced, and the speed of his decisions reflected that. In those Cassini days, Tom would revisit a decision if presented with a compelling argument or if the data demanded it. He was not at all afraid to change his mind.

But somewhere along the way, Karla came to believe, the longer Tom worked at the lab, the higher his status, the greater his responsibility—no longer for single spacecraft but for every spacecraft and for their reliability—the more rigid he had become in his rulings. And by the time he joined Jupiter Europa Orbiter as advisor, his decision-making style was almost unrecognizable. In meetings, Tom would sit at the head of the table, that position of power. Karla was still the study lead. She was still the top of the chain and had the responsibility for making the calls, but Tom was coming in after being JPL's associate director of flight projects and mission success—a position of astounding import and influence. Tom had fifty years of experience behind his decisions, so nobody was going to second-guess him. Nobody was going to tell him no. Tom, Karla began to think, wanted to make Europa his own, was *not* there to mentor.

The orbiter was always going to change significantly, be almost unrecognizable by launch relative to what won the shootout, but whatever eventually flew would be evolved from a common engineering origin and philosophy. Rob Lock, the lead systems engineer on the orbiter, was still there. The original engineering team was still there. Karla was still there. But the new team members seemed to exert an outsized influence, had their own ideas about the mission they wanted, and had no compunction about consigning three years of effort on the Jupiter Europa Orbiter to oblivion.

In February 2010 the lab formally established the Europa Jupiter System Mission pre-project office and installed Tom as its manager. No one even bothered to tell Karla.

Tom tried awkwardly to find a role for her in his new project, but the positions offered were beyond insulting to her: junior jobs, one under a previous *subordinate* on the project. Finally, Tom—it wasn't magnanimity, exactly, what he did next, because Karla didn't think he really realized what he had even done—helped her find another job at the lab, and it was a good one: manager of mission assurance for flight projects. She would be running the only office that touched every flight project at the lab, from the smallest experiments that would launch to the International Space Station, to the biggest spacecraft ever conceived. But after twelve years of keeping the flame of Europa alive, of wrestling with one of the most challenging problems in all of planetary exploration and eventually *finding a solution*—it was hard at the end. Something dear had been taken from her.

Four months later, Clark sent a letter to select members of the lab science and engineering teams announcing that she had taken the new job—that, "as you know" she had become increasingly frustrated with the way things had been going, and that she felt she had become "totally ineffective, and that I can no longer be professionally or personally satisfied when coming to work every day."[332] In making her new position known, she noted that because much of the lab's radiation expertise resided in her office, she would still be working to set up the orbiter for success. (It soon became clear that even that counsel was unwelcome.) And once Karla left, Rob Lock and most of the project's engineers followed. It was clear what was happening: it was an invasion force, and the invaders had won.

Karla's departure was a blow to the science team, and Louise Prockter really chewed on it from across the country at the Applied Physics Laboratory. The details of what had happened were scarce. She knew it was normal at Jet Propulsion Laboratory for managers to be moved around, but Karla had done so much for the mission that it felt utterly inadequate to just say thank you and sign a card. Maybe, she said to Bob Pappalardo, they could all pitch in and commission an art piece. Even if they collected only five hundred dollars, they could probably still get something pretty magical. They reached out to Monica Aiello, an artist whose work was well known in the planetary community for its fusion of science and the humanities. Aiello was all in for Karla, charging just half her normal rate, and in a matter of days, Bob, Dave, and Louise had raised three times the amount they had expected, from forty Europa scientists and engineers around the world.[333]

Aiello created four textured images of Europa, eighteen-by-eighteen-inch panels depicting Pwyll Crater—the most famous and distinct of the craters on Europa—in blues and browns, materials beneath the paint making the artwork three-dimensional. Karla's dad was Welsh, and Pwyll being a Welsh name—well, it was because of him, the childhood kitchen table conclaves of scientists and engineers from around the world, his inspiration to her as a research chemist, that all of this had come to be.

CHAPTER 10

This Earth of Majesty, This Seat of Mars

THEY GATHERED IN THE MAIN BALLROOM OF THE Woodlands Waterway Marriott Hotel & Convention Center in the Woodlands, Texas, just north of Houston. Outside, sunset was fast approaching, and streetlamps and string lights woke higgledy-piggledy. The commercial center of the Woodlands could feel like a bland movie set: inoffensive, Main Street Modern, urban planning by way of plastic-injection molding, designed for the singular purpose of attracting convention crowds and facilitating their need to socialize and have a bite and booze in locations both picturesque and in walking distance, and whatever happened after, chaste or otherwise, was on the forgathered, because the anodyne polyurethane environs would do nothing to stoke primal urges. The restaurants, built along an artificial waterway connected to an artificial lake, were a mélange of familiar names and themed off-brands (e.g., an English-style pub that served fish and chips and looked exactly like you would imagine), and it was all a bit like an amusement park for business executives, the streets clean, pedestrian friendly, aglow with fairy lights and lined ever with fresh flowers real but not.

The Marriott's main ballroom could hold twelve hundred at capacity, SRO, and it was definitely standing room only that evening, March 7, 2011, at the Forty-Second Lunar and Planetary Science Conference.[334] Bob, Louise, and Dave were there. Curt and Susan Niebur were there. Alan Stern, Ralph Lorenz, Jonathan Lunine, Cynthia Phillips, Jim Green, Don Blankenship—everyone was there, eighteen hundred scientists total at the convention, and though they all wanted to be in that room, six hundred or so would just have to wait until later to find out what happened.[335]

The National Research Council was about to release the results of the next Decadal Survey. The announcement would directly affect the careers of not only the Europa team, which needed only a perfunctory nod, but also everyone in the room, whether she studied Venus or Vesta or Ceres or Saturn. Where the Decadal said NASA needed to go in the next ten years, NASA would work diligently to go. Powerful projectors cast the first slide of forty-seven onto two ceiling-to-floor screens.

VISION AND VOYAGES, it said, FOR PLANETARY SCIENCE IN THE DECADE 2013–2022.[336]

Though the Decadal was dated 2013, Ed Weiler had wanted it in hand by March 2011 so that agency planners might use it in 2012 to build the budget for the 2013 fiscal year.[337] Upon commissioning it, he also had a few requests for the steering committee:

1. This time, integrate Mars recommendations into the overall priorities of planetary exploration.

2. Incorporate technical feasibility into your recommendations, and be sure to get credible budget appraisals. Find an independent organization to calculate the costs of candidate missions (i.e., *do not trust the numbers from Jet Propulsion Laboratory*).

3. Do not recommend that NASA increase the planetary science budget. We have already thought of that. There is no more money.[338]

Steve Squyres, still at Cornell, chaired the steering committee. The National Research Council chose him because, in addition to being highly regarded and accomplished in the field, he was the sort of steady-handed workaholic necessary to pull this thing off.[339] The committee decided early on that given Weiler's requests, rather than build on the 2003 Decadal, it would reset the report, recommendations, and reasoning entirely. It would start over. Tabula rasa.

THE JUPITER EUROPA Orbiter as presented to the icy-satellites panel of the Decadal Survey looked like this.[340] It would launch on a medium-lift Atlas V rocket and take a VEEGA trajectory to the Jupiter system: fly to Venus, back to Earth, around the sun, and back to Earth again for a final gravity assist propelling it to the target. When it finally arrived six years later, the spacecraft would spend thirty months in orbit around Jupiter, conducting science on the planet and the Galilean moons: four flybys of Io, six flybys of Ganymede, six of Europa, and nine of Callisto.

The spacecraft would then enter orbit around Europa and spend nine months swimming in radiation, doing intimate science, and answering questions both vexing and mundane. What was the thickness of the ice shell? What was its composition? What caused those weird surface features? What was its interior shape and how did the ice interact with the ocean? And the ocean itself: How big was it and what was the water like? The science would be astounding—the best ever conducted on any moon save our own—and would be completed in ninety days. The extra six months were a bonus. Ideally, Jupiter Europa Orbiter would launch in 2020—eleven years away—and end in 2029 by crashing into Europa, where it would be vaporized on impact. (This made spacecraft decontamination before launch paramount, so as not to feed any stray Earth-born microbes to Europan fish.)

The Europa mission was, of course, but one half of the Europa Jupiter System Mission. One month before they checked into rooms at the Woodlands Waterway Marriott Hotel & Convention Center, Ron, Bob, Louise, Dave, and Tom Gavin flew to France to help the Europeans get their side of the mission approved, too. Unlike the Jupiter Europa Orbiter, which had won its war at NASA, the Jupiter Ganymede Orbiter still needed formal selection by the European Space Agency. And while it wasn't looking grim, it wasn't looking good, either. The prospective project was in competition against two formidable proposals: the International X-ray Observatory and the Laser Interferometer Space Antenna, or LISA. The first would study the evolution of the universe by way of black holes and supernovae. The second would hunt for gravitational waves—ripples in space-time.[341] These were enormous astrophysical questions with supporters spanning the continent and, like the Ganymede orbiter, both competitors had promising prospects for international partnerships.

The Europeans' general selection processes ran parallel with those of NASA. While the Quad Studies were under way, the Europeans held the Cosmic Vision 2015–2025 mission competition, which produced proposals for the X-ray observatory, the Laplace Ganymede orbiter, and TandEM, the Titan and Enceladus mission.[342] (LISA had already been selected for the final showdown.) When the shootout saw the American Europa mission defeat the one to Titan, TandEM, too, took a round to the chest—there was no point in ESA even entertaining that one. And though NASA had made its decision, the Europeans had one more selection cycle to go before settling on a flight project. The interferometer, X-ray observatory, and Ganymede orbiter thus entered an assessment phase during which—as had been the case for the Europa orbiter—

scientists would write inches-thick reports for follow-on feasibility studies. The final step now, before the three were narrowed to one, was a formal presentation to the scientific community. If the Ganymede mission was selected, it would put Alan Stern's insurance policy into full force: it would be difficult if not impossible, politically, for NASA to abandon the Europa mission. But first the Ganymede mission had to be selected.

The culminating event was to be held on February 3, 2011, in the grand amphitheater of the Institut Océanographique de Paris—la Maison des Océans—on Rue Saint-Jacques, just up the hill from Notre-Dame. It was the hundredth anniversary of the building, the century of progress reflected in the contrast between the art in the back of the room—frescoes of sailors pulling sea life onto boats for study—and projection screens in the front, on which were the contesting scientific rationales for exploring the farthest reaches of outer space.

The American contingent came to bring an outside perspective to Ganymede's final overture.[343] For about a week, they did dry runs, polished presentations, tweaked weaknesses—it was a dream collaboration, a happy hint of the partnership to come. When the Ganymede team gave their closing arguments, the Europa team was there for moral support, and in the off hours, the teams joined for evenings of enjoyment in the City of Light. Everything was coming together.

In Washington, meanwhile, in anticipation of the Decadal's endorsement, Curt Niebur finished work on what NASA called an instrument "announcement of opportunity," or AO. This was a solicitation for organizations to propose the science payload elements of the Europa mission. Each scientific tool carried by the spacecraft—things like the camera, the radar, the magnetometer, &c.—was considered a separate experiment, a device distinct from the spacecraft itself. They would be built to specification by the winning teams and integrated later once the spacecraft was ready. To win an "instrument selection" was a huge deal and would bring millions of dollars to an institution, funding its scientists and engineers for years. Accordingly, these were prized contracts, and organizations across government, industry, and academia would move with alacrity to answer the announcement with proposals no less detailed than the very spacecraft on which they would ride. The AO listed the rules of the competition and included a detailed description of things such as power availability, mass constraints, communications methodologies, data return rates, and what an instrument would have to look like physically to actually fit on the spacecraft. Instrument teams would need to know what shielding the spacecraft could provide versus what would have to be radiation hardened internally. Curt got started on the announcement just after the shootout, partnering with Karla to get the details right. When she left the project, the orbiter was still being mutated, and Curt coordinated further with Tom Gavin.

Once the AO went out and instruments were chosen, NASA was committed. The agency could technically stop the train at any time, of course, but there would be legal implications for doing so. An institution spends a million dollars developing, say, a camera for a mission, and the mission is deleted, and that institution is going to want its money back, and then lawyers are getting involved and precisely zero people at NASA headquarters wanted to hear the words *subpoena* or *pretrial hearing*. So this would be the most significant milestone any Europa attempt had ever yet achieved. This was the moment Curt had long toiled toward. It would drop the day after the Decadal endorsement.

MAJOR TALKS AT the Lunar and Planetary Science Conference were spoken practically ex cathedra, no notes, no prepared texts,

PowerPoint projections more than sufficient. Steve Squyres paced the stage that night like Steve Jobs unveiling the best iPhone yet. He explained the whys and hows of the Decadal. He described the whos and the wheres. He really drove home that, look, we didn't do this thing alone or overnight. Squyres explained the easy things: the ongoing missions, the things already flying, the things being built even as he spoke. The Decadal, Squyres stressed, was all about keeping NASA firmly on the side of research. And the Decadal was solidly protechnology, both to bring down mission costs and to do better science.

Then it was time to discuss mission priorities.[344]

First up was the Discovery program. The Decadal endorsed keeping its small missions at current funding levels, adjusted for inflation—about a half million dollars per—and to get those Discovery spacecraft flying every two years or sooner. But the Decadal panel was not asked to recommend or prioritize missions, and it did not. Next were the New Frontiers missions, and the Decadal recommended that the program be boosted to one billion dollars per spacecraft, excluding the price of launch, and that NASA choose two missions to fly in the next decade, among the options: 1. a comet surface sample return; 2. a sample return from the moon's south pole; 3. a small probe to study Saturn's interior; 4. a visit to a "trojan" asteroid (i.e., one that lives not in the asteroid belt but elsewhere, and usually in Jupiter's orbital path); and 5. a Venus probe. These mission concepts were not prioritized. Scientific institutions and NASA centers could propose any type of mission they desired so long as the mission addressed one of those five themes, and they would slug it out on merit and be judged by NASA headquarters. If NASA went wild and found funds for three New Frontiers missions instead of two, the Decadal recommended that the agency also consider an Io observer and a network of small lunar probes.

If you were Louise Prockter, you sat there taking notes in your spiral-bound Levenger planner because you were the supervisor of the Planetary Exploration Group at the Applied Physics Laboratory, and there was a good chance your lab would win one of these mission competitions. You were seated next to Bob, and Curt was right there, and Dave, and your notes were just so. And when the next slide came up, your eyes sped across it, line by line, the way the typewriters you sold twenty-five years earlier raced across the page. *Ding.* Next line. Across the page. *Ding.* And when you finished reading it, you read it again, you sat upright and immobile. Your face impassive, your demeanor stolid. But your eyes betrayed you—it was subtle, but they spoke loudly and in perfect paragraphs, and you made eye contact with no one, which was good, because if looks could kill, you'd have taken out half the ballroom.

Bob sat there, too, his face with pond-like placidity, and Curt as well. Curt already knew what was coming.

FLAGSHIP MISSIONS

(IN PRIORITY ORDER)

I. BEGIN NASA/ESA MARS SAMPLE RETURN CAMPAIGN

. . . and nothing beyond that line mattered.

In the ten years since the first Decadal, no flagship mission had been approved for launch. And now that the NASA planetary science program was broke, it would be a long shot even for the flagship recommendation of *this* Decadal. So whatever came after that line may as well have been *lorem ipsum.* Meaningless.

The decade-plus of Europa mission studies at Jet Propulsion Laboratory no longer mattered. The Quad Studies. The shootout. The JPL-APL partnership. The Ganymede presentation in Paris—none of it mattered. The years thinking through every aspect of a Europa mission, of conquering the radiation belt. The discovery of the saltwater ocean and the likelihood of life. The geomorphological mysteries desperate for data. The much debated thickness of the ice shell. The alkalinity of the subsurface ocean. Even the basic stuff: What does its surface look like? The composition of Europa, its interaction with the other Galilean moons. Its geologic evolution. They'd worked out ways to answer all of it. Knew how to communicate with Earth, how much power they would need. The instruments and measurements and speed of the spacecraft and altitudes necessary to get every byte of data imaginable. The rocket that would carry them, the launch window they'd work from, the trajectories from here to there—where around the solar system the spacecraft would swing, and, once at Jupiter, what it would observe during the slow roll to Europan orbit. They'd figured it all out, the whole thing. Defeated Titan and Enceladus in direct competition. Had been endorsed by the whole of the planetary science community, itself unified by OPAG. And the ten torturous years of study after study after study—try it with solar power instead—try it with a nuclear reactor—try it with a lander—try it without fission—try it with—and none of it mattered.

And then somehow, as the presentation slides progressed, it got worse.

The Mars Sample Return mission—the new consensus priority of the planetary science community—wouldn't *technically* return a Mars sample. Not in this decade. No, it was explained, the Mars Sample Return mission was "multidecadal in character." The mission endorsed by the 2013 Decadal would only cache the samples, i.e., it would scoop dirt into jars. There would need to be another flagship in a decade's time to launch the jars to space, and perhaps another the decade after that to grab the space-borne jars and carry them home.

Mars! Which launched a new mission every two years—had so many spacecraft working actively on or around the planet that it would soon need its own air traffic control system—had so much unexamined science sitting on file servers that even if all work stopped tomorrow and every Martian spacecraft self-destructed, scientists could still spend a decade studying new data. Mars. Steve Squyres—he of the Mars rovers Spirit and Opportunity—spoke so gleefully, and Louise was just blindsided.

As if it mattered, the Jupiter Europa Orbiter was announced as the second-highest priority mission.

The Mars Sample Return mission had been studied only since April 2009. Less than two years. It relied on the landing system of the Mars Science Laboratory (now named Curiosity) to keep its cost low, and Curiosity had not yet launched, let alone landed! Europa, meanwhile, had been studied and reviewed twice now by independent boards going back to 2007 and had been in development in one form or another since 1996! But what really baffled Louise—well, one of the things—was the absurd price tag assigned to the Europa mission by Aerospace Corporation, the private firm handling the independent cost analysis: four-point-seven billion dollars. Where did that number come from? What assumptions did they make to get such an unfathomable figure, and why didn't they ask anyone on the Europa team for a little guidance? I mean, the Europa mission could have cost just over two-point-one billion, but NASA wanted something bigger—the sweet spot!—and so the Europa team made a mission that did more science at Europa—and at Ganymede and Callisto and Io and Jupiter. *That* raised the cost to three-point-five billion. So where did that extra billion come from? And while Europa had been instructed to think big, Mars Sample Return—which, again, *didn't actually return a sample*—sliced itself into thirds and still came in at three-point-five billion, not counting parts two and three and the thirty years the endeavor would entail. The Decadal, for what it was worth, ordered the Mars team to cut another billion from the price, but what did any of it matter? Europa was over.

Bob, Dave, and Louise slipped silently from the ballroom, rode down a crowded escalator that seemed endless and laggard, and walked out of the Marriott and into a fraudulent Woodlands evening. The string lights along the fake waterfront were in full incandescence, and the three just absolutely could not stomach any of it, could not be around any of them—just needed to be away from . . . everyone, everything—and so they picked a direction decisively opposite the perfect palm trees (in Texas) and the specious flowers and frivolity, and walked until freshly painted lampposts no longer lined pressure-washed sidewalks, and still they walked, until they were submerged in the balm of a cool darkness. With every step they proceeded through the Kübler-Ross stages of grief. Anger. Denial. Bargaining. And still they walked, leaving sidewalks and stepping now across manicured fields (for even in the wilderness areas of the Woodlands, the grounds were well maintained), and it was so dark that when they came to a culvert, before crossing they first discussed whether there might be alligators in there, and were there alligators in Texas, but crossed anyway, and found finally a Thai restaurant, because Bob and Louise were vegetarians and its menu offered the most promise, and there was no way anyone else from the conference would be there, and they sat down and ordered from the menu, and but for brief interactions with the server, occasional murmurs about what are we going to do now, they ate almost in perfect silence.

Louise tried to be stoic. I mean, she was still English, after all. She had MESSENGER. In a few days, it would finally enter orbit around Mercury. She had been promoted to deputy project scientist, which, well. It was huge. She was like the vice president of planet Mercury. And she was the group supervisor of the planetary exploration at APL. Dozens reported to her. She had . . . meaning? Purpose? She would survive this. She had been through so much, I mean. Come so far. And Dave—among other worlds in the solar system, he was also a Mars guy, was even the deputy project scientist for Mars Odyssey, still circling the Red Planet, returning spectra and thermal images. And he was manager of Jet Propulsion Laboratory's Office of Science Research and Analysis in the Solar System Directorate. He would be fine.

But Bob. He had Europa. And now he didn't.

CHAPTER 11

E Pur Si Muove

JUST BEFORE THE DECADAL TURNED EUROPA TO crushed ice, the core Jupiter Europa Orbiter team at JPL had relocated from a handful of buildings spread across Saint Gabe to Building 321 (Flight Projects), fifth floor. In the aftermath of the Martian triumph, its members should have packed up and limped back across campus to their previous haunts because Europa was no longer a flight project, but rather, yet again, a lowly study.

Two things kept Team Europa in its new digs. First, the senior leadership at Jet Propulsion Laboratory was thrilled with the results of the Decadal. The two top priorities in planetary science were thirty years of Mars sample return missions (so many spacecraft, so much to do) and a Europa orbiter, both of which belonged to the lab. Something would fly to Europa eventually, regardless of how many decades it might take. Second, the now-ensconced lead of the Europa effort, Tom Gavin, was the one who built the building in the first place, and so he had a smidge more say in who would stay and who would go, and, oh, by the way, we are staying.

The Flight Projects office space was raised fourteen years earlier, in the aftermath of the Cassini launch. The lab, which had previously mounted only a handful of projects simultaneously, soon saw a serious business boom with the onset of Faster-Better-Cheaper and attendant missions to Mars and "small bodies" (comet, asteroid). Because of this, Gavin, then a mere thirty-five-year veteran of the lab and newly minted deputy director of space and earth science programs, saw a problem. Historically, new projects came into the lab, and you found space for them just, you know, wherever. There weren't that many, so it didn't really matter. But now there were teeming teams planted willy-nilly along the San Gabriels. So he called the deputy associate administrator for science at NASA headquarters and catalyzed a revamp and expansion of the JPL campus.

In the new regime, there needed to be flow, logic, and clear lines of communication. Henceforth, key personnel on projects (e.g., the project manager, financial manager, instrument manager, spacecraft manager, mission assurance manager, science leads, &c.—thirty-five people or so in total) would now be kept together by their project's pipeline position: Building 301 (Mission Formulation), where the mission would be conceived; a new Flight Projects building, where an approved project would be developed; and then Building 264 (Space Flight Support), where missions were flown.

Tom wanted the Flight Projects building to be utilitarian, something able to hold about six hundred fifty people and just brimming with meeting space: three major conference rooms per floor, at least, and he wanted each floor to be identical, with no corner offices and thus no fighting over said prestige. He wanted a dedicated floor for design reviews (they chose the basement), because reviews were conducted, presently, at local hotel conference rooms, and the packing of people, possessions, and paperwork to and fro was not worth the lost time. Tom's boss, Charles Elachi, meanwhile, wanted a proper auditorium—the podium, the big screens, the stadium seating with lap desks that unfolded from armrests—just the whole thing—for team-wide all-hands-on-decks. There was also a small gym in the basement for team members (though where the gym came from was a mystery—it wasn't in the plans, but, Tom had to admit, it also wasn't a bad idea). The sixth floor of six total would be saved for the program offices of the Mars and Solar System Exploration Directorates, and also up there would be a patio overlooking all of the laboratory and parts of the cities of Pasadena and La Cañada Flintridge, both of which claimed the lab as their own, the latter pointing to the land on which the lab was built, and the former, the lab's zip code.

The Flight Projects building took just under a decade to move from cocktail napkin to ribbon cutting, opening in October 2009, just after Tom retired. Some called it, informally, Gavin Tower, but because it was a mission's last home until the spacecraft launched, it was better known by its building number: 321—as in: *3-2-1 liftoff*.

Now, on the fifth floor of 321, the Europa team had to regroup. Things post-Decadal moved quickly and not. Tom had already set in motion a series of internal studies of alternate mission concepts, and from headquarters, Curt Niebur encouraged the lab to reach consensus with Aerospace Corporation, which had done the bonkers cost analyses for the Decadal. Regardless of which institution had hit the target more cleanly, JPL or Aerospace, it was clear that the Decadal steering committee wanted a Europa mission that was simpler and cheaper, so do that.

The Decadal.

Bob Pappalardo was OK with the Decadal. The committee had, he felt, acted in good faith, and Steve Squyres, who led it, was the reason Bob was a Europa scientist in the first place. It was Squyres, all those years ago, when Bob was an undergrad in Carl Sagan's course, who gave the guest lecture on Europa that so ensnared, enthralled, enraptured Bob. How ironic that Squyres would be the one to announce the end of Bob's life's work.

In fact, Bob had had time to get his spleen in check, because he knew the Decadal recommendations before the unveiling. Not *far* in advance, but enough that it softened the blow psychologically when the results were announced at the Lunar and Planetary Science Conference. It still hurt badly, though, like knowing in advance that he was going to be punched in the face.

His first hint had come about three months earlier. Bob was at the Moscone Center in San Francisco for the American Geophysical Union Conference when his phone rang. He stepped outside to answer. The comings and goings of taxicabs—their double-tapped horns, those slamming doors, the internal combustion, and persistent exhaust—were nothing next to the cacophony of scientists between sessions, bottlenecked in hallways. You get geologists talking about rocks and they just won't shut up about it. California traffic was quieter.

It was Fran Bagenal on the phone. She was an official reviewer of the Decadal Survey, had scrutinized the flagship mission recommendations, and was miffed straightaway that the steering committee had not pushed the proposal teams to come up with viable, variable price points.[345] Here is the small option for a Europa mission (or a Neptune mission, or whatever). Here is a battlestar. Here is the sweet spot. *We can do any of them.* There were options, and the steering committee could have requested them, if not insisted upon them. Mount the orbiter to a Delta IV Heavy rocket, and you could shorten travel time and add more dumb radiation shielding, the first cutting personnel costs, the latter mitigating the need for custom, radiation-hardened components—it would get rid of the million-dollar computer chips, in other words.[346] Those sorts of simple changes would have chopped a bundle from the bottom line, and the Europa people knew that. But they submitted a flagship in the vein of Cassini and Galileo, a flagship proper, designed to deliver mind-bending science that would keep a generation of researchers busy—not just in the Europa community but also in the wider giant planets community. Given the evisceration of the NASA planetary science budget, Fran felt the Europa team should have been sent back to the clubhouse straightaway to pare down this thing before final submission. And don't get her started on the preposterous sample return sequence proposed by the Mars community.

Of course she couldn't come out and explain any of this to Bob. The Decadal Survey was conducted in secrecy and embargoed until its unveiling at the Lunar and Planetary Science Conference. What she said to Bob was that the community was not likely to come out in support of an overly expensive outer planets flagship mission, and that it was imperative he understand that. Think about how a smaller mission might be designed—a New Frontiers–class mission, for example—and how ruthless their principal investigators—their mission leads—were in cutting away spacecraft elements in order to control costs. So it was a work call, but she was calling also as a friend. She all but scrawled for Bob the word *REDRUM* in crayon.

But Bob stood behind the mission. It was the best mission, he said, the best developed and certainly the one with the best science, and we intend to fly it.

I think you need to prepare yourself for a range of outcomes, she said.[347]

When Bob learned at last where in that range Europa fell, there were two lines in the four-hundred-page final report that he grabbed hold of as though they were floating fragments of debris from a sinking ship. First, "Mars Sample Return was thus prioritized above [Jupiter Europa Orbiter] not primarily because of its science merit, but for pragmatic reasons associated with the

required spending profiles."[348] You take dollars out of the equation, in other words, and the Decadal steering committee was basically suggesting that the science was equal. Bob could work with
that! And though Mars emerged as the highest priority overall,
in the same section as the Decadal committee endorsement came
a caveat: despite being broken into multiple Mars missions, the
caching component of the sample return campaign (i.e., the Mars
lander that would scoop up dirt) was *still* a billion dollars too expensive to fly as proposed. It would need to be reformulated to
square with budgetary realities (i.e., there is no money). The Jupiter
Europa Orbiter, too expensive as well, would similarly require a
descope. And if the budget improved—if NASA somehow found
a chest filled with rare gems and gold doubloons—then, sure, Europa: Knock yourself out. Build your ship and separate it from the
Earth. To lay the groundwork for such an opportunity, said the
Decadal, "NASA should immediately undertake an effort to find
major cost reductions for JEO, with the goal of minimizing the
size of the budget increase necessary to enable the mission."[349]

Well, there you had it! The science was equal, and the Decadal
wanted NASA to keep plugging away at Europa.

Look, Bob was not unflappable. He was fully capable of being
flapped, and flap he did. He found himself sleepy all the time, for
some reason, just swallowed by this incessant fog of fatigue. But
this was his life's work—what was he supposed to do except *keep
going*? So going he kept.

Which is what Curt Niebur wanted—and not only Bob, but
also and equally Louise and Dave—the lot of them. He knew what
was coming before any of them because headquarters had also been
briefed of the results, and it was awkward keeping that secret from
these absurdly smart, steadfast scientists—his friends, with whom
he had been through so much for almost a decade now—but he
had to keep quiet about their coming career catastrophes because

it was his job. Headquarters had to remain an honest broker, and he believed strongly in that, in the process, and in the Decadal. When at last he could talk with them the day after the unveiling in the Woodlands, Curt had things he knew he could say and things he knew he could not say. COULD NOT SAY: He was disappointed in the large mission recommendations. The notion that Mars sample return was on firmer footing than Europa as a mission concept? That the Mars Sample Return study team had a better understanding of the cost and challenges? Well, it made his eyes twitch. Europa had been through the Quad Studies and the shootout—both costed and reviewed rigorously and independently—before the studies of the endorsed Mars mission *had even begun.*[350] COULD SAY: Get the disappointment out of your systems because this is our Decadal, and we all need to get behind it. If you didn't get the mission you wanted, in order to keep planetary science as a whole alive, we have to stick together. COULD NOT SAY: Ed Weiler at headquarters was ready to pull the plug on Europa studies. Zero it out. Shut it down. It's done. The Decadal has spoken! And Ed was serious, and it was nothing personal, but space science as a whole, and planetary science in particular, had been kneecapped by President Barack Obama's proposed budget, and look, how many times were we going to study this thing? How much money did you intend to give Jet Propulsion Laboratory to keep developing these losing ideas, Curt? Because, Curt, *there was no money.* COULD SAY: Keep going. Look, this is new. We've lived with the previous Decadal for a long time now, and that one was our friend, but that Decadal doesn't matter anymore. We all need to digest this one. We've got to think about it. We've got to discuss it, and not just at headquarters, but here too, and at the Outer Planets Assessment Group meeting next week. COULD NOT SAY: He was going to make a concerted effort to sway Ed to, if not fully embrace Europa (wouldn't happen), then at least let it stagger into the tree line, find a nice cave, and recover quietly.

The thing that worried Curt most was that if Ed truly zeroed it out, all the momentum built during the last four years would be lost. That's what happened with JIMO, and Europa was able to recover only because of the previous Decadal endorsement. There was also this Texas congressman named Culberson who wouldn't stop talking about Europa, and he sat on the House Appropriations Committee, and you ignored the guy writing the checks at your peril. But if development died on a Europa mission yet again—just halted entirely until Mars Sample Return started moving—in the six years or so eaten by Mars, Europa would essentially start from nothing: not only as a concept, not only at headquarters, but also in terms of wider support in the community. Science moved quickly, and Uranus and Neptune scientists had been pretty patient so far. That was unlikely to last. (Alas, poor Titan, which was so thoroughly atomized by Aerospace Corporation's cost estimate that it merited not even a mention in mission recommendations large or small.)

Deep down, Curt was convinced that given the maturity and scientific value of the Europa mission, things were not as dire as they seemed. There was a good chance that in the next two or three years, Mars Sample Return would suffer some setback, some overwhelming obstacle, and NASA would need a hot backup. If such an opportunity presented itself—if any chance at all existed for Europa to get into the queue—the outer planets community needed to be ready to slap a plan on Ed's desk, unroll it Normandy invasion–style, jab a finger at the icy space eyeball, and say, "Have I got a deal for you." But to be ready, you had to keep working on it.

The message received by Bob, Dave, and Louise from Curt's boss was less encouraging. Jim Green, an erstwhile Europa advocate, pulled the trio aside after the announcement and told them that it was over. It was a whole production. Green found an empty table in the back of the conference hall, sat them down, and spent a long time, what felt like *admonishing* them? Lecturing them? For Bob, it was like being in the principal's office. And it was so not like Jim that Bob started to wonder about motives. How much of this was Jim and how much was it the direction of Jim's boss, Ed Weiler?

One month after the results were announced, Bob got a taste of Ed's opinion directly while in Washington, DC, to attend the fall meeting of the Space Studies Board, the focal point of space research within the National Academies.[351] Bob was a board member, one of only two or three planetary scientists out of twenty-two at the table. The meeting that day concerned two issues affecting space science: the president's budget and the Decadal.

The lobby of the Keck Center was lined with murals from floor to ceiling, engravings of four thousand years of scientific achievement, among them: the breeding of corn; Galileo's star map; a silicon wafer for the manufacture of microchips. You walked down the hall, and you just *felt* the possible. Room 100, site of the meeting, was of standard size and layout: chairs, tables, wood-paneled walls with twin white screens for PowerPoint presentations. Bob was tired when he entered. It had been a busy month. But he came loaded with questions.

Representatives from the White House Office of Management and Budget had stood at the podium a day earlier and given their take on the president's proposed budget for fiscal year 2012, and things, according to them, just could not have been better.[352, 353] Sure, the next NASA budget request was flat, but you should be thanking us! Flat was the new up!

That was NASA's *overall* budget, however. Planetary science was set to lose money over time, as opposed to the flat (and thus up!) shape of the other science divisions. And not just a few dollars here and there: the proposed budget projections for fiscal years 2013, 2014, 2015, and 2016 would cumulatively carve nearly twenty percent from the planetary science budget. This was an extinction-level event. Already the division was eating ramen to pay its bills. Members of the Decadal steering committee were no fools in their recommendation, foreseeing a doomsday budget scenario. But what the Obama administration was proposing was so much worse than what the pessimistic Decadal committee members had foreseen.

When Ed Weiler gave his talk to the board, he laid out the issues, financial and otherwise, affecting NASA science. In his opening, he said that NASA would *try* to follow the Decadal, budget permitting, adding later—twice—that Europa had an estimated cost of four-point-seven billion. Sure, he conceded, although Cassini—a comparable mission to that of Europa—would cost similarly if inflated to 2011 dollars, the climate had changed. And anyway, he said, there was now an agreement with the European Space Agency to do a joint first phase of Mars sample return (the sample cache mission) in 2018, meaning NASA would have to pay only one-point-five billion.

When the floor opened up, Bob, having Ed pinned down and in public, decided to get some answers. Look, he said, on the Europa side, we get the message, and we are looking hard at less expensive options for a mission.[354] But if we managed to slash costs, how could we be sure that, yet again, Mars mission overruns wouldn't devour the funding that might get Europa going? How could we be assured that the Mars mission in 2018 wasn't yet another case of— and here he was diplomatic—cost over-optimism?

Ed promised to put independent cost reviews in place.

Would there be a public airing of the Mars 2018 mission concept, asked Bob, and would the community be kept involved?

Of course there would, said Ed.

Then Bob pressed harder: It's just that we need to ensure that this airing would include community commentary on whether the Mars 2018 mission would get through the gates set by the Decadal.

And here Ed went into full Ed mode: What gates? They've cut costs considerably, two-point-five to one-point-five. The gates are crossed. The gates are past tense. There are no gates.

Ah, said Bob, but what about *cost realism*? (That is, though he didn't say this, the fictional fiscal promises that led to Curiosity consuming the budget of an entire Discovery mission.)

All of this was starting to really annoy Ed. Bob, said he, you are coming at this as a Europa advocate, and, oh, by the way, even if you get the Europa cost down, we're not going to start a Europa mission until 2030!

TWENTY THIRTY!

Well—and Bob was getting terse and testy, too—are you really suggesting that there won't be a flagship mission until TWENTY THIRTY?

No, said Ed. The Mars sample return spacecraft sequence *is* our flagship mission now, and yes, they are sequential, and no, they will not finish the job until 2030. You do the math, Bob.

Bob was stunned. Ed saw in the Decadal what he wanted to see: that it was Mars missions all the way down. He seemed to overlook the part in the Decadal about getting a Europa mission in the cost box and thus on the flight manifest, and so of course Bob pushed back, and later that day (in an aside), Ed threw Bob a bone and noted that an inexpensive Europa mission would be a good thing, because budgets changed all the time. But it was clear where Ed's heart was, and it wasn't the Jovian system.

Over time, others in the planetary science community pushed back against Ed's interpretation of the Decadal as well. Every time Ed came back from a trip, he would lament that Europa Squad—

both the uniformed regulars like Bob and the underground resistance in the community—had not gotten the message. So he went to Curt Niebur with some instructions.

Europa is dead, Curt! It is your job to carry that message to them, and YOU ARE NOT DOING A VERY GOOD JOB OF IT, ARE YOU?[355]

But Ed, said Curt, Europa is not dead. You keep misquoting the Decadal. If you want to drive a stake through Europa's rocky, saltwater-wrapped heart, you need to say, Yea verily, I, Ed Weiler, hereby slay the icy enigma circling Jupiter. Because you have that power, Ed. But the Decadal cannot be your justification.

And Ed did have that power. But he never killed Europa. Ed was Ed, doing what Ed did, always did, did best: speaking loudly, in very declarative sentences designed with astrophysical precision, to challenge, rile, or annoy—with ninety-five percent of the time, his words not matching his *actual* thinking on the matter. Ed was making a point, gauging his opponent's reaction, poking a stick into the soft spots and waiting for the opposition to fly off the handle. It was all by design and intended to take them off-balance.

It was exasperating but Curt loved it. From his little glass office at NASA headquarters, his bookcases filled with binders of mission concept studies for outer planets exploration, Curt watched associate administrators come and go, some effective, some not, some professorial, some political, but Ed was the only one who would walk in, sit down, look you in the eye, and say: Let's debate this. And Ed never phoned it in. *I am going to push all your buttons and force you to consider every single way of looking at this. That's my job, and we are going to hash this out right now. Without heat there can be no light.* The discussions would come just short of name calling, but Ed would change his mind when presented with a credible argument, and either way, by the time he left, you would think differently about the problem, and progress would be made.

It was the kabuki dance that Ed had once described to Todd May. Ed slayed Gravity Probe B repeatedly and with mirth and quiet calm. But the probe did launch in 2004. Ed killed Alan Stern's Pluto mission in September 2000, banned scientists from even *studying* the mission, vowed publicly that Pluto was "Dead! Dead! Dead!"—his actual words!—his actual exclamation marks!—because of Pluto's price problems, and two months later, Pluto's partisans having found a fiscally feasible way forward, why, Ed hopped on board, and New Horizons launched in 2006.[356] Similarly, Ed killed the *entire Mars program* back in 1999, before presiding over five consecutive Mars mission successes, culminating with the Mars Phoenix landing in 2008, and there were possible sixth and seventh successes in the pipeline—Mars Science Laboratory and the orbiter MAVEN—both wrapping up development.[357] Numbers eight, nine, and ten—the multidecadal Mars Sample Return campaign now embraced by the planetary science community—would follow thereafter.

So when Curt and Jim Green came pounding on Ed's office door, saying no, Ed, you cannot end Europa entirely under the guise of a Decadal decision—there is more work to be done, as expressly, unambiguously described in the Decadal, more studies needed to bring down costs, Ed said: OK. You have a budget, Jim. You find the money for another Europa habitability mission study and knock yourselves out. And Jim found the money.

THE RECONSTITUTED EUROPA science definition team, as directed by Curt, was thirteen strong and had eighteen months to do its job. After that, the money would be gone, and none would likely follow. In the official paperwork, Bob called them the Europa Tiger Team to really drive home the point that this was an agile group of lean and hungry scientists.[358]

Most of the Tiger Team met for dinner on May 2, 2011, at a place called Il Fornaio in a part of town called Old Pasadena. The restaurant was in Smith Alley, named for Willis Smith, a small-time businessman who, a hundred years and change before, had sold saddlery and harnesses to Pasadenans when saddlery and harnesses were what Pasadenans needed. When a three-hundred-room hotel called the Raymond opened a few miles down the road, Smith went into the laundry business. (Someone had to wash those sheets.) The resort catered to wintering well-to-do northeasterners, and with the influx of wealthy tourists, the city boomed and bustled, renovated and thrived. The Raymond attracted presidents and silent screen stars alike, and eventually, residents of obscure stock who knew a good sunset when they saw one.

The hotel didn't survive the Great Depression, but Old Pasadena did, as it always did, from the moment Spanish soldiers stacked the first two bricks that would become San Gabriel Arcángel, the now ancient and still photogenic mission of mortar and stone. Everything proceeded from there.

In a private room in the back of the Smith Alley restaurant, the Europa team was at a standstill. A bottle of white selected by Don Blankenship was passed around. After a day of discourse, the scientists wore rims under their eyes like it was the new black. For four years, Bob and Co. had built the perfect mission for what they wanted at Europa. Their orbiter was derived from the Europa Explorer, which won the Quad Studies in 2006, which in turn traced some of its heritage to the Europa Geophysical Explorer (2005), an internal lab study. The Europa Geophysical Explorer drew much of its science from the Jupiter Icy Moons Orbiter (2005), which drew much of its science from the Europa Orbiter, itself canceled formally in 2001. Europa Explorer became the Jupiter Europa Orbiter, which became one half of the Europa Jupiter System Mission (deceased, March 2011). This didn't even count the dozens of internal variation studies of these missions. The point is: it had been a long time since a Europa mission started from a blank sheet.

Which is how this one would begin.

The Europans had begun the day atop Saint Gabe, on the fifth floor of Gavin Tower, Building 321 at Jet Propulsion Laboratory. Bob welcomed them, dimmed the lights of the conference room, and projected onto a screen a PowerPoint slide, solid white and free of text. They were starting over.

The science side of the team included the usual suspects: Bob, Louise, and Dave, the three of whom would run the project science; Don Blankenship from the University of Texas at Austin was there and would lead ice; Jeff Moore from NASA Ames Research Center was in charge of the geology working group; Bruce Bills and Diana Blaney from Jet Propulsion Laboratory would handle ocean and composition science, respectively; and Melissa McGrath from Marshall Space Flight Center would run the local environment group.

Because Ron Greeley, the maestro who had led the previous studies, was now chair of the Planetary Science Subcommittee of the NASA Advisory Council, and thus unavailable, Curt tapped Bob to stand up the team, with some strings attached: first, invite all the critics (though that's not the word Niebur used)—the members of the community who had real problems with the cost or ambition of the Jupiter Europa Orbiter, or even the orbiter concept itself. So Bob brought on board Fran Bagenal, who was hostile to the ossified structure of flagship-class missions, preferring instead the nimble New Frontiers model of small, agile missions; Bill Kurth and Amy Barr, who had been part of the competing Titan Saturn System Mission and the Enceladus study, respectively; Jack Connerney from Goddard Space Flight Center, who seemed to hate everything Jet Propulsion Laboratory did (too expensive, too big); and David Smith of the Massachusetts Institute of Technology, who was listed as lead author of a white paper submitted a year earlier to the Decadal, asserting that Europa exploration should be divided into smaller, more manageable missions and implying that science definition teams were clueless to the programmatic and political realities of space exploration today.[359] The British-born research scientist spoke from authority, having placed scientific instruments on nearly every major object in the solar system. At the time, Bob and Louise mocked Smith's paper as ridiculous, given the challenges of selling a *single* mission, but he was on the team now, and no one was mocking. No idea was ridiculous anymore.

Next came hours of recap, debating head-on the problems of cost, mission design, and the general malevolence of mighty Jove. Earlier, the team had invited speakers to give insights from other missions and the successful strategies for separating a spacecraft from Earth.[360] Alan Stern talked about how he helped get New Horizons in the cost box and off the ground. Connerney described the spacecraft Juno, set to launch three months hence, and the lessons learned along the way. Alfred McEwan of the University of Arizona described his idea for an Io mission that, rather than orbit the moon, would circle Jupiter nine times or so, capturing a little more of Io with each successive flyby.

A procession of engineers at the meeting described the problems with Jupiter Europa Orbiter, and they discussed mission options: orbiters and multiple flybys and trajectories and what science we could do on the cheap. A lot of ground was covered that day, and by the close of business, everyone needed that wine. Once it was procured in the little room in the back of Il Fornaio, Bob asked what everyone was already thinking: What are we doing here? And they filled glasses.

In retrospect, it was the inclusion of Jupiter research that likely killed the orbiter's chances to get the Decadal flagship endorsement. The Europa Jupiter System Mission went out of its way to incorporate as much Jupiter science as possible, which should have garnered the support of the giant planets panel, thus tipping the results to the Europa mission over Mars. The icy satellites panel was, of course, on board (Europa was an icy satellite) but the giant planets people went their own way, literally choosing *nothing* over a Jupiter system mission. A Uranus mission, which they wanted, wasn't going to happen.

Accordingly, Curt directed the new science definition team that, when developing their mission, in accordance with the recommendation of the Decadal Survey, remember this: We will *never* point the cameras of this spacecraft at the planet Jupiter. Period. It wasn't pettiness. *It was following the Decadal Survey,* hip and thigh, chapter and verse. The Giant Planets people didn't want this mission, and so we would not spend a dime to burden them with it.

Over that bottle of wine or three—Don kept ordering them, and Bob forgot even the color imbibed by evening's end—they came back to Dave Smith's white paper. Smith had proposed three small spacecraft concepts: an orbiter, a multiple flyby, and a lander. At the table that night, they dismissed the lander out of hand as too expensive and unable to meet the science objectives set down by the Decadal. But what would a *small* orbiter at Europa look like? How much would it cost? What about a multiple flyby spacecraft, orbiting Jupiter rather than Europa, and just encountering Europa at intervals, one slice at a time? What science would one do that the other couldn't, and vice versa? The orbiter would handle remote sensing. The multiple flyby would do the geophysics. You break up the mission, a flagship no more. And Bob would captain not some mighty vessel splitting shimmering trails of light on an infinite eighth sea—the HMS *Enterprize* of the eighteenth century or the

USS *Enterprise* of the twenty-third—but rather . . . two somethings smaller.

They had laughed bitterly at this idea before the Decadal was released, but (and maybe it was the wine) the way they were now describing it, the orbiter would be the best little orbiter it could be, and doing only the science that you absolutely needed an orbiter to do, and nothing more. And the multiple flyby would be the best multiple flyby spacecraft it could be, doing only what multiple flybys do best, and nothing more. It could circle Jupiter scores of times, maybe hundreds (why not, after all—this was still a thought experiment), and in its hundreds of flybys, capture Europa one strip at a time. You didn't have to launch the two spacecraft simultaneously. You'd launch one, get your science, and later, when the budgets were favorable, send up the second. Collectively, with this "split mission" concept, you'd get everything the larger Jupiter Europa Orbiter would have done.

The more they talked, the better it sounded. Blankenship, as leader of the ice working group, thought a multiple flyby mission might do a *better* job with his radar instrument than an orbiter. Don first came to the project in 1998, when he chaired the original Europa Orbiter radar study.[361] He had by then been flying terrestrial radar in aircraft over Antarctica for fifteen years, a science he pioneered, and he recruited colleagues from Europe and the United States to converge on JPL sporadically and offer counsel regarding how best to use a radar at Europa. Not everyone in the earth science community jumped at the chance to help, of course; competition in Earth's polar regions was steep, and academic rivals had no interest in sharing their secrets, seeing it as career suicide, an insane act. By the dinner at Il Fornaio, Don was the singular ice scholar who hadn't given up, Ishmael of the cryosphere. He only was escaped alone to tell thee. The surfaces of a planetary body, he knew, were inadequate to the task of assessing whether life arose and flourished on the body. That secret could be found only in the subsurface.

On Earth, if you didn't understand the ocean, you wouldn't understand our planet's ability to sustain life and evolution. The ocean interior—Earth's subsurface—was *everything* when you were talking about habitability. Which is what made Europa special. For celestial objects generally, and Mars being the obvious example, the most a spacecraft instrument could see were the top few microns, the width of a human hair, or, if you were absurdly lucky, a millimeter or two. For all the orbiters and rovers NASA had there, scientists didn't understand the subsurface of Mars—not even a single inch beneath its veneer. Mars is not made of stuff you can see through. But Europa? Its surface was made of ice. And Don's job—his calling—was to see through ice.

He was born and raised in Decatur, Illinois, spent his childhood and adolescence on the till plains, which were as boring a geology as a human being could imagine. To find a single rock in his whole county, you'd need a shovel and patience enough to dig down three hundred feet. But his dad built factories for a living, and in the summer, they would travel, and when Don saw mountains, he would just become transfixed. As a freshman in college in the early seventies, he took every senior-level geology course at the school. It was there that he discovered he was really into ice and glaciers, and when he told his father about it—"I'm really into ice and glaciers"—his dad didn't even blink. Well go figure out glaciers, said the elder Blankenship, reaching into his pocket, pulling out the car keys, and tossing them to his eighteen-year-old son. Go for it.

That summer, Don took the car to Glacier National Park in northern Montana to see what he could learn, spent weeks skulking around, hiking every glacier he could find, and when the season ended, he came home and promptly quit the university because nobody there could teach him what he had just taught himself.

Pocketing his newly acquired knowledge, he then focused his efforts on racing motorcycles.

He first got into dirt bikes as a kid. His part of Illinois didn't have much, but the till plains did lay claim to the best motocross tracks in the world. Don was moderately good at the sport, raced on the weekends, and took jobs here and there to keep the gas tank full. He tried working in factories but found almost immediately that you couldn't race hard and then work at a factory—the human body could take only so much. So he found a job at a bank, worked the drive-up windows. Ahhh—it was heaven. You could sit down all day, like he was being paid to recover from his weekend exertions. After working there for a while, he ran a bank vault, which taught him that giant stacks of money were meaningless (an important lesson to learn if you were in science, as he came to discover). He raced for four years—was a shade away from going pro—and had plans to try for two more years to see how far he could take this thing.

Then young Don Blankenship, not long after turning twenty-one, crashed and suffered a severe knee injury, was told he might never walk again. He spent six weeks in recovery sitting on the sofa next to his mother watching daytime soaps, and when at last he found he could hobble about, he went back to school with vim and verve. This time it was Eastern Illinois University. He was a little more mature, and the geology program was as serious as he was. He hit it hard. To rehabilitate his leg, he hiked tirelessly in the mountains. Don eventually hitched a ride to the Appalachian Trail, hiked a hundred miles in the winter cold. His shanks held up, and it paid off academically: he earned a pretty solid reputation as a field geology guy. After taking a sequence of physics classes, however, Don realized to his shock and horror that he liked physics more than geology—that maybe his future was in geophysics, which in the rock community was heretical, like renouncing one religion for another. Ever the infidel, however, graduate school in geophysics it was.

Six months after he arrived at the University of Wisconsin–Madison, his advisor, the famed glaciologist Charlie Bentley, had him in the back of a C-141 cargo plane—landing gear stuck—as the aircraft went in for a final approach at McMurdo Station, Antarctica. (They survived the landing.) The continent, he found, was a land of punishing extremes: cold, obviously, blizzards occasionally, but also bright: the sunlight from open skies reflected on pristine white below, and without sunglasses, it was utterly blinding, like staring directly at the sun. You went out in the field, and you were in the last real wilderness on Earth, and there was such solitude that only astronauts knew, perhaps, or ships lost at sea. Yet after all Don had done, Antarctic expeditions were fairly inside the box. An essential element of the productivity of a geophysicist working "on the ice," he soon learned, was the ability to tune Rotax two-stroke engines in snowmobiles—a skill he'd picked up in his dirt bike days. This made him really, really effective in the field. He'd be out there, another scientist staring at a snowmobile, lamenting: It won't run anymore; I can't do my work. And Don would walk over, rip apart the engine, fix it, make it go. Thus did motorcycle racing prove critical to the success of his Ph.D.

That doctoral thesis was on mobilized till beneath ice sheets in West Antarctica. At the time, mountain glacier people thought that the glaciers moved on highly pressurized water; that the ice just slid across like a puck on a hockey rink. Don conducted geophysical studies of the bottom ten meters of sheets a thousand meters thick using acoustic waves. It was like doing a high-resolution sonogram of a child, and he discovered that glaciers moved quickly not because of pressurized water, but because they were sliding on mobilized, saturated sediments—real sloppy stuff. Those findings were an absolute affront to classical glaciology. Sliding along mud was just not attractive to the orthodoxy, and they confronted Don at conferences, but he stood his ground. And years later, the part of Antarctica he had studied was drilled. He was sitting in a galley at McMurdo at two thirty in the morning one day when a driller walked in, sat down at the table, and said, I have something for you.

He pulled a baggie out of his pocket. It was the mud!

What's the porosity? asked Don.[362]

Forty percent!

And the pressure?

One hundred kilopascals!

It was exactly as Don had predicted. Poetically, the Illinois till plains that so bored him all his life were created by mud on which glaciers glided. Blankenship grew up on the sort of mud that he discovered in West Antarctica beneath fast-flowing ice streams.

He spent the next twenty-five years studying how the geology of Antarctica controlled the evolution of its ice. He did this concurrently with his Europa work, and by the time he was uncorking a bottle at Il Fornaio, he and his team at the University of Texas at Austin had completed the geophysical mapping of most of the continent through a project he founded called ICECAP. He was ready to get to work on Europa as well. His instruments could see through kilometers and kilometers of ice on Earth, and they could do the same at Europa to reveal the secrets of the ice-ocean interface—a key to understanding habitability. If Earth was any indication, if Europa's ocean floor was populated, the ice-ocean interface would be populated as well. Though the ice on Antarctica's surface was paper white, the underside over water was pretty gnarly, rich in biology, coated in brown webs of algae—entire ecosystems of microorganisms, really—with fish swimming up, nibbling away. Europa's ice might be no different.

But the cost. An orbiter radar would have to be radiation hardened, running about a million dollars per chip to build. And radars were data intensive, which meant a lot of chips. Not so on a multiple flyby mission, however. You wouldn't have to soak in thermonuclear bathwater; you could dip in and out and do so for years. So it was clear which of the two spacecraft the radar belonged on. And by removing the pricey radar from the small orbiter, that spacecraft was suddenly a bargain!

It was a breakthrough—not technical, scientific, or even conceptual—but psychological. That night, there was a clear severing of the past, and the next day, back astride Saint Gabe, the science definition team of this Europa habitability mission started parceling the thing out. Bob literally drew a line on a whiteboard, and on one side scrawled ORBITER and on the other, MULTIPLE FLYBY, and they worked it through, wrote which instrument fit better on each side. The payload practically split itself in half. Each scientific instrument just swooned naturally to the left or right, and with the division, Bob ballparked prices in his head. Billions of dollars had tumbled away from the price tag overnight.

The next day, he emailed Ron Greeley with breathless delight:

The Europa SDT just had its first meeting, which resulted in a radically new approach toward achieving the Europa science objectives and goal. Even the floor ("core") Europa payload that we have discussed in the past cannot be done cheaply. But after several presentations and much discussion on what can and cannot be done from flybys, it became clear that the floor payload splits itself based on what needs to be in orbit, and what could be done via flybys. By splitting the payload into 2 different mission elements, flyby and orbiter, we are now optimistic that each might be done for ~1.5 to 2 B$. We would recommend that the less expensive option go first. It seems probable that this would be the orbital element, which we hope will fit into a $1.5B cost.[363]

He went on to detail what would do what, and where, and who would be doing it, and ended the note by saying, "Let's stay in touch as this idea develops."

The Jupiter Europa Orbiter—indeed, the very notion of *any* outer planets flagship mission—was dead. But the possibility of actually going to Europa? It suddenly seemed as alive as any mission in NASA's portfolio.

CHAPTER 12

The Baltimore Gun Club

THREE DAYS BEFORE JOHN F. KENNEDY COMMITTED NASA to "landing a man on the moon and returning him safely to Earth," Lori Beth Garver was born in the city of Lansing in the state of Michigan. She grew up in nearby Haslett, a quiet, unincorporated community a few miles from East Lansing (not the same as Big Time Lansing, the state capital), and like every green-and-white-blooded East Lansingite–adjacent was fiercely proud of Michigan State University, and when you pedaled your bicycle into town, the city limits sign reminded you just how proud you should be: EAST LANSING, it said, THE HOME OF MICHIGAN STATE UNIVERSITY.[364] Everyone in her family going back generations went to MSU—parents, grandparents—it was just the cat's meow in the family Garver.

Lori hailed from a family of public servants. Her parents both grew up on Depression-era farms. Her dad did a stint in the Marine Corps. He and her mother first met at Michigan State (of course), and when the two graduated, both became teachers. While expecting their first child, her dad realized that the family would need a little more money, so he took the brokerage exam, passed it, and was hired as a stockbroker for Merrill Lynch. He would remain there for the rest of his career, helping people invest successfully in one company or another. Her mom's degree was in home economics, a field that meant a lot to her, and after a couple years of teaching it, she was thrilled to live it, to manage the household. It was a modest lifestyle, but a good one, and Lori and her sisters knew they could one day do anything they wanted, which was not a guarantee for everyone in her community.

Lori got a job at sixteen at McDonald's, and it was just a dream—a thrill—*exhilarating* to work there, two dollars and thirty cents an hour, and she worked hard, got a five-cent raise right away, and that was a pretty big deal, I mean, and she got a car—a VW Jetta (a scandalous purchase in Michigan)—and she toiled under the Golden Arches all through high school, which she loved—school, was successful at it, was small-town super popular, did everything extracurricular: performed in plays, ran track, was on the basketball team, played tennis, was in the band (flute, oboe), did a decade of ballet, was a cheerleader. She was voted Most Versatile her senior year, and that's exactly what she wanted because it's exactly what she was.

Her family on her mom's side was political. The farm had been in the family for one hundred seventy years, and her grandfather was in the state legislature in the fifties and sixties.[365] He was a good Republican and remained a full-time farmer because the legislature met only three months out of the year, and you still had to cudgel out a living when the sessions ended. Later, her uncle was elected to the same seat, and served in the state house and senate in the seventies and eighties.[366] Before Lori could crawl, she appeared in campaign brochures and parades, and once she had mastered standing and walking, the little girl went door-to-door in support of the family. Her uncle ran for Congress in 1980 and then again a decade later, seeking to be the Republican representative of the tenth district of Michigan, and she was absolutely devastated when he lost that first time.[367] But she learned in the process that she loved politics; that if you wanted to make a positive, meaningful difference in people's lives, public service was the way to go.

When Lori told her high school guidance counselor that she wanted to take the SATs, the adult was stunned and bewildered: But why would you want to do such a thing, Lori? Michigan schools used ACT scores! But Michigan State had so suffused young Lori's life that it seemed natural to go elsewhere for college; she had, it felt, already gone to Michigan State from grades one through twelve. She went to Colorado College for her undergrad. She studied political science and economics, and while there, did Semester at Sea spending time in developing countries, seeing with her own eyes levels of poverty previously unimaginable, and it was there that Lori Garver, clinically raised Republican stock—she's seeing this and learning that Ronald Reagan was cutting foreign aid to these people, who had nothing, who were harming no one, and she was aghast and she came home and sought out the person most likely to defeat Reagan, and she asked for a job on his campaign, and got it.

Concurrently: in 1983 Sally Ride was selected for flight on the seventh space shuttle mission, and with the launch of *Challenger* became the first American woman in space.[368] (*Challenger* disintegrated shortly after launch three years and eight missions later.[369]) Until then, Lori was cognizant of the space program—I mean, she knew we had one. She was an eight-year-old when the *Eagle* landed, and it was exciting as far as Big Moments in an Eight-Year-Old's Life went, but right away she just took it for granted that she would one day walk on the moon, too, so in a way it was like watching the ribbon cutting for a supermarket or stadium: it's new and neat, but you know you're going to go there one day, and that'll be even better, and you file away the memory and go about your life. Space was a thing the United States did, but she didn't pay that much attention—until Sally Ride circled Earth. And, sure, it was extraordinary, but it was also ridiculous. This wasn't the fifties. Russia had flown a woman twenty years earlier![370] What took NASA so long? But after that, she was locked in.

The person most likely to dethrone Reagan in 1984 was Senator John Glenn of Ohio, who demonstrated his righteous stuff in 1962 when he became the first American to orbit Earth.[371] Lori got a job on the Glenn campaign as a receptionist and later moved into scheduling. It was an entry-level job out of college, and it was interesting and engaging, and she'd been doing politics her entire life so she knew a thing or two about it, and her higher-ups noticed. Senator Glenn did not defeat Ronald Reagan in the 1984 election. He did not, in fact, even win the Democratic nomination. When he fared poorly in the Super Tuesday primaries, that was that.[372] But Glenn and his team looked out for the young go-getters on the campaign, and when it came time for everyone to go forth and find work elsewhere, the head of personnel told Lori that there was an opening at the National Space Institute, a space advocacy group founded in 1974 by Wernher von Braun. Apply for it, Lori, she was told, and Lori did, and was hired.

It was just a great time in her life. She worked full-time at the institute by day, and at night attended classes studying space policy at George Washington University. Between the day job and night classes, she came to know everyone in the relevant circles. Her professor was John Logsdon, the noted historian, prolific author, and grandmaster of space policy, and over the course of a semester, he would have everyone in the business drop in to give guest lectures. It was a fine advantage if you were willing to work harder than nearly everyone else, and in 1987 the National Space Institute and the L5 Society, a fellow advocacy group, merged to form the National Space Society, and the new organization hired Lori to be its executive director.[373] She was twenty-six and on her way.

The National Space Society was a small outfit relative to the steamroller lobby firms in DC, but they were doing important work at a time when space exploration had reached a nadir. There just weren't that many space advocacy and policy analysis groups, except for maybe the Planetary Society, but it was a relative newcomer, founded only in 1980 by Carl Sagan and two California mountain people: Bruce Murray and Louis Friedman. So here was Lori Garver sitting at von Braun's actual desk, sleek and huge. (He died in 1977, so it was available.[374]) When the spacecraft Galileo and Ulysses launched in 1989 and 1990, respectively, the National Space Society fought the good fight, making sure those missions got their licenses to launch. When you've got screwball activists chaining themselves to fences, you need to deploy rational minds to balance things out, and there she was, Lori Garver of Haslett, locked in debates on the steps of the DC courthouse and testifying on the Hill—executive director of the National Space Society—it certainly sounded official, anyway, and she was the only woman doing *anything* visible for space policy at the time (Sally Ride excluded, who did *everything*, already a walking, eternal echo in history), and the community embraced her.

In 1996 Dan Goldin, the administrator of NASA, called Lori and offered her a job, and that was something. He made her the associate administrator for policy and plans—basically the head of policy at NASA—and from then on, she was a politico. It was never the plan, but there she was. And it was a risk, because once you stepped into the political arena, you would be considered a combatant evermore, but it was worth it, because working with Dan Goldin, my goodness: now *there* was a great American, and a true believer in what the space program could achieve. Faster-Better-Cheaper got a bad rap, but Goldin thought deeply about the problems facing the agency, and Lori's job was to help him. Goldin started each day by picking up the newspaper, and there might be an article on how, say, the medical community made all sorts of mistakes during surgeries because of sleep deprivation, and he'd have Lori get on the phone and get those people in here, and let's transfer to them all the knowledge we have about sleep cycles for astronauts. He also had ideas that were pretty radical at the time about where NASA needed to be, and Lori bought hard into them. Goldin was a visionary, wanted to turn the keys to the space shuttle over to the private sector and, once the International Space Station was space-borne, wanted to hand it off to industry.[375] NASA had built the infrastructure of low Earth orbit. Its work was done. It was up to businesspeople to take it from here, because NASA's mission was to go farther, always: to Mars and beyond. So Lori added commercial space to her portfolio, soon scoring a real coup when Fisk Johnson of S. C. Johnson & Son (the multinational conglomerate behind Drano, Ziploc, Raid, and Toilet Duck) sought to use a NASA-developed bioreactor to fight liver disease from space.[376]

That wasn't her only encounter with space biology. Not long after she took the job, planetary scientists discovered life on Mars.

WELL, THAT'S WHAT they thought.

They called it Allan Hills 84001, and it was first found twelve years earlier by a team with the U.S. Antarctic Search for Meteorites program. Every Martian meteorite was considered *rara avis* because to get a hunk of rock from planets four to three, you would need a Martian impactor of such astounding force that material would be blown to space, and the orbits and timing would have to be such that the rock would fly for millions of miles, find our pale-blue dot, survive entry, and hit the surface rather than an ocean. It took sixteen million years for the meteoroid in question to make the journey.[377] In 1996 a team led by scientist David McKay submitted to the journal *Science* his analysis of Allan Hills 84001, noting that structures within it suggested fossilized bacterial life that could have formed only *before* departing Mars.[378] The paper enumerated and explained discrete characteristics bolstering this admittedly audacious assertion. "Although there are alternative explanations for each of these phenomena taken individually," the researchers determined, "when they are considered collectively, particularly in view of their spatial association, we conclude that they are evidence for primitive life on early Mars."[379]

Well, there you had it! The first direct evidence of extraterrestrial life ever discovered. And this wasn't guesswork from some crackpot on a NASA grant; McKay was the chief scientist for astrobiology and planetary science and exploration at Johnson Space Center. He was literally the world's foremost subject matter expert and one of the smartest human beings to ever walk the Earth. His work was rigorous and meticulous, every *t* crossed, every lower-case-*j* dotted.

The paper hadn't yet dropped when Lori arrived at Johnson as part of an unrelated tour of NASA centers to get up to speed on what the agency was doing and what her job would entail. When she walked in, sat down, and was presented suddenly with this rock from Mars with *fossilized Martian life in it,* she called Dan immediately and said, Um, you weren't going to mention this? Are you kidding me?[380]

And Dan seemed to her just bonkers paranoid.

Are you on a secure line? he asked. We'll talk about it when you get back.

When she got back, they talked about it, gamed it out, strategized how the agency might advance this discovery. The research had been subjected to full peer review. This was the real deal.

Dan wanted the full lube-and-wax job for the unveiling. He even brought in a skeptic who would sit on the panel during the rollout.

But before they did the press conference announcing the discovery, Lori told Dan point-blank that there was one more thing he needed to do: he had to tell the president. Life on another planet? It's a big deal, Dan.

No, no! said Dan. You don't just go *see the president*. Maybe someone at the Office of Science and Technology Policy . . .

Mm, said Lori, no, I think this goes a little higher than that. But she saluted and went to the president's science advisor and explained what had just been discovered, and right away, Dan received a phone call—was summoned to the White House—swept into the vice president's office—and there he was, Vice President Al Gore, inquisitive as ever, and Dan explained what had been found, and how, and why scientists thought what they thought, and Gore listened, nodded thoughtfully, reflected a moment, stood up, and said, finally, in his distinct southern inflection, all stretched vowels and a mouth wrapped carefully around every syllable: *Well, follow me.*

Where are we going?

Oval, said Vice President Gore.

Later that day, Goldin returned to NASA headquarters and sat across from Lori, stunned and starry-eyed. He had told the president of the United States about the discovery of alien life.

And so began the dawn of an exceedingly brief new era. On August 7, 1996, President Bill Clinton gave a televised address from the South Lawn of the White House. In the rearview mirror of history, it might have been the most important address of all time. "If this discovery is confirmed," said the president, "it will surely be one of the most stunning insights into our universe that science has ever uncovered. Its implications are as far reaching and awe inspiring as can be imagined. Even as it promises answers to some of our oldest questions, it poses still others even more fundamental."[381]

And it was all a horrific mistake!

In the end, each of the life-affirming features of the meteorite—now subject to the intense scrutiny of just about everyone whose name was suffixed with a *P, h,* and *D*—was explained by nonlife processes in such a way to suggest that ALH84001 wasn't a pristine cemetery of Martian bacteria.[382] It was . . . just another rock! Sure, another rock from Mars—and that was exciting, because there were then only a dozen identified Martian meteorites in all the world—but ultimately, went the consensus, just another rock.[383]

In 2000 George W. Bush was elected president, and a new NASA team transitioned into power. By the end of her time at NASA headquarters, Lori was most proud of her work developing commercial space. The bioreactor demonstration system was completed just before its 2003 launch on the shuttle *Columbia.* The experiment was destroyed, however, when the space shuttle disintegrated in February of that year.[384]

When she left the agency, Lori worked as a consultant primarily for the big aerospace companies, did proposal development, competitive analyses, acquisitions, and just taught them in general what good NASA contracts looked like and how to win them, which was exactly the sort of consultant work you did when you left government. And then she became "AstroMom," which was not.

One of her clients not of the aerospace industry wanted to fly to space and do experiments there, and he retained Lori's talents to make that happen. So she helped negotiate a great contract with the Russians for the flight, and things moved in Swiss watch fashion. He was soon in Russia doing the necessary medical training for a stint on the space station, but for business reasons, he suddenly found that he could not go up. This did not go over well with the Russians, who absolutely needed the money, and they made it very clear that you (i.e., Lori) needed to find someone else to buy that seat, *srochno.* We have bills to pay! So Lori called Tom Hanks,

James Cameron—everyone she could think of, really, who might want to go to space and had the funds to foot the bill. No dice. So her consultancy came up with an even better idea: Why not send Lori—LORI GARVER OF HASLETT—to space. He was her client, after all. She was the one with the NASA experience. She led the space group at the consultancy. What they would do is get an agent and find sponsors to pay her way, and they would make a big splash of it. And if you're trying to sell product, well, women made seventy percent of household purchases in the United States. Lori was perfect for the job!

She hesitated. Her kids were four and six, and she would have to leave them for months, and that would be hard. There was the danger, too. It was Mary Ellen Weber, her favorite astronaut friend (as one has), who advised Lori to think of her own mom. What if *she'd* had this opportunity? And when you were six, she'd turned it down because of you? Would you have wanted her to have done that?

It was—God, Lori wanted her mom to be able to do *everything*! It was the sort of advice only an astronaut, wise and brave and practically mythological, could have given. And Lori talked to her boys about it, and they *were all in, Mom.*

And so it was set. Lori flew to Russia to undergo the requisite medical tests and training, and she was staying with a Russian family (her translator's mom), and she had negotiated a great seat price, variable, depending on the number of sponsors she could get.

And then her office called.

Hey, Lori, this is hilarious, but Lance Bass from the band 'N Sync wants to fly to space.

And, I mean, to a consultant that was great! Which one was Lance? Was he the cute one? she asked. He was! Her favorite! Nice guy, he flew out, and he and Lori were fast friends, did press conferences. Lance had a manager, agents, and apparent backers, and the

Russians saw serious coin in their future and good night and thank you, Lori, but we've found our real meal ticket.

It was clear to Lori almost immediately, however, that Lance Bass would never leave the Earth planet. Lori had gone after this thing the way a hard-nosed veteran of DC did things: meticulously, strategically, logically. Discovery Channel was her media partner, and it was in for one million dollars. Lance Bass, on the other hand, had MTV, but it wasn't clear to Lori how much, if anything, MTV was willing to pay.[385] She had Visa on board, and it was going to kick in three million dollars, because she was set to fly in November, and here was a mom *in space,* and she had two boys back home. She had to shop for the holidays! It was going to be the first credit card purchase from orbit. She was set to buy something from another of her sponsors, Radio Shack (promising a disappointing Christmas for one of her kids). She had Sudafed in her portfolio. (Astronauts relied on it while on board the space station but could never *say* they used it, because NASA didn't want another Tang—referring to the orange drink mix popularized when word got out that the Mercury and Gemini astronauts had enjoyed it on their flights.) One of her kids played soccer, but Major League Baseball expressed interest, so guess who dropped soccer and took up the bat and ball? It was all set. Disney, when she landed: "Lori Garver, you just returned from the International Space Station—what are you going to do now?"

"I'm going to Disney World!"

The story wrote itself.

Lori was on her way—even had her gallbladder removed because the Russians found a gallstone and refused to put her on the centrifuge. Gallbladderless and so close to launch that she could taste the rocket's dimethylhydrazine, she spent her fortieth birthday doing eight Gs.

But Bass and his entourage of representatives . . . Lance was clueless—a nice guy—but just oblivious to the way things worked.

There was no way, she felt, this guy would find the money to cover a twenty-million-dollar seat on Soyuz. The ruble conversion, however, was just too tempting for the Russians to see the obvious: that the singer's representatives lacked the resourcefulness to get the money. Amazingly, Lori wasn't bitter about any of this because her motivation in part was to get word out about the space station, which Americans seemed to know nothing about. She had been at the agency at the time of its launch, and the station was important to her. And having embraced its commercialization, she wanted to see the money materialize, and she knew now that the advertising dollars were there, just waiting to be spent.

Anyway, Lori went home. Lance Bass could have her seat because he was going to popularize the station, too. But then his agents, as expected, came up short, and the Russians replaced him with a cargo container.[386]

Though she never left terra firma, her AstroMom experience made Lori perhaps the most visible face in commercial space, and she returned to consulting with new ideas for how to do business in the void.

LORI WAS WALKING through Dulles International Airport in Virginia, autumn 2008, when her phone rang.

Said the voice on the other end: I don't know NASA and you don't know me, but I'm Tom Wheeler.[387] I'm supposed to select the head of the NASA transition for the Obama administration. I know exactly one person in the space program, and I called him. He said it should be Lori Garver. The person I called was John Glenn.

Well, of course she said yes.

Barack Obama had not yet been elected president. His transition teams were assembled in advance and briefed so that the day after the election—assuming it went his way—members could parachute into their respective areas of responsibility. In the seventy-seven days between Election Day and the inauguration, the president would need to be fully briefed on the affairs of state so that on day one he could run his new government effectively. On November 5, 2008, President Elect Obama's transition team got to work.

Lori felt the burden of the job straightaway. Here she was, a Haslett dreamer who had worked campaigns since she was in diapers, now helping to facilitate democracy at its finest and most magical moment—the peaceful transfer of power—and she was in charge of space! She dropped her entire portfolio of consulting clients to avoid conflicts of interest, and embarked on a cross-country tour of each NASA center to understand ongoing activities, learn where they were prospering, where they were struggling, what they needed for success, and where they intended to go from here. Her favorite meeting was with Charles Elachi at Jet Propulsion Laboratory. There was no subterfuge in the director's presentation, no obfuscation. He declared that Mars Science Laboratory was over budget and behind schedule, and while they could make the mission launch date if told to do so, there would be a very high probability of failure. Elachi listed the technical reasons for the delay and key things he needed for success, chief among them: four hundred million dollars and two more years. She believed him. I mean, NASA was almost two billion dollars in at that point. You're telling me you need more time? OK. You'll have more time.

For every one of those seventy-seven days, she met with center directors and mission directors in back-to-back meetings. NASA rank and file, overall, were very accepting of Lori and her team. She had been there before, had by now clocked a career in space spanning twenty-five years, was a prominent space advocate, and had a lot of friends. She encountered occasional pushback from senior officials who were nonplussed with a new administration coming in and asking questions, but there was no time for political posturing. Personnel changes could be made soon enough, and she had weekly reports due to Rahm Emanuel, the abrasive incoming chief of staff of the Obama White House. You did not want to disappoint Rahm Emanuel.

Of all NASA's programs, the one that soon clarified in Lori's eyes as the most troubling was Constellation. The program began in response to the Vision for Space Exploration, a long-term plan for human spaceflight initiated under Sean O'Keefe and announced by President George W. Bush in 2004.[388] Broadly, it shared the same goals as George H. W. Bush's Space Exploration Initiative: station-moon-Mars.[389] It was billed as Apollo on steroids, which was almost an understatement. NASA would build Orion, a new category of crew capsule capable of connecting to the space station or carrying astronauts to the moon. The agency would also build a new lineup of rockets, called Ares. (If there was any doubt as to where Constellation intended to fly, Ares is the Greek equivalent of the Roman god Mars.) The smallest of the Ares rockets would launch Orion to the International Space Station. Ares V, the gargantuan heavy-lift variant in the lineup, would send astronauts to deep space. There would be transports between Earth and the moon, lunar landers, rovers, Mars transports, Martian landers— the works. NASA would work out all the kinks of serious settlement on the lunar surface and apply the lessons learned to the Red Planet. The program was possessed of boundless ambition and engaged the entire agency. It, at last, was a vision: a clear, no-hedging answer to What Are We Doing Here, Anyway? Why do we have a NASA? To build the infrastructure and equipment to send astronauts to Mars. And we are going there to stay. We have already started building the equipment. *It is going to happen.*

The problem was paying for the thing. Congress demonstrated little interest in appropriating the numbers NASA needed to get

Constellation off the ground. Had this been a piecemeal effort, the agency might have been able to work around the flat funding, but by the time Lori was running the transition team, billions had been spent on the program, and . . . it just wasn't going anywhere. Its prerequisites were untenable.

The more Lori and the transition team heard, the less sense it made. According to the current NASA plan, Orion and Ares would be paid for by mothballing the shuttle fleet in 2010 (saving three billion a year) and tearing the International Space Station from space in 2016, literally dropping it into the ocean (even though NASA was, at the time, still building the thing).[390] This would save fourteen billion dollars otherwise spent extending it through 2020.[391] But Orion was designed to *go* to the space station and was not scheduled to fly until a year *after* said station had become the world's most expensive habitat for startled fish. Which meant . . . the Orion plan only worked if it assumed that deorbiting the space station wasn't really going to happen. The entire program was like a political sleight of hand.

This did not go over well with Lori Garver, who knew, among other things, that YOU DO NOT START LYING TO THE PRESIDENT ON HIS FIRST DAY IN OFFICE.

Four months after his inauguration, Barack Obama nominated Charles Bolden, a former astronaut and Marine Corps general, to become the administrator of NASA. Obama had previously put forward other candidates, but Senator Bill Nelson of Florida didn't like his first or second or third choice, and Nelson had the power to kill said nominations dead, and did.[392, 393, 394] Nelson was up-front about it: he wanted Bolden as administrator, because, at least in part, he and Charlie had flown together on the shuttle *Columbia* in 1986. Nelson, a payload specialist on the mission, liked Charlie, trusted Charlie, and that was that.

Just before she and Bolden were confirmed by the Senate, Lori set in motion an independent commission to figure out the state of the American space program, and chiefly to get to the bottom of this Constellation business. Led by Norman Augustine, the former chairman and chief executive officer of Lockheed Martin, a panel of ten luminaries in space policy met seventeen times by teleconference and at various NASA facilities between June and October 2009.[395] Sally Ride served on the panel, as did Christopher Chyba, who chaired the first Europa Orbiter science definition team in 1998; Wanda Austin of the Aerospace Corporation (the same that had so vexed the Jupiter Europa Orbiter team in the Decadal Survey); and a mix of astronauts, engineers, scientists, established players in the aerospace industry, and upstarts in the burgeoning New Space movement of private cosmic explorers. At the end of the commission's study, it issued a one-hundred-fifty-five-page report titled *Seeking a Human Spaceflight Program Worthy of a Great Nation*. The set of findings was extensive, examining every aspect of space exploration: where we go first (moon? Mars? elsewhere?); how we get there; why we go there; what we do there; how we pay for it; and how we balance human and robotic exploration. The report presented five options for how NASA should proceed. Each option, in Lori's estimation, was excellent. And every single one of them canceled Constellation.

Cancelation, though, didn't come easy. The White House was ready to delete it, and Lori doubly so. Private sector rocket companies such as Space Exploration Technologies Corporation (SpaceX) and Blue Origin were promising to deliver the holy grail of exploration: cheap access to space, with costs falling by tens to hundreds of millions of dollars per launch. This was so clearly the future, so obviously the way to go. What were we even doing building big government rockets? Why not do what Dan Goldin had wanted to do all those years ago? Hand over the keys!

NASA, though, was not ready to change. Down at Marshall

Space Flight Center, engineers wanted to build giant rockets. That is what they did, what they had always done. Change was not so much anathema as antithetical; why *wouldn't* you build giant rockets? I mean, what else was there? After the Augustine Commission nuked Constellation from orbit, Lori was part of one meeting where a NASA official said, in solution to the problem, *Maybe we could just . . . change the name?* That was his proposal!

That official, knowingly or not, called exactly what would happen. In 2010 Bolden reluctantly (he was a former astronaut, after all) followed through with the wishes of the White House and announced the cancelation of Constellation.[396] And for a brief time, the White House stood by his side. But meanwhile, the administration was waging trench warfare to reform the American health care system and needed all the allies it could gather. When faced with a Senate that wanted Constellation, a NASA administrator who deep down wanted Constellation, aerospace contractors who wanted Constellation, and NASA center directors who wanted Constellation, the White House simply lacked the stomach for fighting on all fronts, and especially on something as peripheral as space. The president would burn no capital on a rocket program, no matter how wasteful it was. It was, after all, just a rocket. Health care was at stake.

Lori was in a car with Charlie when the president's deputy chief of staff called to let them know that the president had decided to cave on Constellation. Lori, he said, make the most of it. Make it a win.

Thus began for Lori Garver the sort of headache that left even the most excessively strengthened aspirin cowering in a corner. Under protest, she made what she knew was just a terrible deal, but the best that she would ever likely get: She would give Congress the Orion crew vehicle and the Ares V heavy-lift rocket.[397] In exchange, the administration would get money for general technology development, the James Webb Space Telescope, and a robust "commercial crew" program—that is: the private sector would one day take over all launches to low Earth orbit. The smaller Ares I rocket was dead, and we were not going back to the moon. (We couldn't afford a lander!)

The White House and Congress were thus agreed. And then the Senate went to Crazytown, dictating in the NASA Authorization Act of 2010 a litany of rocket requirements so specific that you'd think the entire chamber had completed coursework on propulsion engineering.[398] They wanted a rocket capable of lifting seventy to one hundred tons of mass into low Earth orbit, and one hundred thirty tons when married to an integrated upper stage. They wanted liquid fuel engines and solid rocket motor engines. They wanted Ares I and space shuttle technology reused wherever possible. And they wanted it flying by December 31, 2016. It was like an aerospace industry wish list—because it was.[399] Prime contractors pushed hard for the legislation so that existing partnerships might be preserved. Consequently, not only did the Senate want NASA to Frankenstein a rocket, but they weren't even going to allow engineers flexibility to build the best one possible.[400]

From Lori's perspective, it was still Constellation. All they did was . . . change the name! And worse, with the terms dictated to NASA, they had passed an impossibility into legislation. There was no way that rocket would launch on time, and the rocket scientists said so, but the Senate said: No, you have to. It's *right there in the law*![401, 402] Well, said Lori, you could pass a law that says the sky is purple, but that would not make it so.

In August 2011 the space shuttle program officially ended. It was time. Two out of the five had been destroyed during operations, killing all aboard in each instance. They were too slow to launch. Too expensive. There were no emergency egress systems: the light came on that said WE ARE ALL GOING TO DIE and . . . you died.

And the things were not getting any safer with age. Yet when the shuttle *Atlantis* touched down for the final time, an odd, resonant sadness seemed to sweep the nation. It was like the government had decided to dismantle the Statue of Liberty. Despite its costs and setbacks, the shuttle program—formally called the Space Transportation System, or STS—was persistently and now perniciously popular.[403] And with the grounding of the fleet, NASA found itself for the first time in fifty years unable to launch astronauts into space—a significant problem for an agency whose chief purpose was launching astronauts into space. And so—oh, this one hurt—the only way Americans could get to the International Space Station would be to fly from Russia, on Russian-made rockets, in Russian-made capsules.

A happy ending, said Orson Welles, depends on where you stop your story.[404] Because of unrelenting solar radiation, the American flags on the moon had long faded to solid white. The Americans won the space race, it seemed, only if you stopped telling the story in 2010.

IN HIS TWENTY-ONE years since joining the agency, Todd May had learned a lot. He had learned that management and leadership are two completely different things; in some ways, opposites. If you're a manager, you have formal responsibilities, formal accountability, and formal authority. You have a set of resources within which to work. You have constraints and processes. Leadership, though, is the art of gaining willing followers. Your sense of constraints can be less about strict adherence than supporting irregular ingenuity; in some ways, you're trying to take people *outside* of the box, you're taking them to places they didn't know existed. Management is saying we've got only this much money—how do we work with that amount? Leadership might ask how we

spend money differently so that we still achieve what we're trying
to do.

Todd had done over decades the sort of tough, tedious grunt
work materials engineering that, it seemed now to him, young engineers didn't do anymore. Everybody wanted to launch something,
and when Todd first set foot on Marshall—not even a badged
employee of the National Aeronautics and Space Administration
but a bright-eyed contractor—they weren't asking him to launch
anything and would have laughed if he'd suggested that he be put
on flight hardware. First, he had to make his bones, and he did.
He learned later to be a junior manager, and would learn to lead
particular parts of projects, and then projects themselves, and then
entire programs, and in his years, his hard-earned experience had
helped separate a good number of NASA's greatest achievements
from the Earth. But there were tradeoffs. It oftentimes left him
removed from his family by thousands of miles for months at a
time and bleary-eyed because of this emergency or that, solving
problems with Russian engineers who spoke fragments of English
at best and who only months prior had been part of a program to
erase the United States from the map in a total, terrible light show.
He had setbacks, his projects were sometimes canceled before he
started—good projects, worthy projects, like the habitation module
on the International Space Station. Yet he persevered. And he was
really good at what he did, but—and this thinking infected half
of the space program—you build spaceships for a living, and you
sometimes wake up in the morning and think: you've been lucky so
far, but one day the meeting will come when everyone asks: What
are you doing here? Why are you in this room?

By 2011, and despite the insidious doubts that plagued him, May
had done a few things, but with every major role, every promotion,
every ambitious endeavor of which he had ever been part, he had
been asked by someone higher up to give it a go. Todd, we'd like you

to lead this space station module. Todd, we'd like you to lead the Discovery and New Frontiers program at Marshall. Todd, I'd like you to come with me to NASA headquarters. And now there was his boss—everybody's boss—Robert Lightfoot, the director of Marshall Space Flight Center, sitting across from him and asking him to lead what would be the most critical program in the center's portfolio: a program upon which the very survival of Marshall might depend.

At the time of Constellation's cancelation, Todd was associate director of technical on the ninth floor at Marshall. Constellation employed thousands, both directly and through contractors, and Todd worried first about morale. The contractor workforce was about to be gutted. These men and women had been working really hard on this thing, and its eighty-sixing wasn't their fault. And though Marshall wouldn't be chaining the front gates, how could you keep the team primed for whatever came next?

Todd immediately tried to figure out how he might leverage his Discovery–New Frontiers background and started digging into robotic precursor missions—"precursor" as in "precursor to astronauts": i.e., just about every science mission that went to the moon or Mars. A lander, in his mind, was just a launcher in reverse, and launchers were Marshall's thing. And though Constellation was deemed unsustainable in the Augustine Commission report, the document was categorical in stating that the United States still needed a heavy-lift rocket.

Sometime between the teleconferences, the technical meetings, and the flurry of fact-finding trips to Goddard Space Flight Center in Maryland; Langley Research Center and Wallops Flight Facility, both in Virginia; and NASA headquarters—Lightfoot pulled Todd into a one-on-one and offered him the project manager position for NASA's next big rocket program and the centerpiece of any mission to the moon or Mars. They called it the Space Launch System, or SLS.

Todd declined the offer. He said—it was selfish, he knew—I will do this only if it is real. I do not like paperwork exercises. And, May reflected, there were a lot of reasons he was *not* the person for the job, chief among them: he was not a rocket guy. There was no institution on Earth with a rocket culture so strong as that at Marshall. Huntsville even called itself Rocket City. How would the rocket folks take the news that an outsider, even if hailing from within, was now the boss?

Lightfoot dealt with that question head-on, holding an off-site with his direct reports, the executive leadership of Marshall (minus Todd), where he explained that this guy, May, *everything he touched went to space.* He was Marshall's best shot at getting Marshall flying rockets again. Lightfoot asked for one hundred percent of his people to be on board with this. They were.

Todd was not a propulsion engineer, but he was an Auburn engineer. To quote the creed, he "believed in work, hard work." He knew how to take intractable problems and tract them. Puzzles interested him, and especially challenging ones. You're a NASA engineer, and you solve hard problems for a living. It's what made the job fun, exciting; the harder the problem, the greater the thrill. And suddenly Todd was a child again, on his bicycle without handlebars, hands in the air, centrifugal force and his guardian angel keeping everything lined up like a gyroscope. So Todd said yes. And right away, he got to work.

MAY AND THE CORE SLS leadership met first at a lake house owned by Jody Singer, his deputy, and came up with the broad strokes of what they wanted to do. The organization of the program, its management philosophy, the short-term and long-term goals, and one-page plans to get to the next step. The first major milestone on the deep horizon was a mission concept review in which an independent panel would evaluate the plans for SLS and offer an assessment of risk. But to get that far, there would need to be trade studies both internally and with contractors.

A "broad area announcement" went out to industry: anyone who wanted to play—Old Space or New Space (businesses such as Boeing or Blue Origin, respectively)—was invited to contribute. More than a dozen reports came back from industry, and—surprise!—given the parameters, every industrial partner recommended a rocket that looked exactly like the thing they were already selling.[405] Alliant Techsystems, a company that built solid-fueled rocket boosters (i.e., using a literal solid fuel, similar in principle to bottle rockets), proposed a rocket that used . . . solid-fueled rocket boosters! Aerojet Rocketdyne, which built rocket engines, proposed building a rocket with lots of engines! Boeing, which built enormous boosters and cores for its rockets, proposed using an enormous core. And so on. The ideas from industry sifted themselves into general categories, and the SLS management put together three Requirements Analysis Cycle teams—pronounced by their abbreviation, "rack teams"—each of which optimized the problem in some different way. Given the launch date (2016), the launch capability (seventy metric tons), the budget (flat), and the requirement to reuse as much of the space shuttle and Ares I "to the extent practicable" (a glorious term of art!), design the best rocket you can.[406] It must be affordable. It must be safe.

When you build a rocket from scratch, step one is to know your destination.[407] Low Earth orbit versus, say, the moon, requires different rocket sizes entirely. Pickup truck versus big rig. Mars? Even bigger. SLS didn't have a destination—the idea was flexibility. But the ambiguity didn't help. The space shuttle taught NASA that . . . you didn't want to build a shuttle. An in-line vehicle—that single spartan tower, with maybe some boosters on the side for extra get up and go—was just safer (von Braun figured this out in the sixties) and made more sense aerodynamically. So now you have the shape of your rocket. Then you're thinking about trajectory: How do you want to fly the vehicle? What kind of loft do you need, what kind of Gs do you need to maintain, or to stay under: i.e., do you need a gentle ride, astronauts sipping from teacups as they clear the tower, or do you just want to throw something to space with everything you've got?

The destination determines the amount of fuel you'll need to carry. You have *n* number of engines, and they're small relative to everything else. The main engine used by the space shuttle was about the size of a Volkswagen Beetle. The tank attached to it was the length of a football field. So you had to handle that fuel and design the tanks in such a way to feed your engines efficiently. And what *are* the best fuels, anyway? What gives you the best blast for your buck? What's plentiful? What's available? What has the impulse you need when you burn it?

And now you have a broad vision of your rocket. The engines are on the bottom, those big bells pointed at the ground, the core feeder on top. If that doesn't give you enough thrust, then you can up it with smaller boosters—"half stages"—on each side of your prospective launch vehicle. The space shuttle solid rocket boosters, the most famous of these, were the cat's pajamas because they gave you incredible lift off the pad. Overcoming that first moment of gravity was everything; your rocket at that moment would never have more fuel or be heavier. Solids were the supreme and undisputed empresses of lift, intense, unbridled—you *will* separate from the Earth when they ignite—but they're not controllable, you light that candle, and that's it: up, up, and away. So to build a smart rocket with navigation and guidance, you want liquid-fueled rockets that you can control; you want to be able to throttle them this way or that and keep things oriented just so. A liquid/solid combo was a pretty good deal if you could get it.

But where do you place your crew? You hang a vehicle from the side of your rocket and run the risk of, say, foam falling from said rocket and causing a space shuttle *Columbia*-type accident. So you put your people on top of the rocket. But how many people? The size and destination of the crew capsule—but more pressingly, the size and rock-solid reliability of the heat shield at the *bottom* of the capsule—also determined the rocket's configuration. Astronauts didn't go on one-way trips; you'd need that shield for reentry. When seated, the crew's feet were only inches above the heat shield. So during reentry, the shield on one side might reach four thousand degrees.[408] On the other, your astronauts' feet would be cool and relaxed.

The three rack teams had about six months—an unreasonably short amount of time when it came to engineering—to develop their rocket proposals. The teams were necessarily small, each member having an area or two of expertise, and the work was divvied up along those lines: You're great on engines. You're the person for schedule. You're the risk expert. Structures, vibroacoustics—they would come together and mutate the rocket design. There were thousands of variables and equations that, when explained, were the length of wedding toasts. It was hard. It was rocket science! Still, there was a carnival atmosphere at times. Beer and pizza were on the line for whichever team's design came out on top.

None of the Rack Study rockets, in the end, was the most perfect rocket ever; that title belonged to the Saturn V. But they found one that worked.

Rack One was a shuttle-derived rocket. It was always their race to lose, and they were quite confident they wouldn't lose it. They had access to the sixteen space shuttle engines for their design and were thus best able to take advantage of the existing supply chain and knowledge base. So—if you're building a rocket with shuttle-derived technology, and you've got a guy down the hall who's been working on, say, some valve on the shuttle for the last seventeen years, you had an easily checked box without the need for an arduous R&D process. Moreover, the design took advantage of the facilities at Marshall, Michoud Assembly Facility in New Orleans, John C. Stennis Space Center in Mississippi, and Cape Canaveral in Florida.

Rack Two was the emotional favorite. It was the more Ares V–like of the proposals, huge—thirty-three feet in diameter (almost six feet larger than Rack One) and towering two hundred seventy feet in the air.[409] But what won the hearts of all who heard about it was the rocket engine they planned to use: four Rocketdyne F-1 engines, the very same engine that powered the Saturn V. On PowerPoint presentations, when listing the points in favor of the rocket, it just said "von Braun," no explanation needed. And when you looked at its design—well, there was just no doubt about it, the thing was like the love child of Saturn V and Ares V. How the engineers longed to hear the F-1 roar once again! Roar it would: a single F-1 engine was as powerful as three space shuttle main engines, and the Saturn V used *five* F-1 engines per launch. The power output of the first stage of a single Saturn V launch could have powered all of the United Kingdom.[410]

Indeed, if the schedule and dollars had not been constrained—if all that mattered was elegance and engineering—Todd would have just rebuilt von Braun's Saturn V. It was a technically perfect rocket. In the Apollo years, it had all the money in the world. There was never a request for funds that was denied; NASA spent whatever it took to take that hill. But those days were over.

There was another problem: there was no one left alive who really knew how the F-1 worked. Engineers would have to find an engine, disassemble it, and figure out what Wernher knew that they did not.[411] Even for rocket scientists, the Apollo program was like magic. So Rack Two carried a learning curve as steep as El Capitan.

The third team worked off the shelf. It set no one's heart aflutter but was a pragmatist's dream. We have Atlas V rockets. We have Delta Heavy rockets. What can we do to package them? Just keep tying 'em together until you met the power mandate. From a scheduling standpoint, it was the best option: just go to a contractor, buy the rockets, and piece them together like Lego bricks—I mean, problem solved. But the more the Rack Three team worked the rockets to achieve the desired performance, the uglier things got. You needed a cluster of boosters around a core rocket, and you just didn't have enough room for all of them. How did you feed the different engines? Impinged flumes of fire from each of the boosters would melt the bottom of a rocket. And, oh, by the way, with all this *stuff* sticking out of your rocket, how are you going to assemble it with existing infrastructure at Cape Canaveral? In the end, you just couldn't get from existing assets the performance necessary to reach the desired targets. Getting astronauts halfway to the moon would not be ideal.

By far, the hardest problem was cost. The SLS budget was flat. Todd was told to expect about one-point-six billion a year for the first three years.[412] Marshall's original estimates, however, called for more each year: one-point-eight, two-point-six, and two-point-six.[413] But then it got worse: the actual budgets ended up much, much less: around one-point-three, one-point-four-ish, and one-point-four-ish.[414, 415, 416]

The eventually chosen design of SLS—Rack One—used the same engines pulled from the now-mothballed space shuttle fleet. There were sixteen of them, pristine, powerful (if rocket engines were cars, they would be Ferraris), just waiting in a warehouse in Hancock County, Mississippi. They were already rated for human spaceflight. They were reliable; they hadn't been the cause of the two shuttle disasters, and had gone up one hundred thirty-five times total.[417] They were also made for reusability, though SLS as eventually designed would make use of each engine only once. And since you didn't have to worry about reuse, you could crank them way, way up in terms of throughput, just give them the business.

Like the space shuttle, SLS would use two five-segment solid boosters, one on each side of the core stage. Those boosters had completed most of their development work as part of Ares and had thirty years of shuttle heritage, making them practically free. Though the core had the same diameter as the shuttle external tank, it was a new design overall: taller, for one thing, and where the shuttle external tank fueled three engines, this one, as designed eventually, would feed four. It would still be expensive to build, but you had enough engines for four launches.

To further reduce costs, management eliminated a lot of oversight procedures and documentation requirements. May simplified the government interface with contractors who were building things: stop with the thirty checkboxes and multiple levels of approval, everyone; when you find an issue, talk directly with the contractors responsible. Do what you say you're going to do and focus on results. This was a lithe, sinewy rocket program, and a cultural change for Marshall. To help, May ordered the creation of a series of project management courses for the thirteen hundred engineers working on SLS. The rocket would fly, but it would also leave an institution transformed.

TOM GAVIN CHAIRED the mission concept review board that examined the three Rack Studies. Tom and Todd had worked together previously. Not long after Tom had helped spearhead the successful Flight Project Practices and Design Principles initiatives at Jet Propulsion Laboratory, the chief engineer at NASA headquarters contacted him. The agency already possessed its own guide for management and design called *NASA Procedural Requirements 7120.5*, but it was getting long in the tooth. Tom, I'd like you to help review and revise the policies of the entire agency—including human spaceflight—to reflect how projects might be managed going forward. We'll give you a cross-agency team and eighteen months to do it.

Tom agreed, and asked Todd May to be his deputy.

From the start, Todd marveled at Tom's ability to lead. There was never any doubt in any meeting who was going to sit at the head of the table, who was in charge. Tom, meanwhile, knew that Todd, too, had a strong personality and was an effective leader, and drew him close. Better to have him a trusted deputy than the guy on the other end of the table fighting back. They kept each other honest and became close friends.

Ultimately, the *7120.5* review worked out smashingly. When the revision was released in 2007, project managers across the agency were suddenly talking incessantly about which Key Decision Point they were at—and understood what each other meant. But, more important, it was absolutely embraced by the White House Office of Management and Budget, and by Congress. All of this made Tom the ideal choice to lead the SLS mission concept review, and would help Todd demonstrate to human spaceflight the broad strokes of how mission concept reviews—a big part of the work he and Tom had done on *7120.5*—actually worked.

After the SLS review was completed, the decision made, Tom and Todd met at Tom's beach condo. The two of them sat in the courtyard, beers in hand, and Tom talked about the Europa study. Because of the SLS mission concept review, Tom was fully immersed in the capabilities of the rocket. He turned to Todd and asked: Have you ever thought about flying an outer planets mission on SLS?[418]

Todd hadn't, why?

You ought to look at it, said Tom. I think you can get us to Europa a lot faster. On a conventional rocket, it'll take us six or seven years because we'll have to do orbital assists.

May later had his team run the numbers. He called Tom, and asked: How much do you spend on the operations phase while you're cruising out to Jupiter?

Forty million to sixty million dollars a year.

Well, instead of spending that for seven years, we can get you there in two and a half.

So the two men formed a quiet alliance, and the Europa team at Jet Propulsion Laboratory and a contingent at Marshall opened an unprecedented dialogue. No science mission had ever directly influenced the design of a new launch vehicle—that had previously been the exclusive domain of human spaceflight.[419] They couldn't lock in anything, of course. Europa wasn't an actual mission, and they were years from committing to a launch vehicle. But Gavin and May, Europa and SLS, could certainly talk design, refine and harden their engineering, keep each other in mind, continue having the conversation. And from the start, it was clear the two needed each other. Moreover, if the Europa mission didn't have to foot the bill on the rocket—if launch were funded separately—and if they could trim three hundred million dollars from cruise phase operational expenses, Europa would have a stronger case at NASA headquarters and in Congress.

The Obama administration, meanwhile, had yet to commit to a target in deep space. The agency talked humans on Mars, but that was an aspirational goal—not misdirection exactly, but not a real plan. There were no crewed Mars missions on the books, or formal funding to make that happen, and even if there were, Mars would be twenty years away at best. There was a lot of infrastructure yet to build to get there, and no money to do it. Once built, SLS would be a rocket with nowhere to fly.

Europa was a somewhere.

CHAPTER 13

Clipper

DESPITE THE DELETERIOUS DECADAL DECISION, IT never even occurred to Bob Pappalardo that a Europa mission wouldn't fly until he was having lunch with Ron Greeley a few months before completing the split-mission report. The two scientists broke bread in 167, the main cafeteria at JPL, called the Red Planet Cafe.

> RON: So what are you going to do now?[420]
> BOB: What do you mean?
> RON: Well, what are your plans now that Europa isn't happening? Are you going to go back to academia?

Bob didn't know what to say. He—I mean—what? His thoughts collided and folded inward like a derailed train. We're going to do Europa—we've got these two great studies we are doing—what are you talking about?—come on, Ron, this isn't over—you can't stop now!

Bob said aloud, however: I don't know—maybe I'll go back to academia.

And he felt suddenly exhausted. Ron Greeley had given up. The maestro who had taken this thing so far, who had first revealed enigmatic Europa on Galileo. Ron, who co-led the science definition team for the battlestar JIMO! Ron, who was *on the inside*. Who was chair of the Planetary Science Subcommittee of the NASA Advisory Council. Who knew everything, could do anything, going back to the Apollo program. If Ron had resigned to the inevitable . . . then that meant it was inevitable. That maybe it was over—and if it was, then what was I *doing* here? It was their last long visit together, their last long good talk, mentor and protégé, master and apprentice.

Not long after, on October 27, 2011, Bob received what was obviously a hastily dashed message from Ron's assistant, Stephanie Holaday. "Bob," it said. "When you are available to, will you plesase call me . . . I need to talk to you about Ron."[421] Bob went to his office. It felt wrong, the cadence, the typo. He closed the door. He called. Stephanie answered. Earlier in the morning, he was told, there was a teleconference scheduled for the Planetary Science Subcommittee. Ron didn't dial in. Ron, Bob knew, *always* dialed in, was never late—ever. He was too organized, too gentlemanly, to just skip a call. And he *ran* the subcommittee, so it was especially unusual.

Ron's wife, Cindy, was out of town. He and Cindy always traveled together, had seen the world, mountains, valleys, and cities with buildings that scraped the sky, met armed riflemen in the Arctic whose job it was to protect them from polar bears, and met by happenstance the pope in a private audience at the Vatican Observatory. But on this trip, Cindy had traveled alone. Her mother, who still lived in Gulfport, had fallen and broken her hip. Ron and Cindy had just returned after three weeks in France, and Ron wanted to join her but he first had to play catch-up at the office and had a mandatory meeting in Flagstaff.

Cindy tried calling their house. There was no answer. She telephoned her neighbors across the street. They had a key to the house, just in case, and Cindy asked if Ron's truck was in the driveway.[422] Yes it was. OK, said Cindy, something is not right, and she asked them to take the key and go inside, and they did. They found him on the floor in the bedroom. Ron Greeley was dead.

Everything moved so fast. It was a heart attack. Fragments and flashes connected that call and the funeral. The response by the science community—it moved Cindy. It was touching. All the senior officials at Jet Propulsion Laboratory attended the memorial service. She was so surprised by that. She found out that there was a memorial service held for Ron in China! In Wuhan, at the China University of Geoscience, Ron had worked with a professor there and had made an impression.[423] She received a letter of condolences from Sir David Wallace, the master of Churchill College of the University of Cambridge, where her husband had once been a visiting professor.[424] She learned later that a plaque was hung in Ron's honor at the Smithsonian National Air and Space Museum. Ron had always taken his job as a mentor so seriously. He wanted his students to know how to do things correctly. And his postdoctoral and graduate students went on to become the leaders in the field. Ron Greeley had changed the world. And not just this one. He had changed the solar system. The Regional Planetary Image Facility at Arizona State University—one of seventeen such NASA data centers around the world—was renamed the Ronald Greeley Center for Planetary Studies. Ron had been a key scientist on most missions the agency flew, and on every Mars rover to have pressed tracks into rusty Martian soil. Just after he passed away, the rover Opportunity settled in for the winter on the rim of Endeavour Crater—a sweeping, stunning Martian vista—and NASA and the Mars team named it Greeley Haven.[425]

Shortly before he died, Ron had completed the manuscript for his seventeenth book, *Introduction to Planetary Geomorphology*.[426] It was to be the definitive text on the subfield, and the publishers called Bob and asked if he would do the revisions, and—well, of course Bob would.

THE EUROPA HABITABILITY Mission report arrived at NASA headquarters on May 1, 2012, a single seven-hundred-page volume describing the potential missions to Europa in Caro-esque detail.[427] The Europa science definition team had spent one year developing it: a split-mission concept composed of a Europa orbiter and a Jupiter orbiter, two relatively simple spacecraft, each built to suit its strengths and not a single atom more. JPL took the lead studying the Europa orbiter, while APL handled the Jupiter "multiple flyby" option. For the concept to work, you would fly one and eventually the other, or both simultaneously, but the point was, you needed both for a full and accurate accounting of the ocean moon. The Europa orbiter would live hard and die young inside of the Jovian radiation belt, which would poison it fatally in thirty days.[428] The Jupiter orbiter would encounter Europa repeatedly over two years, each time taking some specific, sui generis slice of the icy moon, meticulously mapping it in patchwork. The orbiter would be better for the ocean science; the multiple flyby mission would be better for the ice shell science. Geology was a split decision.[429]

Bob Pappalardo celebrated the delivery of the science definition team's report by getting the hell out of Pasadena.[430]

When he was first hired by the lab, Bob was told that research sabbaticals were a Thing That Existed, but that no one ever actually took them because no one had the time. Well, Bob did, now, and he filled his forms and bid them good day. He needed some space between himself and these papercraft robots. He needed to figure out if it was worth it anymore—the work at the lab, the whole messy business, this Sisyphean effort to get something going. Every day, he sat in his office, at his desk, in his chair, surrounded by *Star Trek* figurines and awards and plaques and textbooks, a globe of Europa and a view from his window of Saint Gabe's geology and flora and sometimes fauna, a few lab facilities here and there dotting the distance. He worked at the world's preeminent institution for building spaceships—and he's in his chair, and every few moments, a gentle chime from Microsoft Outlook, another note for another meeting, another request for information, another mission study, another consult, another talk, another news article speeding across the wire, the decaying budgets of planetary science, Mars rovers and the James Webb Space Telescope gobbling everything green, AND WHERE WAS BOB'S SPACESHIP? He kept taking these leaps, and it was—what if he landed and just . . . kept landing! Kept writing report after report, and PowerPoint presentation ad infinitum. He had given a talk years earlier on Europa and the mission that could be, and after finishing, a radio astronomer in attendance announced: You know, someone could really make a career doing this![431] Bob could see it in his face. This wasn't some joke that just didn't land. It was a rebuke! And Bob brushed it off—what a jerk—but even all these years later, the comment really gnawed at him. He heard, long after the radio astronomer called him out in person, that someone had called him a salesman behind his back. A salesman! Robert Pappalardo! If only he were! There was this *New Yorker* cartoon he had once clipped and taped to his office door. There were two hamsters, one sprinting on his wheel, the other sitting stunned on the edge of hers. There's this look of total clarity on her face. And she says: "I had an epiphany."[432]

Walking out the door, reflecting on the latest report, he was spacecraft agnostic, if not entirely indifferent, about which probe should be approved, if any. These studies. Seven hundred pages for this one. Four hundred ninety-eight for the Jupiter Europa Orbiter.[433] Two hundred eighty-two pages for Europa Explorer.[434] In aggregate, one thousand five hundred pages since he had come to the lab, not counting internal studies he had led or helped write, and not counting JIMO. Fifteen hundred pages! *The Lord of the Rings* was shorter![435] Frodo and Sam walking from Bag End to Mount Doom, ducking Nazgûl and dodging Balrogs, was now officially easier by page count than a journey to Europa. And Europa's fellowship hadn't yet moved a single inch.

This one, this report, landed nicely. It was a good report. As had been the last. And the one before that. And Bob wondered: Would this have any greater effect than its forerunners? He had come to the lab with a three-year personal deadline: get something funded, fabricated, fit for flight, or go back to teaching. Well, he had lapped that limit and then some, and he was tired. He had no spacecraft to show for his endeavors. Not even formal "project" status. They were the imposter tenants of 321.

Pappalardo's research sabbatical turned out to be a bicoastal hat dance, beginning at the University of California, Santa Cruz, and on to Cornell University and Boston University, and closing out 2012 at the University of California, Los Angeles.[436] It was enough time to get some plates spinning. Enceladus was, as a matter of intrigue, white-hot by then. Cassini was flying ever closer and sampling as best it could the moon's chemistry, and it had found organics—carbon compounds—which didn't necessarily imply Life as We Know It but were a necessary ingredient thereof. Ganymede was also getting traction, known now not only to have an ocean and intrinsic magnetic field but also an approved mission: the former Jupiter Ganymede Orbiter had been given a green light by the European Space Agency. So after NASA's edict in 2010 to keep Jupiter Europa Orbiter and ESA's Jupiter Ganymede Orbiter separate so as not to saddle the agency with *Europe's* pusillanimity, just the opposite would come to pass! The Europeans renamed their mission JUICE—the tortured acronym for JUpiter ICy Moons

Explorer and Bob corresponded with Elizabeth "Zibi" Turtle at APL, who invited him to join her team proposing an experiment to fly in its payload. He did a lot of Europa work on the sabbatical, too. At UCLA, he worked closely with Krishan Khurana, he of the very first paper positing that Europa harbored a subsurface ocean, and who had coedited Bob's book on Europa. Krishan had noticed in the old Galileo data an unusual signature that might suggest transient plumes on Europa, something perhaps like those on Enceladus. Such a discovery could be as transformative as the discovery of the ocean itself.

While bouncing from university to university, Bob gave guest lectures on Europa, and, if he was honest . . . he tested the waters. What if he went back to teaching? He had given up a tenure-track professorship in Boulder to go to JPL, a risk in terms of job security and life stability, but it would have been worth it had he managed to get something flying. That said, if he went back to teaching—say, he found a professorship at UCLA—he might still be at least a peripheral part of a Europa mission. But could he lead it? No, not likely. UCLA would have loved it, provided that Bob maintained his publishing cadence. But would the lab allow it? No way.

ON THE OTHER side of the continent, Curt Niebur at headquarters further considered the orbiter and the multiple flyby missions. It would be one or the other, he was certain, orbiter or multiple flyby. (The team had written a report on a prospective lander, too, but no one took the lander seriously.)

To keep things moving forward in Bob's absence, NASA wanted the science definition team to remain active and led by someone capable of snapping bones if necessary to safeguard the integrity of the mission. Curt asked Louise Prockter to take charge, and she accepted.[437] Her leadership came with the benefit of cementing the mission's marriage to the Applied Physics Laboratory. She would make it work, keep her lab on track, and impede any JPL attempts to abscond solo with the entire mission.

Don Blankenship suggested a nautical theme to describe the multiple flyby mission. He worked best by writing out his thoughts, journaling on legal pads that he stored away in great stacks in his Austin office. "The name," he wrote on March 20, 2012, "must reflect the comprehensive traversal of the ocean to establish its suitability for sustaining life and natural selection. A name with a class of sailing ship would reflect that."[438] He wrote out seven options: caravel, cutter, clipper, schooner, barque, corvette, ketch. He circled the word *clipper*. The name was especially clever in that the Decadal and NASA headquarters both claimed that an outer planets flagship mission was unaffordable right now. In the nautical nineteenth century, a clipper ship was just a *little bit smaller* than a flagship.

And the Clipper concept was turning heads. It was something different, cheaper, doable. In the consensus findings that emerged from its January 2013 meeting in Atlanta and sent out to the wider planetary science community, the Outer Planets Assessment Group reported "its unequivocal and strongest support for the Europa Clipper mission. The ultimate result of more than a decade of ever more detailed study and down-selects, this mission offers paradigm-shifting, flagship-level science at Jupiter's ocean moon."[439] At headquarters as well, Europa Clipper was the consensus favorite over the orbiter, but there was one pressing concern: it broke the established order of scientific exploration.

For any planetary body, the order of missions went: flyby, orbiter, lander, rover, sample return, astronaut. That's how Apollo did it. That's how Mars, in theory, was doing it. Europa had its flyby with Voyager 2. It had several flybys with the spacecraft Galileo. And now the Europa community wanted multiple flybys again? Why would

NASA spend money on another Galileo? It did not compute. So Curt and his boss, Jim Green, had to clock a lot of hours explaining why the Galileo–Europa Clipper comparison was specious. Galileo had eleven flybys of Europa during which its data were constrained. Its instruments were state-of-the-art . . . in the early eighties. The two spacecraft were completely different, and a multiple flyby by-passed the radiation bath. You'd get a low-cost mission that lasted years rather than days. Europa wasn't ready for a lander, the team argued; the moon's surface was still too great a mystery. Under muted protest, headquarters allowed Europa Clipper to stay.

Having now established what the science definition team could do with skeletal funding on two tightly focused spacecraft, Curt challenged them to try to find new sweet spots. The Europa Clipper was ideal for ice science but weak on ocean science. Given a little more money, could you get the ocean science in there? Likewise, the Europa orbiter wowed with ocean science but disappointed on ice shell stuff. And so the science definition team under Louise started study round two: the "enhanced" Europa missions.

The first meeting under her aegis, she felt, did not go well.[440] The scientists did what she had seen them do a hundred times and what she herself had probably done: they ran chaotically off in their own directions, an orchestral scherzo playing out in time. Louise stood there in front of the room, and no matter what she tried, she absolutely *could not* bring this group to consensus. How nice Bob's sabbatical must have been! And whatever words she said aloud, in her mind, she could not stop saying, like a mantra: *Icannotdothis-Ineedtoleavethisroom Ineedtowalkawayfromthisroomrightnowbe-causeIamnotgoingtomakethiswork*, and it was the exact moment when she appreciated, and not for the last time, just how good Ron Greeley really was. But just as it fell to Bob to finish Ron's book on geomorphology, it fell to Louise to finish this one, and she would find a way.

CHAPTER 14

Princess-Who-Can-Defend-Herself

OF THE ONE HUNDRED TWELVE THOUSAND WORDS comprising Public Law 113-6, enacted by the Senate and the House of Representatives of the United States of America in Congress assembled on the twenty-sixth of March in the Year of Our Lord 2013—the short title: Consolidated and Further Continuing Appropriations Act, 2013, the purpose: making consolidated appropriations and further continuing appropriations for the fiscal year ending September 30, 2013, and for other purposes—fewer than forty words, a single clause of a single sentence, mattered to an orb of ice, water, rock, and iron six hundred million miles away.[441]

Customarily, a budget, or, short of that, a "continuing resolution" to keep the government funded and open for business for some set amount of time—a week or a month or a fiscal year—doled out dollars in big bites and with onerous sentence structures. NASA, to name one agency, needs five billion dollars to fulfill its

science mission, from launching telescopes to buying swag to hand
out at school science fairs, so:

> For necessary expenses, not otherwise provided for, in the
> conduct and support of science research and development
> activities, including research, development, operations,
> support, and services; maintenance and repair, facility
> planning and design; space flight, spacecraft control, and
> communications activities; program management; person-
> nel and related costs, including uniforms or allowances
> therefor, as authorized by sections 5901 and 5902 of title
> 5, United States Code; travel expenses; purchase and hire
> of passenger motor vehicles; and purchase, lease, charter,
> maintenance, and operation of mission and administrative
> aircraft, $5,144,000,000, to remain available until Sep-
> tember 30, 2014.

But the will of Congress is more nuanced than just telling a
government clerk she has five bil to blow as she likes. Accompanying every appropriations bill is a guidance report that lays out
precisely what will be done with those dollars.[442] Out of that five
billion, We the People expect you to spend one hundred forty-six
million on the Mars Atmosphere and Volatile Evolution orbiter.[443]
And sixty-five million on Mars Science Laboratory. &c. However,
unlike the appropriations bill ("I give thee $5,144,000,000"), those
precise dollar amounts are, technically speaking, only suggestions.
They carry no weight of law. Oh, you will hear about it if you defy
the stated will of Congress—but you won't hear about it from the
inside of a jail cell.

Which is why John Culberson of Texas, in his wheelhouse as
a senior member of the Commerce, Justice, Science, and Related

Agencies Subcommittee of the U.S. House Appropriations Committee, inserted not into the report, but into *the budget itself*, a single, absurdly specific sentence solving some long-unfinished business of his: "Provided, That $75,000,000 shall be for pre-formulation and/or formulation activities for a mission that meets the science goals outlined for the Jupiter Europa mission in the most recent planetary science decadal survey."[444]

Culberson, meanwhile, made his opinion known to top-level headquarters officials that he wanted a spacecraft to touch that ice.[445] He never kept his desire for a lander a secret—it was the only way to answer definitively the life question—and a rogue group at the lab kept him appraised on a surreptitious lander study they were conducting. (Bob, Louise, and Dave knew nothing of it and were miffed mightily to learn it had taken place behind their backs.) NASA thus directed the Europa Habitability Mission science team to investigate the lander as a third spacecraft option.

Culberson was by then in his ascendancy in congressional appropriations, with increasing sway in Republican politics by way of the emergent Tea Party movement, of which he was a founding member in the House, spurred by frustrations with Democratic success passing the Affordable Care Act. He was philosophically disposed toward low taxes and slashed spending; "Obamacare," as it was sometimes called, offended him.[446]

Still, some things needed a little coin. When John the junior subcommittee member funded JIMO all those years ago, the NASA administrator took a look at the budget (the law of the land) and the report (pretty please do JIMO), and tossed the report in the trash.

Not this time, the administrator wouldn't. Just like that, NASA *by law* had to get going on Europa and spend serious dollars on its development. It was, in fact, the only mission illegal for NASA not to fly.

John Culberson had expanded the Katy Freeway into the widest highway in the world, and one way or another, he would build a much longer highway. He would get his spaceship to the Jovian system.

THAT SEVENTY-FIVE MILLION dollars did not send Curt Niebur skipping to the local liquor store to buy a bottle of sparkles, singing and swinging from lampposts along his merry way. Just the opposite.

Look, the money was wonderful. Jim Green, the head of the planetary science division, had been scrounging for years to keep Europa alive, and now there was a huge pile of coin to press forward on mission development. Jim, Curt, and the Europa team scattered across America could work wonders with it, but the administration's hand had been pushed, and it had no problem slapping back. The White House was adamant: it was not pursuing a planetary science program right now. And unless NASA (via the White House) *requested* money for Europa, Congress could keep cash coming all day long, but the agency would not enter any long-term agreements to spend the money beyond the year appropriated.

The split between the White House and Congress on NASA funding could be measured in Grand Canyons. The Obama administration came into office with a certain set of priorities and stuck to them. Of NASA's four major scientific disciplines—heliophysics (i.e., the sun), astrophysics (i.e., the stars), earth science (i.e., rock no. 3), and planetary science (i.e., everything else in space, dust mote to gas giant)—Jim Green was told early on and point-blank which stood where.[447] Earth science came first; under this administration, climate change would *not* be ignored. Second came astrophysics, but more specifically, the James Webb Space Telescope—successor space observatory to Hubble and billions of dollars in the red. It was being built in Maryland and thus represented by Barbara Mikulski, who was still the ranking member and future chair of the Senate Appropriations Committee. If she wanted James Webb, she would get it, wrapped with a big red bow. Next came heliophysics, because it was small, with few pennies to shake from its piggy bank, and dead last was planetary science. There were only so many dollars, and the money for earth science had to come from somewhere.[448] Why? Because Congress had foisted the SLS rocket on the administration, and there was no more money. From Lori Garver's desk on the ninth floor at NASA headquarters, it was a terrible decision to have to make, but a decision had to be made. Planetary science would just have to take one for the team. The Mars Science Laboratory—the rover Curiosity—had just landed successfully on Mars. This was as good a place to pause as any.

In the president's budget request for fiscal year 2013, Jim lost three hundred million dollars.[449] In one year. One-fifth of an entire scientific discipline: gone! And not *just* in 2013. As the Office of Management and Budget had told the Space Studies Board, the money would not return in 2014. Or 2015. Or 2016. Or 2017. The solar system was now being balanced on mountains of empty piggy banks. You lose twenty percent of your budget, and you tighten your belt . . . around your neck.

On paper, it was the sort of existential science cut not seen since Reagan. And through 2017? Forget launching flagships to Mars or Europa. You're trying now to keep from switching off the spacecraft you've launched. You're trying now to keep from switching off the office lights.

And the Office of Management and Budget was not playing a game. These cuts were real. From Curt's office, although that seventy-five million dollars would enable all sorts of great work, it was a one-time thing. A Europa flagship mission was not a seventy-five-million-dollar mission. It was a two-point-five-*billion*-dollar mission . . . if it stayed perfectly on budget (which was unlikely—just look at Cassini or Curiosity). Europa was still in preformulation. Culberson was able to finagle seventy-five million dollars this year (and yes, Congress did restore much of the twenty-percent budget cut to planetary), but what about next year? The year after? The year after that? Once development really ramped up, there would be consecutive years with nine-figure price tags.

So all things considered, having the money was in some ways worse than Europa's panhandling days, because at least when it lived on a shoestring and a prayer, you could rage like Lear at the heavens, curse the feckless fiscal priorities of the American government. But to have the money and *know* it would amount to nothing?

Shortly after the bill was signed by the president and Europa had sixty-nine million dollars in hand (Culberson's target was reduced by a budget sequestration and federal rescissions), on April 22, 2013, Jim Green directed JPL to focus solely on the multiple flyby concept and to discontinue work on the orbiter and lander.[450, 451] Europa Clipper was not the agreed-upon mission concept across the agency, but Jim was planting the flag, making a call that his superiors seemed incapable of making. "Given the funding for Europa mission pre-formulation efforts recently authorized in our fiscal year 2013 appropriation," he wrote to the Europa team, "we can now move forward beyond the initial study phase. Please have your team focus solely on the 'Europa Clipper' concept, i.e. a multiple flyby mission, and do not continue development of the Europa Lander or the Europa Orbiter mission concepts at this time."

The studies backed up his conclusions. The enhanced Europa orbiter had limitations that the study team could not overcome.[452] To get the data necessary to achieve its new goals, the mission would need one hundred eight days in orbit as opposed to thirty.

But a longer orbit meant more radiation protection, and such shielding was precisely the thing that killed the Jupiter Europa Orbiter. To address the ice shell science and reconnoiter a landing site, meanwhile, the orbiter would need an ice-penetrating radar and a high-resolution reconnaissance camera—which could be done, but both were data-intensive instruments. The spacecraft would have to make the most important observations as quickly as possible and blast them back to Earth immediately, because regardless of shielding, the spacecraft would eventually (or might suddenly) die of radiation poisoning and crash into Europa. But the real problem with the enhanced orbiter was its inability to work out Europa's composition: it could not carry two mass spectrometers, to say nothing of a thermal imager, without compromising its higher-priority payload. Which meant if the orbiter flew, the nature of the surface would remain a mystery—its salts, organics, and chemical makeup would remain blank spots on the map. Moreover, the enhanced orbiter was not "cost neutral," as requested; it would break the cap.[453]

An enhanced Europa Clipper, on the other hand, addressed every stated objective posed by the science definition team going back to the Jupiter Europa Orbiter (and by extension, the studies before that one as well). The spacecraft could accommodate a magnetometer, a thermal imager, a gravity science antenna, a high-definition reconnaissance camera (for a future lander), and a plasma instrument—everything, in other words, you would need to learn about Europa's ice shell, ocean, composition, geology, and eventual landing site. Clever modifications to the model payload kept Clipper cost neutral. The mission would last for years, cost half the money, and do most of the science that the late, great, (allegedly) four-point-seven-billion-dollar Jupiter Europa Orbiter would have done, and because Europa Clipper would orbit Jupiter, NASA could, when the time came, safely dispose of the spacecraft.

Rather than crash into a potentially habitable ocean world, it would dive into the Jovian abyss (or maybe even Io!—the possibilities were endless), where it would be vaporized. There was practically no risk of contaminating the whales of Europa.

The Europa Clipper decision, however, was limited to the planetary science division. The way these things had worked, headquarters might well decide later to fly an orbiter instead. Curt therefore warned the science definition team to be ready for anything. Should Europa Clipper be killed and an orbiter chosen, everyone should be ready to sing with a happy heart the praises of an orbiter and to make the best doggone orbiter mission NASA had ever seen. The idea was to make it hard for NASA to say no to *any* Europa mission—redundancy of the most exhausting and dissonant sort.

With a spacecraft concept now sort of chosen, the science definition team reformulated itself as a Europa Science Advisory Group, paring down its membership to the most essential players.[454] No such group was strictly necessary, but the consensus from everyone who wasn't a Jet Propulsion Laboratory engineer was that Jet Propulsion Laboratory engineers made design decisions that sapped spacecraft science. You needed the tension of a science team to keep the engineers honest. Tom Gavin might have been the best in the world at building a spacecraft, but he was not a Europa scholar, and if he had to make a call that would harm the science in the interest of attaining some elegant technical solution, he would go with the technical solution every time and sleep very well that night. So Louise remained chair of the advisory group, and a reinvigorated Bob was back as study scientist.

Culberson had acquired an almost mythical stature by now among the Europans, who referred to him alternately and endearingly as "Our Benefactor" or "the Buffalo." Headquarters set aside fifteen million dollars of the congressman's appropriation to

further develop and retire the risks associated with the science instruments composing the spacecraft's payload.[455] Curt called the
effort ICEE—Instrument Concepts for Europa Exploration—
pronounced like the convenience store staple frozen fruit drink (and
don't think the PowerPoint slides forgot about that: a frozen Europa in a red-and-blue cup).[456] And he and the Europa leadership
team were insistent that if at all possible, pull from flight-proven
hardware. If somebody else built some specific radiation-hardened
chip, *use that chip*. Don't reinvent wheels and introduce uncertainty.
If at all possible, avoid thaumaturgic solutions to technical problems. The Europa Clipper mantra: No miracles.

THOUGH CURT WAS at the epicenter of the budget clash between
Congress and the White House, it was fine. He would deal with it
one roll of the rock at a time. Two years earlier, he had learned perspective in the worst way possible. And all of this? He kept a clean
desk. He would handle it.

On January 3, 2011, a combination PET-CT scan had revealed
seven small spots on his wife's lungs.[457] Susan had spent the second half of 2010 undergoing radiation and chemotherapy, and the
two had hoped these would be the images giving her a desperately
deserved all clear. Four days later, however, doctors delivered a diagnosis of cancer—her fourth recurrence in four years. She started
a clinical trial of a possible treatment, but Susan was under no il-
lusion about what the recurrence meant. Survival involved sudden
trips to emergency rooms, chronic headaches, sensitivity to bright-
ness, dizziness, and nausea.[458] Respite meant lying on a bed in a
darkened room. Those days were hell.

But she was also Susan Niebur, the woman who had kicked
open the door at NASA headquarters for scientists under the age of
eighty; Curt would still be working on hex bolts for aircraft landing

gear at Raytheon if not for her! She had helped spark what would be an irrepressible grassroots cultural change at the ossified agency nerve center. She was a mom at home, and a mother to so many careers, and as ever, she had work left to do.

Soon after diagnosis no. 4, mornings involved visits with lawyers to get her affairs in order, and afternoons were spent watching PBS Kids with her little ones. But she also found time to organize advocacy initiatives for cancer patients; blog prodigiously on her personal online journal, *Toddler Planet*—hundreds of thousands of words for hundreds of thousands of readers—on parenthood, cancer treatments, reflections on life, love, parenthood, and mortality. Through Women in Planetary Science, which had taken on a life of its own and become a unifying force in the field, online, in the press, in the flesh, she organized talks and meetings, facilitated networking events, and worked to change institutional malpractice that disproportionately affected women. She attended conferences, published papers, spoke on panels, and presented posters. Some days she was bedridden. Some days she celebrated birthdays at bowling alleys with her boys.

One day, her eldest son noticed the Lego minifig in her office, Princess-Who-Can-Defend-Herself, with the eyeglasses and the sword, and exclaimed: "She's you! She's inside you, fighting the cancer!"[459] And so she was.

But time, Susan knew, was short. At the time of her rediagnosis, her youngest had just turned four, and not long after receiving the news, she wrote in her journal: "And every day of my life, I live now for you and your brother. . . . The pain and fear and uncertainty that you're reading about in these archives—please know always that for you—for you—it was worth it."[460]

Her final blog post described an exchange with her husband, her bed "strewn with children's toys, books, and an oxygen tank." We got this far, said Curt, "because of your amazing strength, commitment, and love for your family that you have shown since you were diagnosed." We got this far, she responded, "because of you, always at my side, supporting me, joking with me, taking me to yet another appointment and holding my hand. Kidding me about the speed I drink the contrast shakes, and raising eyebrows with me as the tech's hands jiggle as he tries to place the line."[461]

On Wednesday, October 17, 2012, the Division for Planetary Sciences of the American Astronomical Society awarded, posthumously, the prestigious Harold Masursky Award to Susan Niebur "for outstanding service to planetary science and exploration."[462, 463] She had died on February 6, at just thirty-eight years of age. You didn't even have to know her to feel devastated. A community this small, you felt everything. You lose someone like Susan, and the axis of the Earth felt somehow to wobble slightly. "All that survives after our death are publications and people," Susan wrote on her blog. It was, she said, her mantra. And when she left, science went on, publications and people supported by the shoulders of one giant more.

CHAPTER 15

Ocean Rising

CURT'S INBOX CHIMED AT 6:14 P.M. ON DECEMBER 5, 2013, though he didn't see until later what had arrived. The email was from Kurt Retherford, a senior research scientist at the Southwest Research Institute in San Antonio, the subject line: "Discovery of Europa's Water Vapor Plumes." Curt fumbled with his iPhone, opened the message.

> I'm pleased to inform you about an exciting discovery that will soon be published in Science magazine. Our Roth et al. manuscript is titled "Transient Water Vapor at Europa's South Pole." I'm sure you'll agree that this important finding will have a tremendous impact on the future exploration of Europa as a potentially habitable world.[464]

Attached was a PowerPoint file giving the basics. Curt tapped out his response, his thumbs a percussive roll against the snare drum of his smartphone screen:

> This is pretty damn exciting. Can you get me a
> copy of the paper by any chance? It won't go
> any farther than me. And does anyone else at
> HQ know about this (someone in astrophysics
> division, for example)?[465]

Came the reply:

> No need to curb your enthusiasm! Go ahead
> and run down the halls of HQ.[466]

Curt began reading the ten-page paper: "In November and December 2012 the Hubble Space Telescope imaged Europa's ultraviolet emissions in the search for vapor plume activity . . ." it said; ". . . statistically significant coincident surpluses of hydrogen," it said, ". . . and oxygen," it said; ". . . emissions above the southern hemisphere," it said; ". . . these emissions are persistently found in the same area over ~7 hours, suggesting atmospheric inhomogeneity," it said. "They are consistent with two 200 km high plumes of water vapor," it said.[467]

Curt ran down the halls of headquarters. Jim had to know about this. Then the two of them ran down the hall together. John Grunsfeld, the associate administrator of the Science Mission Directorate, had to know about this. Then Curt corralled the public affairs officer for NASA headquarters, and within one hour, it was set: the *world* would know about this.

Europa had plumes.

Just like Enceladus. Europa was blowing hundreds-mile-high columns or fans of water vapor directly into space. The phenomenon had long been hypothesized (or at least, really hoped for), but no solid evidence had ever presented itself. The twin spacecraft Voyager on flyby, Galileo at Jupiter for nearly eight years, New

Horizons and Cassini during gravity assists, to say nothing of the land and space telescopes looking, looking, looking—scientists going back to the original Galileo had been staring at Europa with one tool or another, from a homemade spyglass of hand-ground lenses pointed from a little patch of Padua, to an orbital telescope with a one-ton mirror—and nobody had ever seen the suspected jets of ocean water.[468] Until now.

The lead authors of the paper were Lorenz Roth of the Southwest Research Institute in San Antonio, Joachim Saur of the University of Cologne in Germany, and Retherford, also of Southwest Research. They had taken the observations using Hubble's ultraviolet camera.

Lorenz led the data analysis that followed.[469] That spring, he showed Kurt what he had found, and what he found in November was . . . Europa. But what he found in December was Europa with a big bright blob above its lower limb.

Well, said Kurt . . . what do we do about that?[470]

The group had studied the disc of Europa previously, finding nothing. There was no "scan for water" button on Hubble, and the data were at the hairy edge of what the telescope could do. Lacking a cosmic divining rod, they were looking for "atmospheric inhomogeneity." What they found this time via Hubble were atoms glowing with light—an auroral process, electrons in Jupiter's magnetosphere smashing into gases coming off Europa.[471] This wasn't in itself a left field finding. The researchers had seen something similar at Io, and the phenomenon was associated with oxygen molecules. What caused Kurt, Lorenz, and Joachim to gasp and sit up in their chairs was the presence and quantity of hydrogen atoms. The ratio of oxygen and hydrogen glowing at their respective, spectrally characteristic wavelengths was a telltale sign of the two elements being split apart from common molecules by projectile electrons. And the molecules were emerging from the same place:

near Europa's south pole. The men were seeing plumes of water vapor being blasted from inside Europa.

Roth, Saur, and Retherford didn't report the results at science conferences, didn't even hint to colleagues over coffee. Just the opposite: they attacked the data, threw stones at it themselves, because if they got it wrong, they would look like lunatics. So they did everything possible to drill holes in their discovery. They drafted their science paper and even added a long, supplemental appendix outlining everything they had done to really undermine their own ideas.

Confident at last in their observations and interpretations thereof—but including every caveat and *more observations needed* disclaimer they could pull from the annals of academia going back to Socrates—they submitted to the journal *Science*. It wasn't lost on them that the discovery would have implications for how one went about exploring Europa. If you could fly through its plumes the way Cassini could fly through those of Enceladus, you would eliminate the greatest barrier to Europa exploration: getting through that ice shell. With plumes, you wouldn't have to find a way to the water. Just as with Enceladus, the water would find its way to you.

Right after they were accepted by *Science*, Retherford contacted Curt, knowing he was the lead person doing programmatic planning for outer planets exploration. They weren't buddies, but, I mean, they had been in the same room before at public meetings and conferences, so he expected Curt would be interested. He didn't know, however, that Curt would be *this* interested.

Here is how NASA alerts the press and informs the public. The agency sometimes does media briefings at science conferences— the Lunar and Planetary Science Conference, the fall meeting of the American Geophysical Union, the winter meeting of the American Astronomical Society—and those talks tend toward the interesting and in extreme cases, exciting, but don't necessarily lead to large words on A1.

Sometimes, however, NASA wants editors at the *New York Times* and *Washington Post* to take extra notice. To do this, they hold a press conference at NASA headquarters. A reporter is invited to 300 E Street SW, Washington, DC, and she knows something big is about to happen. The agency doesn't even consider this sort of summoning unless it sees a reasonable shot at making the front pages of every major paper in America.

Of course, NASA is a government agency, and so there are procedures to follow and a bureaucracy to appease. Division directors, division public affairs officers, directorate public affairs officers, a science mission directorate editorial board (called, informally, a murder board), the head of public affairs for the entire agency . . . all hoops. If the process glided along unhindered, it would take about a month to make everything happen.

Curt had one week. The results, he learned, were set to be reported during a talk at the (coincidental) fall meeting of the American Geophysical Union, the largest annual gathering of geologists in the country. Without NASA, the plumes would get good coverage because it was good science, but Curt wanted to bury the needle on this thing. It was terrible timing. It was ideal timing. Everyone would be there: twenty thousand scientists, plenty of press, as well as Roth, Retherford, and Saur presenting their findings. So while plumes survived the murder board for headquarters press conference approval, the venue of the fall meeting was just too perfect. They would throw the press conference there. They would dial this thing to eleven.

FOR MOST OF Europa's history, the mission died inside the four walls of NASA headquarters. It died, in fact, on the third floor. Ed Weiler had higher priority projects. On the rare occasion that the associate administrator gave ground, however, and let Europa reach the ninth floor for consideration by the administrator, it died there. The mission was too small. Or too big. Or too expensive. Or overly ambitious. Or insufficiently ambitious. Let's go to Mars instead.

Congress, meanwhile, no matter Culberson's appropriations defiance, would never, ever, ever win over this White House, which was still inflexible in its commitment to earth science and James Webb. Unless: NASA headquarters, from the security guard at the front door to the administrator on the ninth floor, had to be solidly, unambiguously, implacably, persistently, irrepressibly committed to a Europa mission. The administrator would need to walk to the White House, march into the Office of Science and Technology Policy and the Office of Management and Budget and not take no for an answer.

But the only way to do that—the only way to win over the administrator—the only way, in fact, to make him willing to throw his star on the table if given a no—was to create a story so compelling that it connected emotionally, cerebrally, and spiritually to the White House, despite its misgivings about prioritizing anything that didn't orbit the Earth. And the story as written, so far, just wasn't enough. The spacecraft was credible, the science superb, the mission concept ready to come out of the oven, but the whole thing was missing . . . something.

Then Curt walked into Jim Green's office with a paper reporting the discovery of plumes on Europa.[472]

Jim saw a story coalescing.[473] The Space Launch System rocket, announced two years earlier, was making headway. It was the centerpiece of human spaceflight—would eventually be the most powerful rocket ever built by NASA. Jim filed that away mentally. The SLS could get Clipper to Europa in less than three years. Clipper gave SLS somewhere to go. Now, seeing those plumes on Europa, Jim knew: the connection was made. You could take full advantage of everything known about Enceladus, all of Cassini's evidence, observations, and experiments . . . and fold it in. Just like that! Better still, Europa was much, much bigger than Enceladus. If Enceladus were a golf ball, Europa would be a basketball. Europa was huge! Something that size blasting an ocean into space—blasting it *two hundred kilometers high*,[474] the height necessary to be seen by a space telescope five hundred million miles away—just imagine how much water vapor was being expelled to hit that altitude? Which meant what was happening at Europa was even *more* phenomenal than what was happening at Enceladus. And Europa's ocean was billions and billions of years old: plenty of time to develop life.

So this press conference would be one of the most exciting things Jim had ever done in his life. Because he knew what would happen when it was over.

CHAPTER 16

Train Driver

JOAN SALUTE CAME TO NASA FROM PURDUE UNIVER-sity, where she worked as a computer programmer for the Laboratory for Applications of Remote Sensing, which studied things from orbit. She was not a space person per se—I mean, when she was a girl, she joined her family and a thousand other people to watch the Apollo 11 landing from the Wright Brothers National Memorial in North Carolina.[475] Park personnel presented the landing on a big bank of television screens outside. Where better to see it? Neil Armstrong even carried on the *Eagle* a thumb-sized sliver of the Wright Flyer's propeller and a small swatch of the plane's wing.[476] Joan, eleven at the time, didn't understand exactly what she was witnessing—the magnitude—but it stayed with her and turned out to be her comic book origin story, her radioactive spider bite. While writing lines of Fortran for the remote sensing lab, she developed a rare skill set: how to process and interpret satellite data. Other universities and government agencies, including NASA, would dial into Purdue's supercomputers to take advantage of their bit-smashing abilities, and before long, she was training representatives from those institutions in the ways of data analysis. When she moved to California in 1982, she reached out to managers at Ames—a Purdue client—and they offered her a job in its earth sciences division. She started as a contractor and was badged as an agency employee six years later. In 2004 she moved to headquarters, where she worked as the program executive for the Lunar Atmosphere and Dust Environment Explorer (but pronounced *laddy*), a satellite that circled and studied the lunar exosphere and looked for moon dust in space. Seven years later, Jim Green added the Europa Habitability Mission to her portfolio.

A program executive was a project's in-house representative at headquarters. If you were a mission project manager at JPL, you didn't want constant calls from congressional affairs, the NASA chief financial officer, or the office of the chief engineer. Enter Joan, whose job it was to run interference for the project and act as its advocate—and enforcer—at headquarters. It fell to her to keep projects in line with agency expectations. She was to Europa management what Curt was to its science team.

Her first Europa meeting at JPL was in 2010. Joan hadn't known what to expect. Engineers are prickly already, and the lab could sometimes instill employee paranoia when it came to the priorities of NASA. But across the agency—at every field center and partner office, from the redwood forest to the Gulf Stream waters—there was one catholic conviction: everyone at headquarters is an idiot.[477] A speed bump. In the way. Not at All Welcome Here. Before she joined headquarters, Joan had spent twenty years equally devout in that belief. And now she was on the other side. Everyone had only just taken their seats in the conference room, Tom Gavin presiding as per custom from the head of the table, when she walked in.

Tom saw her, stood up, introduced her to everyone, and then introduced himself to her.

Can I get you some coffee? he asked.[478]

Joan was stunned. The engineers were stunned. The coffee-maker was stunned. Anywhere else, any other meeting at any other center, and she would have been told where the coffee machine was, and just take your time. But here was Tom—in Gavin Tower, no less—and he didn't just get her a cup of coffee—he found a mug, *washed* it, and made a fresh pot. And this was really happening and the jaws of his staff were slack, their eyes wide. Tom Gavin was—well, he was Tom Gavin! And he was treating this . . . headquarters creature . . . so . . . so *regally*. WHO WAS SHE?

Joan would come to understand that Gavin wanted emphatically to establish her as worthy of extraordinary respect by the team. It made her job going forward—one where she would sometimes be the bad guy—a lot easier. It also made his retirement easier, knowing things would be right with headquarters once he punched out for the day one final time.

Tom retired, again, in 2012. He was seventy-three, and he just didn't have the energy to run a flight project any longer. Gavin told Charles Elachi, the lab director, that it was time to find a real eight-day-a-week flight project manager.[479] During Tom's transition out, Keyur Patel—who Karla Clark believed wanted the job but would be bad at it—stepped in briefly as manager of the Europa Habitability Mission. He didn't last long. Headquarters was happy with Keyur's selection by JPL management, but the choice did not sit well at the Applied Physics Laboratory. From Tom Magner's vantage, to suggest that Patel had thinly veiled contempt for APL would have been to suggest the presence of a veil, and Magner felt like his full-time job had become saving the alliance. Ultimately, Elachi offered Keyur the top spot over the Deep Space Network, and Keyur took it, to the relief of many on the project leadership team.

Asked for his recommendation, Tom Gavin endorsed Barry Goldstein for the job of project manager. Barry had done the Phoenix lander—a hard, major league mission—was a manager of enviable ability, was the obvious choice. He even had familiarity with Europa, having sat on the technical review board that examined the orbiter, lander, and Clipper concepts, and like everyone else on the board, was just besotted by the multiple flyby concept. Of course Barry wanted to run the project!

One person who was not asked for an opinion on the matter was Bob, who, he felt, as project scientist, was supposed to have at least a *say* in who his project manager would be. Indeed, he had been told explicitly that he would be consulted when the time came to choose a project manager. When he went to lab management with this complaint, their response was: Whatever.[480]

Since his days as Cassini project scientist, Bob had tried to assert and protect the authority of the role in lab missions. His job had one portfolio, and the project manager had another, but any management decision affecting science had to go through him. That was the whole point of the project scientist position, and it was the only way to protect the scientific integrity of the mission. Karla got that, and Tom really got that, went so far, in fact, as to take the organizational chart—the hierarchy of who reported to whom—and put his name and Bob's on the same line, with a comma between the two.[481] But Barry . . . there was never any ambiguity in Barry's mind about his role vis-à-vis Bob.[482] There weren't any commas on the new org chart.[483]

And Barry wanted to change some things on Europa Clipper before they became problems. The spacecraft was to use a power system called the advanced Stirling radioisotope generator, which was being developed at the NASA John H. Glenn Research Center in Cleveland.[484] Though the Stirlings promised greater efficiency than traditional thermoelectric generators while using less plutonium (which was ever in short supply), they were still in study and as a result presented ridiculous levels of technical risk to the mission. Barry told the team it could either get solar panels to work—a relatively new concept in outer planets exploration, enabled by breakthroughs in highly efficient solar cell technology—or they could use traditional radioactive power sources, but, listen: Quit trying to make Stirlings happen. They won't fly on this spacecraft.

Then there were Clipper's articulated antennae—a high-risk feature for low-priority science. If it moves, it can break. (Ask the Galileo team.) You bolt down something, you've eliminated a list of lurking mechanical misfortunes. So once again: find a way to do it with a fixed antenna, or it's out.

The most pressing problem, perhaps, was planetary protection. The idea had been to just remain relaxed about the whole thing. Let the spacecraft marinate in microbes during development, and, at the very end, submit the spacecraft for "dry heat microbial reduction," i.e., stick it in an oven and bake at two hundred thirty degrees Fahrenheit for fifty hours.[485] Politely speaking, Barry did not like that idea. It had worked in the seventies for the twin landers Viking, but any time you exposed a sophisticated, sensitive spacecraft to extreme temperatures, you introduced the risk of damage, obvious or imperceptible.[486, 487] When he had managed the lander Phoenix, its sample collection arm—a relatively tiny part of the spacecraft—had to be baked, and those three thousand minutes of terror were enough for one project manager's lifetime.[488] To do it on the entire spacecraft? No, thank you.

These were meaningful changes Barry was insisting upon, and he knew he was asking a lot. But he brought implementation experience to the table, had spent his entire career under pressure, delivering spacecraft on a schedule, and that mattered, went a long way with engineers. It was one thing to make demands, but quite another if your team knew that you had done it yourself, that you had lived through what you were asking them to do, and that you were doing it even now.

BRIAN COOKE ONLY knew he didn't want to drive the train. When he was in the fifth grade, he decided to be an engineer when he grew up. His understanding until then was that an engineer wore the hat, leaned out the locomotive window, and blew the horn. Somewhere in there was this business with a shovel and coal. But no, it was explained to him: an engineer takes things that are sitting around and makes better things out of them. Well, he heard that and he was in! Brian eventually attended Virginia Polytech Institute and State University and in 1995 earned an undergraduate degree in multidisciplinary engineering.

He had at the time a friend who had been working an internship at JPL and needed to return to Pasadena. He didn't want to make the drive alone, so he asked every girl he knew if she might want to join him on the transcontinental trek, VA to CA, and none said yes. Out of options, he asked Brian to ride shotgun, and Brian said yes, but on the condition that he (i.e., Brian) could go rollerblading on Venice Beach. When they arrived in town, however, it was raining—purposefully—vigorously—indefatigably—and the rollerblades sat unused in the corner of his hotel room.

His friend finally took pity on Brian and offered to bring him on a tour of the laboratory. By the time it was over, Brian had four job interviews lined up. (He had that kind of personality.) But because he had come to Pasadena strictly to rollerblade, he had packed only Tevas, t-shirts, and shorts for the trip, and showed up for said interviews wearing exactly that. Still, he was hired straightaway. (JPL was that kind of place.)

Cooke was soon developing command sequences to send to

Cassini for execution at Saturn. Three years later, he was hired to work on GALEX—the Galaxy Evolution Explorer, a small orbital telescope being built on an absurdly small budget: one hundred fifty million.[489] (Its price tag customarily would have been in the large percentages of a billion, but this was the Faster-Better-Cheaper era. You just made it work for less.) GALEX launched, was a success, and filled a critical gap in the observational abilities of astrophysicists, but what its engineers really bragged about over beer was that one of the images taken by the telescope became a stock photo in the space screensaver for Microsoft Windows.

Brian next worked on the Dawn spacecraft, which would fly to Vesta and then to Ceres. It would be the first spacecraft to visit a dwarf planet (Ceres), and the first to enter orbit around two separate objects. Brian was the project verification and validation engineer on that one. He made sure it did what it was supposed to do, to spec—with no surprises come liftoff and arrival—and was even tapped to serve as launch director. Before it flew from Earth, it was he who said, *Dawn is go for launch*, which, I mean, was pretty cool.[490] Twelve years now into his career, he had a good sense of what the lab was doing right, what it was doing wrong, and he had a burning desire to go make his own mistakes. It was the first time he was really able to say, "If I were in charge, I would do things a little differently," while having experience enough to mean it.

He first came to know Tom Gavin while working as project system engineer on the Kepler space telescope. As Brian was the lead technical authority, part of his job involved assessing engineering issues encountered by the project and reporting regularly to a council composed of the head of safety and mission assurance; the head of astronomy; the head of engineering and science; and the head of the council, the associate director of flight projects: Tom Gavin.

Brian called it the four-headed dragon. You would walk to Building 301, go up to the fourth floor, room 427—it was one of those conference rooms with the boardroom layout—and you knew exactly what to expect. The four heads were well known for their even temperament: they were always angry. And there was Tom Gavin at the end of the table, twenty-five feet away, glaring at you, the other three seated adjacent to him, scowling, all totally unimpressed by anything about you. You're presenting opposite them, and it feels almost ritualistic, an offering. The dragon fed on technical risks and issues. Your spacecraft-in-development ran into a potential problem—say, a transient signal was discovered and, under *very specific* conditions, could cause some unintended behavior—and you briefed the council on it, and they evaluated your solution. Of the four members of the panel, Tom was the worst. He would get spun up, and his . . . *enthusiasm* . . . would spin up the other three, and they would start to dig in—hard. You never wanted to go in unprepared. You identified a risk item, and you explained the solution in excruciating detail, and you also brought fifteen pages of backup. And they would go through every single page in your supporting documentation. You were walking them through schematics, through circuit diagrams for your exact fix, what work was left to do, the details of said work, schedules, milestones. Between the four heads, they had seen everything, success and failure, and knew when you were tap-dancing up there. You couldn't just wing it, and you couldn't deliver a "good enough" solution, because they knew what would and would not work. At the end of your presentation, they either gave you a thumbs-up or a thumbs-down, gladiator style. Their collective intuition, Brian came to see, bordered on clairvoyant and was one of the secret sauces of the laboratory that made it work so well.

ON FEBRUARY 10, 2010, while waiting in the recovery room after his wife had given birth to their daughter, Brian pulled out his cell phone. He was standing at the window looking out from the hospital, and there was snow on the San Gabriel Mountains, and for whatever reason, he thought of Tom Gavin and called him. Cooke had heard that the Europa team was restructuring, and he wanted to put his name out there if an opening appeared for a technical lead. Tom was noncommittal. He made Brian sell himself for a few months before finally bringing him on board, and after Karla Clark and Rob Lock, the lead system engineer, were out and the Jupiter Europa Orbiter died with the Decadal, Gavin had Brian restart the study team and get moving on the Europa Habitability Mission concept. From the start, the new spacecraft went through a series of redesigns, each in search of what engineers called "elegance."[491] *Elegance* had no definition, exactly. You knew it when you saw it. Spacecraft feng shui. It wasn't aesthetics, necessarily—pleasing to behold in the traditional sense: sleek lines and sweeping curves—the starship *Enterprise*. Science vessels tended often to look ungainly and hodgepodge; parts stuck together. This was because, once in space, they had no need for aerodynamics, never actually confronting air. Rather, it was engineering elegance: the ability to meet objectives with grace and style. You had to maintain mass efficiency while distributing the science instruments in such a way that they would not interfere with one another. This was all done in computer software. During pre-project planning, you worked only with a straw man payload; you knew it carried a magnetometer, but you didn't know *which* magnetometer. Each would be made to order by some other institution, which meant quirks and necessary accommodations. Beyond that, you had to consider the construction of the thing, and potential problems down the road. For example: an early Europa Habitability Mission concept had radiation-sensitive avionics embedded in a vault that was itself tucked inside the propulsion system and surrounded by fuel and oxidizer tanks. The tanks provided additional radiation shielding. It was brilliant! Unless you had a problem with avionics late in the spacecraft build—how would you get to them?

Each spacecraft concept started with a cylindrical body. The vessel at launch would be affixed atop a rocket pushing it into space from the bottom; a tall cylinder was thus as mass efficient as you could get. Various parts of the spacecraft would be moved around, repositioned, creatively mounted, and tested. Even in the earliest stages of design, the software produced models of adequate fidelity for engineers to conduct structural analysis, with answers accurate to eighty percent or higher—easily enough to make large-scale strategic and architectural decisions. The devil, as ever, was in the details.

When the multiple flyby concept study moved from APL to JPL, Tom had insisted that the spacecraft be made modular. Rather than construct a monolithic vessel—the way you might build a house—he wanted it designed as a series of mechanically separable elements, built independently and in parallel, each tested to one hundred percent comprehensiveness and merged late in the development process with high confidence that the system would work. This was how airliners were built. You would need, then, only limited testing of the fully assembled spacecraft. A further advantage was that if you ran into problems with any one element of the spacecraft, all work didn't stop.

Once Barry came on and the Stirling generators were removed, the team initially went with multimission radioisotope thermoelectric generators, but engineers latched quickly on to using solar panels instead. For something the size of Europa Clipper, solar simplified a lot of things. Just avoiding the nuclear launch approval process made it worth studying. With solar, no one would chain themselves to the fences of Cape Canaveral at launch. Only one other outer planets mission had ever attempted to use solar power, however— the spacecraft Juno—and Juno wasn't yet orbiting Jupiter, so it was a bit early yet to call it a success. Furthermore, unlike the solar cells of Juno, whose orbit avoided the worst of the Jovian radiation belt, Europa Clipper's would have to fly stridently through the belt and survive.[492] And if the radiation didn't kill the cells, the temperature might: a Europa spacecraft would follow an equatorial orbit, unlike Juno, which meant that half the time the largest planet of the solar system would separate the sun and the spacecraft. Not only would you be running on batteries during that time, but while on the far side of Jupiter, temperatures would plunge to cryogenic levels.[493] This would need to be a hearty spacecraft indeed.

Moreover, solar panels introduced complications to the science payload. There were questions about their "magnetic cleanliness." Solar arrays were notorious sources of electromagnetic noise because of their size, composition, and the currents running through them, which could impair the integrity of the magnetometer.[494, 495] Electrostatic discharge could interfere with the plasma instrument. Radiofrequency noise could affect the ice-penetrating radar. And the solar panels would need to be big—very big. The core Europa Clipper spacecraft would be about the size of a small car. The solar arrays would make the structure the length of a basketball court.

The gusto with which the engineers embraced the solar option alienated members of the science team, and Bob in particular.[496] They had been left out of the discussion of the trade to solar. The project scientist had authority in all matters affecting science, and here, Bob argued, was an element of the spacecraft that was *inarguably* affecting the science—degraded radar, defiled plasma instrument, diminished magnetometer—deteriorated data would obviate the point of going. Look, Bob understood the advantages of solar: it cost less and simplified an awful lot. He wanted solar to work. But the decision in his view wasn't ready to be made, and certainly not unilaterally. Until the trade studies were done, Bob didn't want to sign on.

It was a power play—literally—and power being king in space exploration, it pitted Bob against Barry in conversations you could hear down the hall, and not for the last time.

CHAPTER 17

Step Forward, Tin Man

BOB STOOD ON THE ELEVATOR AT NASA HEADQUAR-
ters, the 9 button lit. He was wearing a suit and tie, so everyone
knew: it was serious. January 24, 2014, and he was here at last, in
the Lanyard City. He was off to see the administrator.

NASA headquarters could be any municipal building or corpo-
rate office in America: a little dingy, well worn, too many cubicles,
too little space, posters pasted on surfaces, flower bouquets and
bobblehead dolls on desks, stickers and tchotchkes and tall stacks of
paperwork teetering, tottering, threatening to spill into walkways.
The carpet . . . adhering to the five-second rule would have been
at great personal risk. No one built spaceships at headquarters, and
there were no marble columns out front. Unless you knew exactly
what you were standing in front of, it was just *some place* in DC, and
you would walk by it and never bother looking up. Architecturally,
the Residence Inn across the street was about as impressive. Certainly no one would confuse HQ's ground floor with the Smithsonian Air and Space Museum. There were a few space artifacts (or
at least replicas of space artifacts) in display cases in the lobby, and

off to the side there was a little gift shop where you could get your official NASA gear, t-shirts and ball caps, and postcards of Mars. On the floors above: Things getting done. Employees in business casual racing to meetings; half-eaten birthday cakes on breakroom tables; coffee mugs in sinks; fridges filled with Tupperware lunches; people talking in hallways; ten thousand chimes from ten thousand simultaneously open copies of Outlook for every agencywide memo; the clacking of keyboards, the clicking of mice.

You didn't brief the NASA administrator every day. Most, even at headquarters, would never brief the administrator in their entire careers. The JPL delegation to headquarters (of which Bob was a part) left nothing to chance, wanted things just so, because this might be their one shot, and they didn't want to throw it away. And so for this Europa presentation, every slide title and bullet point was iterated by committee until every word had been drained of meaning, sentences stripped clean to the bone, so simple in structure that Cormac McCarthy might have suggested sprinkling some semicolons in there. Pictures were used in lieu of words whenever possible. It's not that anyone thought the administrator was an idiot; rather, everyone recognized that he was neither a scientist nor an engineer, and furthermore, the man attended briefings eight hours a day, five days a week, for years now. When he left the Europa briefing, you didn't need him reflecting later about Europa's surface albedo of zero-point-six-four. You wanted him to remember from the slides: "Very high science merit" and that Hubble "Indicates water plumes!" And, of course, a great big image of Europa Clipper astride the SLS rocket in its Apollo color scheme.

A few days before giving the talk—far too late in the process for the customary eighty edits and heated discussions over whether an image's border should be thickened by two pixels or one—Bob surreptitiously slipped in a slide. It was a shot of Europa's chaotic terrain taken by the spacecraft Galileo, with the caption WHAT IS

THE ROLE OF LIQUID WATER IN SHAPING EUROPA'S BIZARRE SUR-
FACE?[497] After weeks of discussion and rehearsal, Bob's role in the
presentation had come down to explaining what excited him about
Europa. He realized that the NASA administrator probably didn't
even know what Europa's face looked like—why would he? So
maybe this picture of the chaos region, rugged and crystalline, so
alien from anything else in the solar system, might get his attention. Why do we need to go to Europa? To figure out things like
this.

Bob didn't know if Administrator Bolden (or Mr. Administrator, but never, it was drilled into him, Charlie, or Mr. Bolden, or
General Bolden) would be a friendly audience. One month earlier,
during a break in a separate meeting, the administrator had popped
into a gathering of the NASA Advisory Council to make a few
brief remarks.

We have to stop thinking about flagships, Bolden told the
council. "The budget doesn't support that."[498] The agency wanted
"more, less-expensive types of missions," and even *attempting* to fly
a flagship would mean "eternal battles" with the White House. In
the eyes of many at headquarters, Bolden apparently among them,
after the punishing experiences launching Galileo, Cassini, and the
rover Curiosity—each in the course of its development a lightning
rod for criticism by presidents and Congresses alike, all over budget
and each ever at risk of cancelation, hundreds of millions of dollars
therefore wasted—*flagship* was the literal f-word. You didn't say it
in polite company.

Bolden's declaration was seen generally as a preemptive strike
against congressional micromanagement. Scuttlebutt had it that
Frank Wolf, chair of the Commerce, Justice, Science, and Related
Agencies Subcommittee of the House Appropriations Committee, was set to announce his retirement at the end of the next session.[499] Which meant twelve months separated Charlie Bolden

from *Chairman* John Culberson. If this Texan interloper was dictating terms now, what would he do when handed almost unlimited power over the NASA budget?

Bolden got his answer two weeks later, when the congressman stated his plans outright to the *Houston Chronicle:* "I'm certain that there's life elsewhere in the universe. And I'm also certain that the first place we will discover life on another world is Europa," he said.[500] "If I'm successful in becoming chairman of the subcommittee, that's going to be right when the Europa mission will need its maximum funding. It needs to be a flagship mission. The biggest and best we've ever flown." Culberson raised the stakes weeks later in the Consolidated Appropriations Act of 2014, again inserting into law: "Provided further, That $80,000,000 shall be for pre-formulation and/or formulation activities for a mission that meets the science goals outlined for the Jupiter Europa mission in the most recent planetary decadal survey."[501] If you were Joan Salute at NASA headquarters or Barry Goldstein at Jet Propulsion Laboratory, you were doing the math, carrying the one, and smiling. The Europa Clipper still had a surplus of cash from the seventy-five-million-dollar infusion in 2013. Plus the eighty million, it literally could not spend money fast enough.

Though Europa now had in the bank *twenty times* what had been spent on the studies Quad and shootout combined, Bolden still held the trump card for any prospective mission.[502] Until Europa appeared in the NASA budget request—until the agency made it Facebook official, asked for the money, committed in word and deed to Doing This Thing—it was virtually impossible for the lab to do any long-term planning, or, more urgently, to sign any long-term contracts to turn hopes into hardware. The money might be in the bank, but the Europa team would have no withdrawal slips, no ATM card. Without Bolden's signature, John Culberson was paying for the most elaborate paperwork exercise in the history of space science. Without a formal agency buy-in, Europa was completely Culberson contingent and one heart attack or falling piano away from vanishing entirely from NASA's radar.

And so, yeah, when Bob pushed 9—goin' up!—he was a little bit terrified. The elevator carried him from floor three, where he had met with Joan and Curt for a pregame conference. The two were not invited to the meeting and wanted to brief Bob on any last-minute business and the leadership disposition. Of the twenty-eight slides today, Bob would speak for only a few of them. He knew what to say—had given ten thousand talks on Europa—but this ten thousand and first was the one that mattered most. Charles Elachi, Jim Green, and John Grunsfeld, the associate administrator for science, would be there, maybe a few others Bob knew. But they would be background. He would focus on the administrator.

Bob had been on the ninth floor before, but still, it caught him off guard. The elevator doors slid apart, and *Whoa—this place is a little nicer than the others. Am I supposed to be greeted by someone? Am I allowed to walk down this hall?* There were fancy photos on the wall: from Apollo, the shuttle, the space station, Hubble. It was eerily quiet, clean, bright, and unnerving.

Conference room ACR2 was smaller than he expected, and darker.[503] It was an interior room, no windows, and there was a small conference table around which only a handful of VIPs sat. It was positively intimate, and practically one-on-one when you zeroed in on Administrator Bolden—much less formal than Bob expected. It was not a presentation setting with big screens or a podium. You didn't have to project or wear a microphone to reach the audience in the back of the room. (You could lean forward and touch them with your fingertips.) Bolden was known for his breezy bonhomie and approachability (rank-and-file NASA adored him), which was both welcoming and, for those used to more . . . clinical . . . feedback, disorienting.

OK, thought Bob, *now I am going to talk to the administrator of NASA.*

And when came his slides, he did. He made the science case for Europa. He described the whys and wherefores of the ocean and ice shell, the rocky interior touching water, the plausibility of habitability, presented the chaos terrain, and—

Sometimes you gave these talks to agency brass, whether at the center- or headquarters-level, and you wondered if your audience artificially amped up their interest—Oh, wow! Wonderful! Unbelievable! Amazing!—but meanwhile were wondering in the backs of their minds what they might have for lunch later that day, or whether they had left the stove on, or whether the sitter knew to pick up their kid from school today at one thirty rather than three, because it was early dismissal—but Bolden seemed blown away by it. He leaned forward, narrowed and then widened his eyes, really, really absorbed what he was seeing, and seemed truly transfixed by that parting view of Europa, the blocks and the crazed etchings in chaotic terrain. Bob didn't push forward. He let the slide linger, let the man at the head of the table take it all in, offered a cursory description of what the administrator was studying, but mostly let the image speak for itself. Next slide.

When his few slides were up, Bob handed off the presentation to Barry Goldstein. His relationship with Barry remained less than ideal and on an errant trajectory, but they gave great talks together, a perfect doubles tennis team. Barry really, seriously knew his stuff, and Bob admired Barry's brain, a vault of mission minutiae and spacecraft specifics. Barry, joined by Firouz Naderi, head of Solar System Exploration at JPL, explained the business case for Europa: why the Clipper concept enabled the lab to fly a successful mission for less money ("Ninety percent of the science for fifty percent of the cost," Curt had drilled relentlessly into them to tell anyone who asked, ever) and precisely why Europa Clipper was the unexpected dream payload of the centerpiece of human spaceflight, the SLS rocket. Charlie Bolden, a human who had experienced spaceflight, remained mesmerized.

The meeting lasted an hour. It was conversational in tone, and Bolden interjected when he had questions or just wanted to make general comments. Charles Elachi, afterward, congratulated his team from the lab, and Bob left feeling very good about it.

As he should have: the talk worked. But being Europa, there was a catch.

Curt and Joan learned later that Bolden, though supportive of a mission, wanted it done for a billion dollars.[504] Someone had told him that it could be done for a billion dollars, see, and he wanted everyone to get on that. Europa. One billion dollars. Everybody got that?

EUROPA COULD NOT BE DONE FOR A BILLION DOLLARS. Curt had literally spent the last ten years proving that Europa at one billion was an absolute metaphysical impossibility. It didn't matter if you were talking about orbiter, lander, or multiple flyby—no way, no how. But some bureaucrat at the White House Office of Management and Budget had heard from someone at some university that, Oh, fellas, Jet Propulsion Laboratory just isn't trying hard enough—we have a miracle cure for the billion-dollar barrier—tiny spacecraft called "cube sats"—and give us the chance (and money!) and we will prove it to you. And thus, Europa at one billion: do it.

Thankfully, David Schurr, the deputy division director for the Science Mission Directorate, stepped in. He had been briefed by Curt a thousand times on the state of Europa and what could be done for how much, and convinced the White House budget examiner that rather than simply *direct* a billion-dollar mission, it would be prudent to request a white paper on how Europa was costed, and why the only way to achieve Decadal-compatible science was to fund a mission at the magic number of two-point-five billion, which, if that congressman from Texas was to be believed, Congress was willing to do.[505]

It fell to Curt to write said paper. It was due in six weeks.

Ten days later, February 3, 2014, Curt left headquarters for the airport on an itinerary taking him from National to LAX, and then by car to JPL. The lab was hosting a Europa science team meeting and a workshop on icy world habitability. But when he got to the airport—and, look, he had checked on this, confirmed this, had been assured that, yes, Curt, it was done—his training request had slipped through the cracks, leaving his plane ticket unpurchased. Unlike most of his colleagues in the Europa effort, Curt was not a million-miler with United or American Airlines—he had a family who needed him and better ways to spend the taxpayers' money, and, at most, he took eight flights a year on business. So these things happened. He rolled his carry-on back to headquarters to get the paperwork straightened out and to catch a later flight.

Back at the mother ship, he went through security, called the elevator, doors slid open, entered, lit 3, goin' up, 2, third floor (*ding!*), the doors slid open, and he stepped into the nerve center of American space science, the cube farms, the posters, the birthday cakes, and there was David Schurr, who gave him a friendly hi, followed by: Europa is officially in NASA's fiscal year 2015 budget.

It was like a piano had fallen on Curt's head, but in a good way. A good piano. How much? he might have asked, if he could speak at all.

Culberson's, ah, strong messaging did the trick, Schurr explained.[506] That white paper for the Office of Management and Budget on why a billion dollars is impossible? Forget the six-week deadline; I need it by Friday.

Curt canceled his trip, swept his desk clean of competing items, and set about drafting the best white paper in the history of paper, or of the color white.

THEN HE WAITED.

On February 20 he and Joan holed up in her office, picked up the phone, and dialed. Ten days earlier, Bolden had given word that he wanted to take ownership of Europa and, during the budget rollout on March 4, hoped to announce that the agency was officially pursuing a mission there. But no decisions had been made, and if the White House pushed back, Bolden would definitely take no for an answer. A strategic implementation planning meeting was called, and after frenzied email traffic, Curt and Joan were told to prepare slides for it. The slides were due in three days, and neither of them would be allowed to attend the meeting.[507]

SIP meetings were, at best, an ambiguously defined concept. The American space program distilled to an endless gauntlet of meetings and telecons, with the occasional foray into space. At headquarters, the SIP was among the most baffling of those meetings, because nobody really knew what it meant or what it did. Some higher-ups weren't even sure what the abbreviation stood for, exactly. So planning for a SIP was like packing for a mystery vacation. Maybe you were going to Tahiti. Maybe you were going to Denali. Good luck!

Four days later, John Grunsfeld, the associate administrator, gave guidance to Jim Green, who gave guidance to Curt. The presentation to the administrator and whoever else would be at the meeting was to focus primarily on the science. Curt had a number of problems with this, not the least of which was that he would be preparing science slides for an unknown presenter with, quite possibly, far less knowledge than him, and, oh, the meeting would decide the fate of a project he had spent ten years working on. He spent the weekend immersed in the pending presentation, shaping slides, and the following Tuesday, he and Joan sat down with Grunsfeld and walked him through them. Grunsfeld had a few ideas for how this mission ought to go and wanted them included in the presentation. Specifically: NO SOLAR. He wanted this thing to fly for at least eight years, and nuclear power would allow that and then some. Curt explained the benefits of solar: that there was a plutonium shortage, which might lead to delays, and the headache of the Department of Energy and nuclear launch approval, the doomsday press coverage and the people chained to Cape Canaveral fences, and Grunsfeld's response: That's fine, Curt. I want nuclear.

Nuclear it was!

Curt included also in the draft presentation a slide discussing the release of an announcement of opportunity for instrument development, allowing organizations to develop detailed designs of the scientific payload the mission required. Hundreds might be submitted, and NASA would choose the best magnetometer, the best radar, the best camera, and so forth, and sign contracts with the winners to build the things. An actual payload of instruments real and true would allow project engineers to develop spacecraft subsystems at a granular level.

For Curt's purposes—as with the ill-fated AO (as they were called) for Jupiter Europa Orbiter—the very act of releasing a multiple flyby AO would commit NASA to a Europa mission in a way that a speech, Decadal endorsement, community insistence, congressional pressure, or a line item in the budget would not. Once NASA solicited proposals for a science payload, institutions would start pouring their own dollars into developing their instruments. And after spending hundreds of thousands of dollars developing the best imaging instrument in the history of space exploration—an instrument *made to order*, designed to survive the Jovian radiation environment—NASA, at its peril, might cancel said mission. Though there were caveats in the announcement designed to protect the agency should this thing go south, those who had spent all that time and money would have *just enough ground* to sue the U.S. government to recoup the costs incurred to prepare their proposals. Still, NASA's lawyers could handle that. What the agency could *not* handle were institutions crying foul—that NASA had pulled a fast one, a bait-and-switch, was a poor steward of taxpayer funds—anything, really, that the press could latch on to and make messy, and suddenly you have associate administrators called to testify before Congress and getting grilled on C-SPAN. It would be ugly, and NASA hated ugly more than anything else.

Grunsfeld saw Curt's and Joan's AO slide only as he was standing to leave for another meeting. He was running late, but he stopped, stared at it for five seconds or so, and said, finally: Leave it in.

He also allowed another line to survive his edits: one designed with surgical precision to solve the agency leadership's inability to make decisions on this thing. Curt called it the Joan Bullet. "Decision on class of mission needed by Fall 2014 to feed Budget submission for FY16 overguide for new initiative," it read.

The way she worded it, and the way she brought it up with Grunsfeld, explained it (Lord, was she smooth about it), and, most important, *got him to support it*—well, Curt was in awe. The Joan Bullet put a countdown on the thing, entered headquarters leadership into a tacit agreement: NASA would explore smaller missions, but if it could be proven to the ninth floor that a small mission was not worth doing, then agency leaders absolutely had to make a decision by the fall. No more delays. No more bouncing from one billion to four billion to three billion to two billion to—Pick one. Because once NASA committed to the *size* of a Europa mission, everything would fall into place. The agency would stop looking at other options when the best one was staring them in the face.

Everyone had been scared to force the issue. But Joan didn't scare so easily.

BEFORE THE SIP telecon, Curt ran to the local liquor store to buy a bottle of Veuve Clicquot, in the event things went their way. Not only were he and Joan not invited to the meeting, but they weren't even invited to the teleconference! The thing had been reserved for center director-level leadership and above. Still, if you read the email just so, there was nothing in there that expressly *forbade* them from joining. So they got their hands on the dial-in information and website feed, and Joan and Curt met in her office, huddled over the telephone, made the call, hit Mute, and waited.

As was procedure, the start of the meeting involved a roll call of persons who had dialed in. Each one sounded off until all were accounted for. Curt and Joan looked at each other and Curt felt his panic rising. The system knew how many people were there. What do we—

Joan unmuted the phone.

Joan and Curt are here, she said.

Seconds that could be measured in hours elapsed as they waited for someone to call them out and forcibly kick them from the meeting. Before anyone had an opportunity to send them on their merry, however, the director of Goddard Space Flight Center pressed on, said, "Chris Scolese here," and the roll call continued, and . . . they were in!

At the start of the meeting, Bolden announced that he wanted to approve a Europa mission before the end of his tenure at NASA. It had previously been word of mouth, but to hear it from the man himself made it feel suddenly legitimate—a solid fact.

John Grunsfeld gave cursory opening remarks about a Europa mission, about the urgency to choose a mission class, about the instrument announcement of opportunity, and he set things up for Jim Green, who presented Curt's slides. It didn't escape Jim that in all these years, this was the first time he had ever been allowed to go beyond a white paper and present Europa on the ninth floor.

And so after decades of false starts and millions of words across thousands of pages and hundreds of meetings and scientists growing older and engineers growing despondent and uncomfortable hours spent looking at data and budgets and with the help of a Tea Party guy in the Houston suburbs who had seen the light and was ready to harvest alien lobsters from the Ocean of Eden, Jim began.

He opened by speaking briefly about the landmark discovery of Europa's plumes and the NASA press conference to come out of that. Next slide. He went back to the basics. He presented the Galilean moons and how to make an ocean. Io: This one's too hot. Ganymede: This one's too cold. Europa: This one's just right. Next slide. He described Europa: the ice shell and its terrain; the ocean; the ocean floor and its hydrothermal vents blasting out hot water where—if its ocean was anything like that of Earth—life would be teeming, and if there was life *there* [points to ocean floor], there would be evidence of it up here [points at surface]. Next slide. What are we talking about when we talk about life on another world? He explained the habitability triad: water, energy, and chemistry, one at each corner, and how only with all three could you get Life as We Know It, and, by the way: Europa has all three. Next slide. All of this is directed by the Decadal Survey—the most recent and the one ten years ago. Scientists were *desperate* for this mission. Next slide. The Decadal wanted a less expensive mission, and we listened. The post-Decadal mission concepts that have been under intense development by Jet Propulsion Laboratory: an orbiter, a lander, and a multiple flyby mission. We did the studies. The lander was too expensive. We did more studies. We looked at the orbiter that would circle Europa, and the multiple flyby—Clipper—that would orbit

Jupiter. We did more studies and found a way to enhance Clipper to do the job of both.

The reason why this [points at Clipper] is better than *this* [points at orbiter] is because we took a page out of Cassini's book! Next slide.

Planetary science has been very methodical in how it explores the solar system. I bet you didn't know that! [*laughter*] It's sometimes hard to see. Why will we fly by Pluto next year? Why don't we land on it? Why don't we bring back a sample? Because we are methodical. When studying another world, we fly by, we orbit, we land, we rove. Flyby is for reconnaissance. If we like what we see, we come back and orbit. And when we orbit, we are looking for an overall view of the body, and we find a place to land. Once we get ground truth, we send a spacecraft to bring back a sample. It's what we did at the moon. It's what we are doing at Mars. And we will do that with every object in the solar system that is important to us. Next slide.

We know that the radiation belt at Europa is so crippling that no spacecraft in orbit can survive there for very long. We've been studying orbiter concepts for fifteen years. We found that the most recent, cost-effective orbiter would last ninety days there, tops. Mr. Administrator, that thing was going to have a heck of a time. See, we need an ice-penetrating radar, but the problem is that radars produce an enormous amount of data, and with an orbiter, we can't send all that data back; there's just not enough time. So the spacecraft computer would have to do onboard processing to figure out the important things in the radar data, and send home only the important stuff. And to execute it in ninety days? That's tough to do. Next slide.

Orbiters are all about getting a global view. Here is where Cassini comes in. Each time it orbits Saturn, Cassini swings by Titan to change its orbital plane—its angle of travel. Titan is an enormous moon and has the perfect gravity for that. So Titan lets Cassini fly higher or lower over Saturn and see different parts of its rings. And every time Cassini flies by Titan, it keeps its science instruments switched on and gets some new slice of the mysterious moon. By orbiting *Saturn*, we have been able to capture eighty-five percent of *Titan* . . . because of all those flybys. So if we want a global view to understand Europa at Jupiter, we can do it just like Cassini: with multiple flybys. Next slide.

We orbit Jupiter, fly by Europa, collect data, zip out of the radiation belts and away from Jupiter, and send *all* that data home. Once that's done, we come back in and go back out. Send the data home. Right back in and right back out. Send the data home. We won't have to survive for months in the radiation belts, won't need all those chips for onboard data processing or all that radiation hardening. The price plummets and satisfies the demands of the Decadal Survey—both in terms of cost and science. Next slide.

And now, here comes the SLS. Because instead of spending eight years flying to Europa, we can get there in two and a half. Instead of spending a fortune building a spacecraft capable of surviving both the intense heat of a Venus gravity assist and the cryogenic cold found at the far side of Jupiter, we can just go there in a straight shot. Only the SLS can do that. Only the SLS can open up the outer planets, make them as accessible as a trip to Mars.

The final slide was a list of the big bullet points. Grunsfeld had told Curt and Joan to include everything here—even the mission class decision—but changed his mind about the announcement of opportunity. After the bullets were covered and everyone had a chance to soak them up, the screen switched from PowerPoint to Word, with the full text of a memo stating the meeting's outcomes and decisions. It was already written, the goal being to print the thing and have it signed by all those present *before they left*, lest it never get signed.

After summarizing things, someone asked: *Does the memo cover everything we discussed?*

Curt and Joan inhaled sharply, held their breaths, leaned into the phone.

Wait, said another. *There's nothing about the announcement of opportunity in the memo.*

Here is where everything would fall apart. Curt felt—

Well, let's just add that in there, said someone else. And someone started typing words about the AO—right there, right on the call—into the shared decision memo on the screen. David Schurr helped guide the wording based on his previous conversation with Curt on the subject.

And just like that, fifteen years after the agency made its tentative first steps toward Europa, and for Curt Niebur, after a full decade of trying to get a mission going and an instrument announcement out the door . . . it was done.

When the meeting ended, Curt realized that no one had uttered a single negative word throughout. All his years of arguing with Ed Weiler, of funding studies from coins found on sidewalks, of begging, persuading, challenging, driving fists into tables, and meetings coming just short of name-calling—after years of feeling as though he were hammering his head into a brick wall, unsure if anyone ever heard a syllable he had ever uttered . . . they had. They had heard it all, understood what he had been saying, what the mission needed, why Europa mattered. And no longer was he or Joan or Jim alone in carrying the flag for Europa at headquarters. The National Aeronautics and Space Administration was carrying it now. There were still decisions to be made—mission class, mission concept, price—but as preposterous as it sounded, the decision had been made to make those decisions. All else would follow.

Curt and Joan popped the champagne.

CHAPTER 18

One Inch from Earth

BACK TO *2001* AND *2010*. THOSE WORKS OF KUBRICK AND Clarke, jointly and respectively, are about humanity's place in the universe. They explore our origins, the workings of the cosmos, God, the discovery of the first evidence of extraterrestrial life, what we do with that knowledge, and how it leaves us transformed. This is what science fiction does best, and it is entirely fitting that a thoughtful collaboration between geniuses would be the first to ask serious questions about Europa specifically. As Philip K. Dick wrote, "It's not just 'What if . . .' It's '*My God; what if . . .*'"[508]

A sequence of missions beginning with a spacecraft Europa Clipper might provide an answer. What if there is life elsewhere? How would the human psyche handle its discovery? And if we find it, *what do we do with it*? Thus far, the whole history of space exploration has been one of disappointment as it relates to the life question. Probes stole from the human imagination brontosaurs wading through Venusian swamps, civilizations of Martians with their great carved canals, and, with each successive spacecraft, even lower forms of life. The whole of planet Mars couldn't even give us a shrub. But German philosopher Immanuel Kant, who wrote pro-lifically about life on other worlds, observed in *Universal Natural History and Theory of the Heavens* that barren planets, like barren stretches on Earth, are in fact necessary, for "would it not be a sign of nature's poverty rather than an evidence of her abundance, if she were to display with diligence all her richness at every point in space?"[509] It remained in his lifetime, of course, a problem of technology, and he noted in *Critique of Practical Reason* that "if we could get nearer the planets, which is intrinsically possible, expe-rience would decide whether such inhabitants are there or not; but as we never shall get so near to them, the matter remains one of opinion."[510]

And in opinion, as the popular contemporary response to plan-etary science has suggested, there is great comfort. Though Kant seemed certain that life existed on other worlds—that he "might well bet everything that I have on it"[511]—were Europa Clipper to fly through a plume and directly image some saucer-eyed celestial sea bass blasted into space, the same people who deny the moon landings (despite video footage from the moon, the laboratory of lunar samples at Johnson Space Center, and the ability of the Lunar Reconnaissance Orbiter to image the Apollo landing sites down to their footprints) and who declare Earth's changing climate to be an opinion (despite *personally experiencing* its catastrophic effects) would certainly do the same for life on Europa. It's just easier. Opinion is easier.

A house cat might go her entire life believing she is the only house cat in the entire world. She has her caretakers—her met-aphorical gods (though cats might think this the other way around)—and if she is kind, shows love and devotion, there will be food and water and comfort when there is hunger, illness, or injury. The metaphysics of *felis catus* allow it to stretch on the back of a sofa or lie on a stack of papers because it is at the center of the universe.

What more peaceful place could there be? Otherwise, as Kant continues, when describing the awe-inspiring starry heavens, "a countless multitude of worlds annihilates, as it were, my importance as an animal creature, which must give back to the planet (a mere speck in the universe) the matter from which it came, the matter which is for a little time provided with vital force, we know not how."[512]

Humankind has had a good run as the only house cat in the world. But once we learn of others, how would things change? Religion is surprisingly well equipped for the eventuality. At first glance, God's only apparent purpose in the Abrahamic religions is to keep creating until arriving at human beings, at which point the purpose is to tend to our garden. We are the whole point of the universe, the purpose for creating the heavens and the Earth, for giving us light, water, sky, land, plants, and animals. We are even created in God's own image, says the Bible. We are the beginning and the end.

And yet these religions have no problem per se with alien life. One might read the ancient texts as guidebooks for how to deal with extraterrestrials—and not peers in appearance or general ability, but the real scary stuff: completely uncommon, perhaps with ill intent. Angels are not of this Earth. They are, in fact, very clearly creatures of some order higher and more terrible than humanity, and they have dreadful powers. In 2 Kings, a single angel defended Jerusalem by slaughtering an army of one hundred eighty-five thousand: "And it came to pass that night, that the angel of the Lord went out, and smote in the camp of the Assyrians an hundred fourscore and five thousand: and when they arose early in the morning, behold, they were all dead corpses."[513] Moreover, when they are not disguised as humans, angels sound, frankly, terrifying. The book of Daniel gives a pretty good idea of this: "His body also was like the beryl, and his face as the appearance of lightning, and his eyes as lamps of fire, and his arms and his feet like in colour to polished brass, and the voice of his words like the voice of a multitude."[514] In Luke, an angel appears and scares the bejeezus out of the father of John the Baptist: "And when Zacharias saw him, he was troubled, and fear fell upon him."[515]

These are literal extraterrestrials—from the Latin: outside of Earth.

Despite the biblical proclivity for people to worship false idols every instance God looks the other way (see: the Israelites and golden calves), angels, even with their horror-show fire-eyes and splatterhouse reputation, are never worshipped. And when someone looks like they're *about* to start praying to their sudden lightning-faced new lord, the ancient texts go out of their way to put a stop to this angel-worshipping business, every single time. "I fell down to worship before the feet of the angel which shewed me these things," wrote John, mysterious author of the Book of Revelation. "Then saith he unto me, See thou do it not: for I am thy fellowservant, and of thy brethren the prophets, and of them which keep the sayings of this book: worship God."[516]

It is not just Christianity and Judaism, of course. Whether talking about the Jinn in the Islamic faith, the Brahmā in Buddhism, or the Asura in Hinduism, things above humans and otherworldly are not outliers to be explained away or otherwise apologized for. They are foundational elements of religion itself, and thus, if the books are to be believed, of human existence.

"Be fruitful," said God, "and multiply, and replenish the earth, and subdue it: and have dominion over the fish of the sea, and over the fowl of the air, and over every living thing that moveth upon the earth."[517]

Still, God here raises a question worth considering: What of the animals *not* of the Earth?

Higher creatures would arrive on Earth from a position of strength, however benevolent they might be. Like Zacharias, trouble and fear would fall upon us for that reason alone. They know a lot that we do not. But if religion prepared us for the extraterrestrials who arrive in flying saucers, it falls short in preparing us for those aliens arriving in canisters bearing JPL branding. Catholic theologians have wrestled for centuries with the question of whether sentient alien life not descended of Adam and Eve would have been born with original sin—not their garden, not their apples, after all—and thus, would baptism be necessary?[518] But where would we fit in a solar system whose *lower* orders of life had no relation to our own? Who share none of our DNA? Who were grown in their own Gardens and Oceans of Eden unrelated to that of Earth? A microbe, mackerel, mermaid, or monster in the Europan depths doesn't likely share an ancestry with humankind. Would a Europan creature even be *considered* an animal?

What makes life on Europa so compelling a subject to consider is that, unlike what might one day be discovered on Mars, Europan life has a real chance at complexity. Sure, Martian microbes would be thrilling—they would transform biology and finally give astrobiologists something to do at all those conferences. But . . . they are at least slightly disappointing next to a space guppy. Were scientists to discover a Europan creature that wiggles, it would not be long before they would get one in a lab. As John Culberson astutely asserts, no discovery would so mobilize and orient NASA; whichever world offers up life first for our probes would be the world to attract a generation of exploration. But by what right do we proceed? The first discovery of the first living organism off of Earth will also mark the first time that humans—earthlings—present themselves practically to the cosmos. Are we an invasive species? Are we *allowed* to slice open that Jovian jellyfish? Would one deep-fry a Europan squid with the same relish as its terrestrial counterpart?

Europa's proximity complicates everything. In a universe with one septillion stars, sure, even the hardened skeptic of alien life can grant that maybe the set of circumstances that happened here happened also on some other planet in some other solar system. But if whatever happened here happened *two planets over*? And not even on an Earth-like world—that fantastical, waterlogged Venus—but rather, on a little ice ball circling a giant hydrogen hurricane sphere? We would not be the only house cat in the world. We would not even be the only house cat in the *house*. If genesis occurred two times in three planets, then habitability is not likely an aberration, and Earth is not some lonely cactus in a vast, indifferent desert; it is a blade of grass in a sweeping, verdant meadow.

The ambition of the Europa mission studies and the implications regarding religion were not missed by Ron Greeley. In June 2003, while leading a JIMO science definition team meeting, someone asked: "Will this mission answer the question 'Is there a God?'"

"Yes," said Greeley. "But it is priority two."[519]

BOB PAPPALARDO STOOD in the front of the room. One hundred fifty faces stared back. There was a time when everyone on the Europa team could fit in the back of Il Fornaio. There was a time when the team could fit in a single booth in a Thai restaurant. Now they were in the basement auditorium of Gavin Tower, room B20, Building 321, as in *3-2-1-liftoff*, and . . . they had made it. They had grown by an order of magnitude overnight.

August 4, 2015.

Sixteen years earlier, he was a postdoctoral researcher—Lieutenant Junior Grade Spock of the spacecraft Galileo—and published a paper pondering a subsurface ocean on Europa. It would have been illogical to commit wholeheartedly one way or the other, and he ladled the article with lines such as "no definitive evidence" and "viable, but unproven"—was unwilling, even, to say that such an ocean existed *in the recent past*, though he granted that "warm, low-viscosity material at shallow depths" played *some role* at *some point* in shaping Europa's geology.[520] Anything more, and you were just being foolhardy—cavalier—audacious—brazen—snotty—insolent. But while he was hesitant in journals, he *just knew*, and he tipped his hand in interviews at the time. Spock, after all, was half human.

Three years before that paper, Pappalardo published another asserting the geological evidence of "solid-state convection" in Europa's ice shell: that beneath the granite-hard surface, cold ice sank ever deeper in the ice shell, and warm ice rose, in a slow dance over hundreds of millions of years.[521] Maybe there was an ocean beneath it. Maybe there wasn't. Who's to say!

But now it was no longer "What if there were an ocean . . ." It was "*My God;* what if it were *habitable?*" Now he was Captain Kirk of the Europa Clipper, and he was going to find out for sure. They were met in Building 321 for PSG-1: the first project science group meeting of the Europa Multiple Flyby Mission. Not a concept. Not a study. Not a pre-project. It was official. All of it.

Well, not the name. Oh, they still called it Europa Clipper, though NASA headquarters hated it so. It was, officially—for now—the inelegant yet precise Europa Multiple Flyby Mission. But every time Joan wrote a decision memo for the project back at headquarters, she wrote in "Europa Clipper." And every time, someone changed it back: "Europa Multiple Flyby Mission"—what Curt called the Europa mf'n mission. But Joan was possessed of arachnid patience. She wasn't going anywhere, and her plan was this: she would wait out everyone. Agency leadership arrived and departed as though headquarters were Heathrow, and sooner or later, their replacements wouldn't know any better—wouldn't know how headquarters honchos once hated the name—loathed it, really—fought for years to prevent it—and would one day sign what Joan had written right there in black and white: "Europa Clipper."

And once it said Europa Clipper in a signed memo, it would, officially, be called Europa Clipper.[522] Just like that! So let it be written, so let it be done.

Those hundred and fifty faces facing Bob belonged to the project management team and the teams of each of the nine instruments selected for flight. Nine! An actual science payload, contracts signed, budgets written, money spent. 1. An ice-penetrating radar to see what exactly was happening inside of Europa's ice shell, and how it interfaced with the ocean, led by Don Blankenship; 2. a camera suite to image Europa at up to a half-meter resolution; 3. an imaging spectrometer to determine Europa's composition; 4. a magnetometer to figure out Europa's ocean; 5. a plasma instrument that would work in conjunction with the magnetometer in doing ocean science; 6. a thermal imager—a "heat detector" for finding extant and active geologic areas on Europa, such as the eruption sites of plumes; 7. a mass spectrometer to measure Europa's atmosphere and any material ejected therein; and 8. and 9. an ultraviolet spectrometer and a dust detector: two instruments that had not been previously part of the Europa Clipper straw man payload. The latter two instruments were deemed critical by headquarters for flying through prospective plumes and making out their makeup.

In seventeen years, there had been eighty-five people, cumulatively, on six science definition teams: Europa Orbiter through 1999, JIMO through 2004, Europa Explorer in 2007, the Europa Jupiter System Mission through 2010, the Europa Habitability Mission through 2012, and Europa Clipper through 2014. The faces came and went but for Bob, Dave, Don, and Louise. Many were in the room now, but not all of them. Some, like Ron, were there in spirit, but you just felt Ron's crushing absence. If only he could have received those joyful emails, first from Louise and then from Curt, disbanding the science definition team—not because the money was gone or because the study was dead—but because the job was done.[523]

The team members were released so that they might pursue the instrument announcement of opportunity. "This will be the first time in quite a while," Curt had written in farewell, "that NASA is bereft of direct scientific input on Europa mission formulation from the community. . . . This group has faced and overcome significant challenges, among them the sheer duration of the effort as well as the programmatic vagaries that have been repeatedly introduced, and I thank you for your steadfastness and the opportunity to work with you."[524]

Two weeks after the SIP, NASA formally requested fifteen million dollars for Europa in its fiscal year 2015 budget.[525] It was a crack in the dam, and Congressman John Culberson thought that was cute, picked up a sledgehammer, and smashed the sliver into a yawning breach by appropriating—in law—*one hundred million dollars* for Europa mission development. Afterward, as predicted, he was made chair of the subcommittee, and when word went out of that triumph, Bob couldn't help but email some members of the Europa community, "Hang on, it's gonna get fun!"[526]

In January 2015 the ninth floor at NASA headquarters at last committed to the big decisions at the project's acquisition strategy meeting. Culberson had pushed their hands and was clearly not going anywhere, giving the agency assurance that the mission would not soon be abandoned by Congress for some other shiny object. And so it was settled: after all that, the shrinking and splitting and fusing and failing and improvising, Clipper would be a flagship-class, two-billion-dollar (in fiscal year 2015 funds) multiple flyby mission. The effort was directed to Jet Propulsion Laboratory, which would lead its development, in conjunction with the Applied Physics Laboratory.

Europa was made a mission, real and true, on June 17, 2015, with the ratification of Key Decision Point A at NASA headquarters.[527] The meeting itself was highly formal, ceremonial, cheerfully solemn. There were presentations to key stakeholders at the agency on what they were buying, what risks would be incurred, what the Europa multiple flyby mission would learn, and how it would change the course of solar system exploration. After the presentations were given, the room was polled.

—Science Mission Directorate?
—Go.
—Office of the Chief Engineer?
—Go.
—Chief financial officer?
—Go.
—Office of Safety and Mission Assurance?
—Go.
—Office of Legislative Affairs?
—Go.
—Legal counsel?
—Go.
—NASA administrator?
—Go . . .

Afterward, Bob, Barry, Joan, and Curt gathered around the signed Key Decision Point A memo as though it were the Declaration of Independence, took photos of it, took photos *with* it.[528] Eleven signatures later, Bob was no longer a Europa study scientist or the Europa pre-project scientist. He was the project scientist of the Europa mission. Yet another bottle from the good widow Clicquot was uncorked that day.

And now? Now they had to do this thing. There were still so many subtle, specific, detailed decisions to be made. I want to get an image of some section of the surface at a certain resolution—what camera focal length do I need? Do I need a color filter? How close to the surface do I have to be? How slow must we fly so as not to smear that image out? Each element of every little decision affected something else. There was always a trade. During each flyby, the science team wanted every instrument running simultaneously. Could solar power afford it, if the mission went that route? They were great, but my God, those panels were big. They launched folded and had to be deployed. They degraded over time. The mass spectrometer, the particle and magnetic field instruments, the dust detector—they needed an environment of a certain cleanliness—you needed to know that they were measuring Europa and not something radiated by the spacecraft. There were data rate questions: What kind of compression were they going to settle on? (You couldn't forget the lessons of Galileo!) Was the antenna going to be big enough to get those data back to Earth before the next Europa encounter? They had to start figuring out trajectories. Each instrument had its preferred pathway in orbit; some might want the Europan equator, and some might want the poles, and all of that would have to be negotiated across forty-plus flybys. Would those massive radar booms interfere with the field of view of the narrow-angle camera? You can't have a boom in your FOV! They still had to work out how each instrument would react with every other, and with the components of the spacecraft, and how they would interact with the Jovian environment, and how all of those interactions affected the science measurements, and—

Everything had always been notional! This is what we want. They were now, at long, long last, in this-is-what-we-*need* territory.

Six weeks after the mission was made official, the full team, the hundred and fifty, gathered as one for the very first time. Bob had just gotten married, and planning the project science group meeting was a lot like planning his wedding (and included many of the same people). He drew heavily on his brief and rocky tenure as project scientist of Cassini. He knew that the same scientist on two different missions tended to act differently on each, that it was the *mission*'s personality that elicited or stifled certain behaviors, which in turn fed back into the mission, repeating. So it was important to Bob to establish Europa Clipper's demeanor from the first. They were one team.

And since headquarters hadn't yet approved the name of the mission, they needed . . . something. They needed a totem, a symbol. Something around which they could gather, rally. Something to call their own.

And the answer was obvious.

A monolith.[529]

That big black slab of perfect proportions from *2001: A Space Odyssey*. The same mysterious monument that transformed life on Europa in *2010: Odyssey Two*. Five years earlier, Rob Lock, the project system engineer of Europa Explorer and the Jupiter Europa Orbiter, sent Bob an email with a facetious Monolith "action" figure being sold online. ("It has 0 points of articulation, so you don't have to worry about it bending in half or anything. It will help your paperclips, your stapler, even your tape dispenser evolve into sentient life forms. So long as your desk isn't on Europa. 'Cause you gotta leave Europa alone."[530])

So Bob would buy one. Not an action figure (though he did buy one of those), but a life-sized Kubrick-Clarke-beguile-and-evolve-the-caveman monolith. Thankfully, when you need a monolith in a hurry, there is no better place to live than Los Angeles, and you ask friends who work in theater and film. Steve Vance, an astrobiologist at the lab, was married to Sara Fenton, an actress and filmmaker who in turn knew Emiliano "Emi" Rios, a Hollywood production designer.[531] The two discussed it, Bob and Emi, and Emi, it turned out, really knew his monoliths. (If the dimensions weren't exact, after all, it wouldn't be a monolith; it would just be a rectangle.) Emi even researched how Kubrick's original was made.[532] The monolith would be just under 8 feet tall—94.5 inches high, 42 inches wide, and 10.5 inches thick—*the proportions are right, Bob*—made of medium-density fiberboard and painted in low-gloss black with a pewter metallic tint. Aliens couldn't have done a better job.

While that was being built, Bob googled up a 3D-printer mono-lith model and had key chains made. Written on them:

EUR

OPA

PSG1

2015

IT'S

FULL

OF

STARS

. . . the bottom four words being Dave Bowman's final, mysterious message in the novel before being whisked away by some alien force and eventually reborn as a starchild.[533] The words were even engraved in Gill Sans, the same typeface as seen in the film's credits. Bob ordered a cake to match the key chains—solid black icing, and, again, in perfect monolith dimensions—but the bakers refused to cut individual cake slices in 1:4:9 dimensions, so Mabel Young, newly wed to Bob, happily handled confectionary responsibilities.

Look, Bob wasn't throwing a party here. He organized and out-lined to the minute two long days of introductions to Europa, to the Clipper concept, to how changing designs (e.g., going from nuclear to solar) would affect science investigations. They would refine the science traceability matrix and mission science requirements, dis-cuss trajectories, mission operations—it was a crash course in all things ice world science—but there would be years of development before the spacecraft saw space. The earliest it would launch was June 6, 2022. The earliest it would arrive at Jupiter was March 5,

2025. The team stuff, though, could not wait; mattered more, perhaps, than anything at this point in the project.

It had been a long time coming when Bob began the meeting, and he made a proper production of it.[534]

BOB: Welcome, everyone, to Europa PSG number one.

[applause]

Oh, yeah, we ask you to silence your cellphones. We've been working this a real long time.

I appreciate you all being here.

I hope that this meeting is both informational and enjoyable as well. And let's start now.

Standby LX Q1

Standby Video Q2

LX Q1 **GO**
[Health state—black]
LX Video Q2 **GO**
[Video intro]

[Over the sound system: the sunrise fanfare from Richard Strauss's "Also sprach Zarathustra"—famed from 2001: A Space Odyssey. Cccccccccc-Gggggggg-Cccccccccc-EGC-E-flatGCcccccc! Projected on large screen at front of the room, a clip of the film, the sun rising over the Earth. The sun reaches its zenith and . . .]

Standby LX Q3

LX Q3 **GO**
[Spotlight stage left]

[Suddenly, at the front of the room, is an eight-foot MONOLITH.]

The startled room erupted into cheers and applause at their surprising (and quite imposing) mascot—you just wanted to reach out and touch it!—but that wasn't all: Bob was now walking forward holding a femur bone, as seen also in the film. (Where does one get a femur bone? From an eBay storefront called Skeletons and More.[535]) He tossed it toward the screen—the timing was just so—and it was met on the overhead by the femur flung by the ape-man in *2001* . . . only instead of the famous jump cut to an orbital nuclear weapon in space, it cut to Europa Clipper.

There were too many people present for all the scientists there to introduce themselves and say their interests or instruments, or whatever—you'd have a sleeping auditorium by person sixty-three—so Bob found a better solution. He had each investigation team stand up.

Everyone from the radar team stand up . . .

Now everyone from the camera team . . .

. . . and so on, and at the end:

Now *everyone* stand up. That is the last time we are going to introduce ourselves as a member of an instrument team. We are one team. We are the Europa science team.

EUROPA NEVER STOPPED tempting. In 2014 Louise Prockter discovered plate tectonics on the icy moon. It was a spare-time thing. She just saw this section of its ice shell, and it made sense. She met up with a colleague, Simon Kattenhorn, at a conference. He had written the chapter in Bob's book on the tectonics of Europa. She knew he had never worked so hard on anything in his life as he had on that chapter, and he was justifiably proud of it, and he was very emphatic in stating in the text—unambiguously, so there was no doubt on anyone's part—"The upshot is that there is no indication of a terrestrial-like, global plate tectonic system on Europa."[536]

And here was Louise, holding these pictures.

Look at these features, she told him, showing him an image taken years earlier by the spacecraft Galileo. They look like subduction zones on Earth.

Simon's initial reaction was that it was probably a complete coincidence. But then he really, *really* looked at it.

I don't know what's going on there, he said, but it's weird.

They spent the next year corresponding, sending Photoshop-processed images around, trying to work out what was going on with the ice shell. He had a sabbatical coming up and joined her for three months at the Applied Physics Laboratory, where they could figure it out in person. Louise at the time was still leading the Europa science definition team and working as deputy project scientist of the MESSENGER mission to Mercury, and so Simon took the reins on the work. They printed out images of Europa and Simon then literally cut apart its ice shell with scissors, sliding plates around until there was no longer any doubt that large portions of the icy moon's surface were vanishing. And if they were vanishing, there was only one place for them to go: down.

It was exciting. It was terrifying. They developed a litany of lines of evidence. Louise called it "career-ending stuff," because Earth was the only known world to possess plate tectonics, and if they were wrong, the best case involved colleagues calling them deranged and wondering how they got this far without people realizing what lunatics they were.[537]

To find the first non-Earth world with plate tectonics was heady stuff, but more important: if Europa had plate tectonics—if large plates of ice were being pushed beneath other plates—it meant that stuff on Europa's surface was being *pushed into Europa's interior.* The severe radiation environment around Jupiter created oxidants on the surface, which would be necessary for Life as We Know It to take hold in the ocean. But the radiation penetrated only the first ten centimeters of an ice shell twenty kilometers thick. Those oxidants had no way to get into the water—until now. The plates were pushing those oxidants into the ocean; the whole world was a veritable oxidant conveyor belt!

The paper was published in *Nature Geoscience* in 2014 and kept the Europa conversation moving forward.[538] Europa Clipper would provide exponentially higher-resolution images than Galileo and allow for their hypothesis to be tested. The things no one knew for sure about Europa were precisely the reasons that NASA needed to go back.

LOUISE, FROM THE front row of the basement auditorium of Building 321, looked around the room at her fellow members of the Europa mission, and, yeah, this was real. After all this time. Bob was in perfect form—it was a brilliant beginning, the bone, the monolith. There were so many scientists there that day, many of whom had never done anything on Europa before. Maybe related things, but now that everyone was in the room, it felt good, but it felt a little weird, too. Almost like the new faces were intruders in some way—that wasn't the right word—it's just that the core group had been working together for so many years. There was Don Blankenship. He'd been doing this since 1998! She met him on the JIMO study, and they had worked together on every study ever since. All these people—every mission had its own personality, and she was swept briefly with melancholy that this one would soon change. Which was good! The mission would evolve and grow into the future, into what they had all been working toward for fifteen years.

Part of it, I guess—how could anyone ever know what they had gone through? No one would ever know the trials endured to get everyone into this room. I mean everyone knew about the troubles getting a proposal going, but *they weren't there*. They heard updates from the OPAG meetings once a year . . . a couple of slides, and that was it! But it wasn't *their* lives; it was hers, and Don's and Dave's and Bob's. She felt suddenly like the veteran of some horrible war. It had been such a huge part of her career, had taken so much, and would demand so much yet, that even now, it was real—but it wasn't a relief. Until she got those images back from Europa, she might never believe they had done it. Maybe even then she wouldn't believe it. She had internalized the setbacks. And until those data were on Earth, in computers and being processed, so much could still go wrong. For the next . . . ten years? Fifteen? She would never be able to sit back, kick off her shoes, and say, I'm done now. Because she was not. They were not. She did not come this far to only come this far. And this is where it would start to get really, really hard, because now they had to make the difficult decisions. Before, they were trying to move it forward, streamline it, hone it, and sculpt it into this amazing, beautiful, irresistible thing to NASA that was going to go and do unbelievable science at this unbelievable moon. But all that, all the years and tears, it wasn't to get to Europa. It was just to get into this *room*.

Surrounded now by her colleagues in the auditorium, it was good. Now there would be new ideas, better ideas, from some of the best scientists in the world, and they would want to understand why the Europa leadership had made the decisions they had. Suddenly, you were—it was like getting married. You're where you want to be, but now you've got a whole new family to learn about and work with, and some you were going to get along really well with, and some . . . not so much. Everything was different now. It was exciting, though. So exciting, to finally be a real mission. After all this time. It was going to happen. They were going to study Europa, its geomorphology. Its shell. Its ocean.

She had once been to the bottom of the ocean on Earth. Her

initial research at Brown University involved the study of volcanoes on Venus and how they compared with volcanoes on the midocean ridge, mountain chains on our ocean floor. The job required countless hours of mapping sonar data taken from the bottom of the sea and comparing them with radar data from the surface of Venus. One day during a talk, her doctoral advisor, good old Jim Head, let it drop—but casually, because he was nothing if not charming and ever a showman—that one way to study how volcanoes might form in a pressurized environment like you see on Venus is to study their analogues on Earth, and, oh, he had already set it up: they were going to take an Alvin deep-sea submarine to the floor of the Pacific Ocean.

The bottom of the ocean, Louise learned, was like being on another planet, and the dive down was like getting there.[539] They departed from port on the RV *Atlantis,* a research ship with a crane attached, and the Alvin attached to that. Their target was the East Pacific Rise, where met two tectonic plates beneath the surface of the Earth: the Pacific Plate and the North American Plate. They were looking at hyaloclastite flows—something like Washington State's Mount St. Helens, but on the ocean floor. Where magma erupts into the water, it shatters into billions of pieces, and they begin to settle. She was there to observe and drill into one of those hyaloclastite volcanoes.

There were two others on board that day: an oceanographer from Monterey Bay Aquarium, and the pilot. It was the first day that Louise wasn't seasick, thank God, and the way it works is that you get in the submarine, and they lower you into the water, and you just wait for a bit, bobbing around, mostly to make sure you're not claustrophobic or in danger of losing your mind, because there are three people crammed into a few square feet, and you'll soon be surrounded by one hundred eighty-seven quintillion gallons of water one mile below sea level. You're wearing jeans and sweatshirts, because it gets cold down there, but no shoes, because you're not really walking around or anything. You don't drink anything the night before or the day of, because there aren't facilities down there (they really drive home that point in the paperwork), and when you get in the Alvin everyone respects everyone's personal space, limited though it may be.

Then you start to sink. The sunlight can still penetrate the water for the first fifty meters, and it's alien already, fluorescent flecks all over, biota of some sort. You continue sinking for hours, and soon it's pitch black and you're just sitting there, chatting anxiously, and when you reach the bottom of the ocean, the pilot turns on the headlamps, and it is nothing like you expected to see. It's white rolling dunes all around you—more like the desert, really, than anything—and every now and then you see some weird whitish translucent creature creep by. It's not heaving with life, fish everywhere. It's more severe than that, but what you see, suddenly you're playing Twister with your newfound friends to peer from the tiny portholes. And then you're sampling, trying to drill into the hyaloclastite, and having a hard time collecting core samples with your robotic arms. It's lunchtime, and you're eating sandwiches at the bottom of the Earth. Once air runs low and enough is enough, you rise to the surface, sinking in reverse. The pilot let Louise drive the submarine for a bit, but she was rubbish at it. (It looked so easy.)

The bottom of the Europan ocean would likely look nothing like the one on Earth. The fine sandy stuff forming dunes down and along our ocean floor was composed of bits of shell, sand, and sea creature. It's all weathering and decomposition down there. The seafloor of Europa would be rough, more Martian than not. There's no weathering, no ground-up pieces of stuff settling out.

Unless there's life. Then it's anybody's guess.

Acknowledgments

This book could not have been written without the support of the planetary science community, and especially Louise Prockter, Bob Pappalardo, Don Blankenship, and Curt Niebur. Seven years is a long time to answer anyone's questions.

My sincerest thanks go to Geoff Shandler, who acquired and edited this book, working miracles with those blue and red pencils. Thank you for your encouragement, kindness, and craftsmanship.

I am indebted to the team at Custom House—particularly Peter Hubbard, for adopting this book as your own and bringing it to shelves; Ben Steinberg and Kelly Rudolph, who have worked so hard to make it a success; and Molly Gendell, keeper of all knowledge.

Thank you to my agent, Stacia Decker, who made this thing happen, and whose superpower is talking me down from ledges. I am grateful also to Shannon Stirone, for sage advice during composition, and to Kris Gallagher, who has been there from the start.

Lastly, thank you to Kelly, Alexander, and Amelia, for your love and support, and for living with this project for so many years.

Notes

1. K. K. Khurana et al., "Induced Magnetic Fields as Evidence for Subsurface Oceans in Europa and Callisto," *Nature* 395, no. 6704 (1998): 777–80, doi:10.1038/27394.

   Khurana's paper in *Nature* was the first to posit that an odd, unexpected magnetic field found at Europa by the Galileo magnetometer is not intrinsic, or borne from within, but rather is *induced* by Jupiter's magnetic field, in the same way that an airport body scanner induces a magnetic field in the keys you are carrying in your pocket. The only way such a thing could be possible would be for Europa to either be made of copper, which it wasn't, or for there to be a liquid, subsurface ocean on Europa, which, apparently, there was.

2. M. G. Kivelson et al., "Galileo Magnetometer Measurements: A Stronger Case for a Subsurface Ocean at Europa," *Science* 289, no. 5483 (2000): 1340–43, doi:10.1126/science.289.5483.1340.

   Margaret Kivelson, as principal investigator of the magnetometer instrument on Galileo, would later convince the project to fly the spacecraft in a particular orientation in order to test the ocean hypothesis. Because Europa's magnetic field is tilted, by flying on the *other* side of the tilt (versus the trajectory of the initial measurement), the magnetic field would flip if it were induced. They flew on the other side of it. It flipped. The experiment proved successful, and Kivelson thus discovered the second global ocean in the solar system—the first, obviously, being on Earth.

3. R. Pappalardo, telephone interview by author, April 7, 2015.
   See also J. Moore, telephone interview by author, July 17, 2017.

4. G. Vane, email message to R. Pappalardo regarding the lunch conversation at SSES, November 16, 2004.

5. R. Pappalardo, telephone interview by author, October 27, 2017.

6. M. J. Rutherford and P. Papale, "Origin of Basalt Fire-Fountain Eruptions on Earth Versus the Moon," *Geology* 37, no. 3 (2009): 219–22, https://doi.org/10.1130/G25402A.1.

7. W. N. Charman and C. M. Rowlands, "Visual Sensations Produced by Cosmic Ray Muons," *Nature* 232, no. 5312 (1971): 574–75, https://doi.org/10.1038/232574a0.

8. R. Pappalardo, telephone interview by author, April 8, 2015.

9. J. Achenbach, "NASA's 1976 Viking Mission to Mars Did All That Was Hoped for It—Except Find Martians," *Washington Post*, June 18, 2016, https://www.washingtonpost.com/national/health-science/nasas-1976-viking-mission-to-mars-did-everything-right—except-find-martians/2016/06/18/749701f6-2c15-11e6-9b37-42985f6a265c_story.html.

10. R. Pappalardo, interview by author, March 7, 2017.

11. "Longs Peak—Keyhole Route," National Park Service, last modified June 5, 2018, https://www.nps.gov/romo/planyourvisit/longspeak.htm.

12. There was a real art to helping a grad student reach his or her potential in planetary science. A student comes along and says, I am interested in Europa, and Bob says, OK, why? and the response would be his guide. Here is a possible project. Let's see what evolves. Or a student would have a particularly strong skill set (or an interest in cultivating said skills) and that could lead to interesting things, too. He might be at his desk during office hours and receive an email from the Canadian Geological Survey. Hey, Bob, we wrote this paper on the Canadian high Arctic that might be of interest to you. (Frozen parts of Earth being a good place to study frozen worlds less accessible.) A year later a grad student materializes and says, Dr. Pappalardo, I am interested in Europa, and

Bob says, What are you interested in, exactly? and the student says, Well, I'm interested in Earth analogs, and I'm taking a remote sensing course now and I like that. Well, says Bob, there's this weird site in the Canadian Arctic that might be a good analog for Europa—why don't you look at the paper and tell me what you think from a remote sensing standpoint. The student goes on to research, say, the deposits of sulfur present in the Arctic site, and gathers existing satellite data, and one thing leads to another. Bob is writing grants now, to get her out there to do some field geology, and linking her up with the Jet Propulsion Laboratory people who can actually take those remote sensing spectra from satellites and who are working on identifying sulfur-rich deposits from orbit autonomously.

A student comes in and says, I'm interested in Europa, and I'm also interested in programming. Bob thinks for a moment and says, Well, OK, maybe you can write a program to map the structural features on Europa to compare them to the stress patterns predicted from existing models. There's a group at the University of Arizona doing it, but they shouldn't be the only ones in the universe who can calculate Europa's surface stresses, right? Let's see if we can duplicate that model. Once the project is under way, Bob mentions it to a geophysicist colleague who says, THAT IS NOT HOW YOU DO THIS! VISCOSITY IS CRITICAL! IT'S NOT JUST AN ELASTIC PROBLEM! (this is how geophysicists talk) and the student and the geophysicist link up and develop something better than any previous model to come before.

That is how science is done.

13. R. Pappalardo, telephone interview by author, December 29, 2017.

14. J. W. Wood, *Pasadena, California, Historical and Personal; A Complete History of the Organization of the Indiana Colony, Its Establishment on the Rancho San Pascual and Its Evolution into the City of Pasadena Including a Brief Story of San Gabriel Mission, the Story of the Boom and Its Aftermath, and of the Political Changes and Personages Involved in This Transformation: Churches, Societies, Homes, Etc.* (Pasadena, CA: s.p., 1917).

This is a remarkable account of how Pasadena came to be, worth reading for its historical scholarship as well as its literary merits. Concludes the author, J. W. Wood, of his beloved city: "But of the future! It may require no prophetic vision to see it. Invention and genius, well applied, will confer their magic, and we can in our horoscope, discern clearly a rehabilitation that will give to this community a new fame." How right he was! The book is now in the public domain and may be downloaded in its entirety from Google Books. And in case you are wondering: yes, the title of the Wood book—about a Pasadena mission—inspired the subtitle of this book, about another Pasadena mission.

15. The settlers from the northeastern United States arrived on horseback and in carriages, buggies, and wagons drawn thereby, and they sought a new life—amazing that nobody had settled this land before!—and in 1874 established the Indiana Colony on land that had long been parceled from the dominion of the Mission San Gabriel Arcángel. It would be work to make this thing a success, Indiana Colony. Nearby Los Angeles, after all, wasn't much to look at, crude and bedraggled, but my God, this California sun! Have you ever seen anything like it? The colony did well enough that first year, and when at last it petitioned for a post office, the postmaster general cast eye across application and returned it rejected. What kind of name was "Indiana Colony"? You didn't arrive by *Mayflower,* people; try again. So the settlers settled on the name "Pasadena," on recommendation of a friend of a friend, a missionary working with the Chippewa in the Mississippi Delta. We want something that sounds really *Indian,* you know? Something that means "entrance to the valley" or "crown of the valley." Something like that. And the missionary, his hands mostly tied because, I mean, come on, returned the requested translations: Weoquanpasadena or Tapedaegunpasadena. But that wasn't really what the villagers of Indiana Colony had in mind, though the "pasadena" part was quite lovely, wasn't it? Which translated from the Chippewa language more literally as "this is a valley," but I mean it's not like any Chippewa *lived* in California, so who would know? Pasadena, California it was.

16. S. O'Keefe, telephone interview by author, May 31, 2017.

17. National Aeronautics and Space Administration (abbreviated NASA from this point forward), "New Horizons—The First Mission to Pluto and the Kuiper Belt: Exploring Frontier Worlds," press kit, January 2006, https://www.nasa.gov/pdf/139889main_PressKit12_05.pdf.

18. R. Taylor, *Prometheus Project Final Report* (982-R120461) (Pasadena, CA: NASA, Jet Propulsion Laboratory, California Institute of Technology [abbreviated JPL, CIT from this point forward], October 1, 2005), 112, https://trs.jpl.nasa.gov/bitstream/handle/2014/38185/05-3441.pdf?sequence=1&isAllowed=y.

See also M. J. Wollman and M. J. Zika, *Prometheus Project Reactor Module Final Report, for Naval Reactors Information* (Washington, DC: U.S. Department of Energy, 2006), https://inis.iaea.org/collection/NCLCollectionStore/_Public/37/107/37107481.pdf?r=1&r=1.

Much of the information about JIMO and Project Prometheus in this book derives from the final report, as well as from multiple interviews with members of the JIMO science and engineering teams, and from Curt Niebur and Sean O'Keefe at NASA headquarters.

19. National Research Council, *Priorities in Space Science Enabled by Nuclear Power and Propulsion* (Washington, DC: National Academies Press, 2006), 45, https//doi.org/10.17226/11432.

Mike Griffin, the then-incoming NASA administrator, placed the number at "eleven billion dollars and counting" during testimony before the Science, Space, and Technology Committee of the House of Representatives on June 28, 2005. A billion dollars is quite a difference, and NASA is notorious for those sorts of varying figures everywhere you look, in part because projects are sometimes developed across divisions and centers, and sometimes include launch costs or technology development. Pinning down the price of anything done by the agency is more an art than a science. Every project has stakeholders at headquarters, the White House Office of Management and Budget, the House, the Senate, NASA centers, and aerospace companies. Sometimes, numbers get fumbled. In every case, I have attempted to find the most reasonable project cost estimate available from the most reliable sources.

20. NASA, Johnson Space Center, "Space Shuttle Mission STS-31," press kit, April 1990, 37.

See also National Security and International Affairs Division, *NASA Program Costs: Space Missions Require Substantially More Funding Than Initially Estimated—Report to the Chairman, Subcommittee on Investigations and Oversight, Committee on Science, Space, and Technology, House of Representatives* (Washington, DC: U.S. General Accounting Office, December 1992), 8–9, http://archive.gao.gov/d36t11/148471.pdf.

See also P. K. Martin, *NASA Cost and Schedule Overruns: Acquisitions and Program Management Challenges* (Washington, DC: NASA Office of Inspector General, June 14, 2018), 1, https://oig.nasa.gov/docs/CT-18-002.pdf.

See also *James Webb Space Telescope Independent Comprehensive Review Panel Final Report* (JPL D-67250) (Pasadena, CA: JPL, CIT,

October 29, 2010), 30, https://cdn.theatlantic.com/static/mt/assets
/science/499224main_JWST-ICRP_Report-FINAL.pdf.

NASA's Hubble cost estimates are always all over the place. The
NASA press kit from the spacecraft launch claimed that the Hubble Space Telescope itself came in at $1.5 billion, with another "$300
million for the science and engineering operations which have been
supporting both the spacecraft development and the ground science
operations at Goddard and the Space Telescope Science Institute,
and $300 million for the design, development and testing of servicing
equipment to maintain the Telescope's 15-year expected lifetime." So:
$2.1 billion in 1990 dollars, when the telescope launched. But a 2010
independent comprehensive review panel for the James Webb Space
Telescope claimed that Hubble cost $2.8 billion (after adjustment to
1990 dollars) without its costs broken into various elements. A 2018
NASA inspector general report claimed that Hubble cost $1.2 billion,
presumably for the spacecraft and presumably in 1990 dollars. A 1993
report from the Government Accountability Office (GAO) agreed that
the Hubble spacecraft itself (in other words, sans mission operations
costs and data analysis) cost about $1.5 billion in 1990 dollars. Ulti-
mately, the "a third of that" assertion went with the press kit, whose
foundational dollar figure was confirmed by the GAO. But see what I
mean about costs?

21. National Security and International Affairs Division, *NASA Program
     Costs*, 8–9.

     See also "Space Shuttle and International Space Station," NASA
     Kennedy Space Center, last modified August 3, 2017, https://www
     .nasa.gov/centers/kennedy/about/information/shuttle_faq.html.

22. "Weather History Results for Washington, DC (20546), October 22,
     2004," *Farmers' Almanac,* accessed June 4, 2019, https://www.weather
     .org/weather-history.

     There was a high of 57.2 degrees F, with 8-knot winds. It drizzled
     that day, too, but at 0.01 inches, not enough to be worth mentioning.

23. R. T. Pappalardo, W. B. McKinnon, and K. K. Khurana, "Europa: Per-
     spectives on an Ocean World," in *Europa* (Tucson: University of Ari-
     zona Press, 2009), 702–3.

     For an alternate hypothesis on the thickness of the Europan ice
     shell—and, oh my goodness, is *this* a bitter rivalry!—see also the ex-

cellent R. Greenberg, *Unmasking Europa: The Search for Life on Jupiter's Ocean Moon* (New York: Copernicus, 2008).

For the deepest hole ever drilled, see also A. Kröner, "The Superdeep Well of the Kola Peninsula," *Precambrian Research* 42, nos. 1–2 (1988): 208–10, doi:10.1016/0301-9268(88)90019-8.

The exact thickness of the Europan ice shell is unknown, which is one reason planetary scientists are so determined to get a spacecraft to Europa. Multiple lines of evidence (and pretty much every scientist featured in this book) suggest a thickness greater than fifteen kilometers. For the "thin ice" argument, check out Rick Greenberg's book *Unmasking Europa*, in which he also has some strong words to say about the scientists in this book. I'm not able to argue either way for the geophysics of Europa, but I'm definitely right about the people described in these pages.

24. National Research Council, *A Science Strategy for the Exploration of Europa* (Washington, DC: National Academies Press, 1999), https://doi.org/10.17226/9451.

25. C. Niebur, *Investments in Europa Mission Concept Studies,* September 16, 2009. This document was written in response to a request from the 2013–2022 Planetary Science Decadal Survey steering committee.

26. E. Weiler, interview by author, September 4, 2017.

27. K. Batygin, telephone interview by author, February 22, 2018.

This is the simple version of the story of everything, and I am grateful to Konstantin Batygin of Caltech for helping to build the solar system. Thank you also to author and astrophysicist Adam Becker for fact-checking me through the formation of the sun and planetary scientist Kirby Runyon of the Applied Physics Laboratory for getting me from there to the formation of the Earth. Additional thanks go to Alfred Nash of Jet Propulsion Laboratory. Ever want to feel humble? Try to read even the simplest astrophysics paper.

For the basic narrative of the beginning of the universe, see also the following sources: E. M. Burbidge et al., "Synthesis of the Elements in Stars," *Reviews of Modern Physics* 29, no. 4 (1957): 547–650, https://doi.org/10.1103/RevModPhys.29.547.

See also J. Pandian, "How Are Light and Heavy Elements Formed (Advanced)?," Ask an Astronomer (Astronomy Department, Cornell University), last modified June 27, 2015, curious.astro.cornell

.edu/about-us/84-the-universe/stars-and-star-clusters/nuclear-burn
ing/402-how-are-light-and-heavy-elements-formed-advanced.

See also C. Palma, "Nuclear Fusion in Protostars," Astronomy 801:
Planets, Stars, Galaxies, and the Universe, Penn State College of Earth
and Mineral Supplies, accessed October 14, 2019, https://www.e-edu
cation.psu.edu/astro801/content/l5_p4.html.

See also A. H. Guth, "Inflationary Universe: A Possible Solution to
the Horizon and Flatness Problems," *Physical Review* 23, no. 2 (1981):
347–56, https://doi.org/10.1103/PhysRevD.23.347.

See also A. D. Linde, "A New Inflationary Universe Scenario: A
Possible Solution of the Horizon, Flatness, Homogeneity, Isotropy and
Primordial Monopole Problems," in *Quantum Gravity*, ed. M. A. Mar-
kov and P. C. West (New York: Plenum Press, 1984), 185–95, https://
doi.org/10.1007/978-1-4613-2701-1_13.

See also A. Albrecht and P. J. Steinhardt, "Cosmology for Grand
Unified Theories with Radiatively Induced Symmetry Breaking," *Phys-
ical Review Letters* 48, no. 17 (1982): 1220–23, https://doi.org/10.1103
/PhysRevLett.48.1220.

See also J. Pandian, "Could the Universe Have Expanded Faster Than
the Speed of Light at the Big Bang?," Ask an Astronomer (Astronomy
Department, Cornell University), last modified June 27, 2015, http://
curious.astro.cornell.edu/physics/104-the-universe/cosmology-and-the
-big-bang/expansion-of-the-universe/627-could-the-universe-have-ex
panded-faster-than-the-speed-of-light-at-the-big-bang-intermediat.

See also P. Sutter, "How Can the Universe Expand Faster Than the
Speed of Light?," Space.com, last modified July 2, 2016, https://www
.space.com/33306-how-does-the-universe-expand-faster-than-light
.html.

See also K. Spekkens, "How Can the Universe Expand Faster Than
the Speed of Light During Inflation?," Ask an Astronomer (Astronomy Department, Cornell University), last modified June 27, 2015,
http://curious.astro.cornell.edu/physics/109-the-universe/cosmology
-and-the-big-bang/inflation/664-how-can-the-universe-expand-fast
er-than-the-speed-of-light-during-inflation-advanced.

See also M. Strassler, "Inflation," Of Particular Significance, last modified March 17, 2017, https://profmattstrassler.com/articles-and-posts
/relativity-space-astronomy-and-cosmology/history-of-the-universe
/inflation.

See also K. Tate, "Cosmic Inflation: How It Gave the Universe the Ultimate Kickstart," Space.com, last modified March 17, 2014, https://www.space.com/25075-cosmic-inflation-universe-expan sion-big-bang-infographic.html.

See also "The Origins of the Universe: Inflation," University of Cambridge, Stephen Hawking Centre for Theoretical Cosmology online, accessed October 14, 2019, http://www.ctc.cam.ac.uk/outreach/ori gins/inflation_zero.php.

28. W. Johnson and R. Riegel, "Where Did All the Elements Come From?," MIT Haystack Observatory, accessed October 14, 2019, https://www .haystack.mit.edu/edu/pcr/Astrochemistry/3%20-%20MATTER/nu clear%20synthesis.pdf.

29. About the phrase "light as we see it didn't exist yet": Those fresh and free electrons turning heads were doing so in a very real way for primordial photons, scattering them every which way. If you could go back to this period in the history of the universe, it would all appear as a sort of hot, glowing, opaque mass, like looking at the surface of a star (but way, way hotter). Consequently, even the most powerful telescopes cannot see earlier than the Big Bang plus about three hundred eighty thousand years. Once basic atoms formed, however, those photons could travel without hindrance, and the universe became "transparent." Is the phrase "light as we see it didn't exist yet" literary sleight of hand? Yes, because, of course, *light existed*. But I hope sympathetic astrophysicists would grant me a little artistic license here.

30. F. Bagenal, email message to author regarding Jupiter magnetosphere, September 8, 2018.

See also F. Bagenal and S. Bartlett, "Magnetospheres of the Outer Planets Group—Graphics," University of Colorado Boulder, Laboratory for Atmospheric and Space Physics, accessed October 11, 2019, http://lasp.colorado.edu/home/mop/resources/graphics/graphics.

31. J. Sustermans, *Galileo Galilei (1564–1642)*, circa 1640, oil on canvas, 867 mm x 686 mm, Royal Museums Greenwich, accessed May 21, 2019, https://collections.rmg.co.uk/collections/objects/14174.html.

32. G. Consolmagno and P. Mueller, *Would You Baptize an Extraterrestrial? . . . And Other Strange Questions from the Astronomers' In-box at the Vatican Observatory* (New York: Image, 2014).

This is an extraordinary work by two legends in the field of space science. Though the title suggests a focus on aliens, the book is more accurately an examination of how science and religion complement each other, and the role of the Roman Catholic Church in the history of astronomy. The authors discuss at length "the Galileo affair," and common misconceptions. On that subject, the inescapable conclusion is that Galileo was not imprisoned for locking horns with Church dogma. Rather, he was imprisoned because he was just a real jerk about the whole thing, publicly daring the pope to push back—and lost.

33. T. L. Heath, ed., *The Works of Archimedes* (London: Cambridge University Press, 1897), 221–22.

   Accounts of Aristarchus explaining his model of heliocentrism were conveyed by the Greek mathematician and inventor Archimedes in *The Sand Reckoner*. Aristarchus's work is largely lost to time, possibly due to the annihilation of the Library of Alexandria.

34. J. M. Pasachoff and A. Van Helden, "Simon Marius vs. Galileo: Who First Saw Moons of Jupiter (Abstract)," in *Abstract Book, DPS 48 / EPSC 11* (Washington, DC: American Astronomical Society, Division for Planetary Sciences, 2016), 172.

35. D. Lauretta, telephone interview by author, March 5, 2018.

36. R. M. Nelson et al., "Laboratory Simulations of Planetary Surfaces: Understanding Regolith Physical Properties from Remote Photo-polarimetric Observations," *Icarus* 302 (2018): 483–98, https://doi.org/10.1016/j.icarus.2017.11.021.

37. R. Greenberg, "Tides and the Biosphere of Europa," *American Scientist* 90, no. 1 (2002): 48, https://doi.org/10.1511/2002.1.48.

38. H. Melosh et al., "Is Europa's Subsurface Water Ocean Warm?" (presentation, Thirty-Third Lunar and Planetary Science Conference, League City, TX, March 2002), https://www.lpi.usra.edu/meetings/lpsc2002/pdf/1824.pdf.

39. K. M. Soderlund et al., "Ocean-Driven Heating of Europa's Icy Shell at Low Latitudes," *Nature Geoscience* 7 (2014): 16.

   See also K. Soderlund, telephone interview by author, February 2, 2018.

40. J. Green, interview by author, May 6, 2017.

   See also P. K. Byrne et al., "Limited Prospect for Geological Activity at the Seafloors of Europa, Titan, and Ganymede; Enceladus OK (Abstract P21E-3385)," in *American Geophysical Union, Fall Meeting 2018* (Washington, DC: American Geophysical Union, December 2018).

This abstract presents an alternate view to the Io-at-the-seafloor hypothesis.

41. A. C. Clarke, *2010: Odyssey Two* (New York: Ballantine Books, 2001).

42. Ibid.

43. "NASA Anniversary: Apollo 11 Moon Landing, July 20, 1969," NASA, last modified July 20, 2010, https://www.nasa.gov/news/media/audio file/Apollo_11_Moon_Landing.html.

See also "Apollo 11 Mission Logs, July 20, 1969," NASA, accessed May 22, 2019, https://spaceflight.nasa.gov/history/apollo/apollo11 /july20.htm.

Much as Hamlet might have said "Oh, that this too, too sullied flesh would melt . . ." or maybe "solid flesh," or maybe "sallied flesh," depending on the folio, the Apollo 11 transcripts are in a constant state of historical reevaluation. For example, did Neil Armstrong say "One small step for a man" or "One small step for man"? He insists he said "a man," and that the "a" was lost in transmission. In this case, *pick* and *kick* are both possible messages in that second sentence he spoke on the lunar surface. Don't @ me.

44. After taking the job at Jet Propulsion Laboratory, Bob lived that first year and a half in a rented house in Altadena, and the landlord lived next door, and his first summer there it broiled outside, and the landlord explained that sure, this summer was warm, but this was unusual weather, Bob! I mean one hundred plus degrees? Not here in Altadena, not ordinarily, no siree! Not every day for a month like we're seeing now, I mean! So Bob stuck it out, and the next scorching summer it hit one hundred again and that, too, lasted a month, and Bob, not at all cool with this, got in his car and drove west, one eye on the road, one eye on the dashboard thermostat. In the Eagle Rock/Burbank area, it dropped to ninety-two. By the time he hit the 405, it was eighty-five. At Eleventh Street in Santa Monica, it hit seventy-two, and he was resolved: I will live west of Eleventh. He found a bohemian bungalow, Venice beau ideal, just before the real estate boom and not far from the canals or the beach. It got cool in the evenings and was about an hour's drive from the lab each way, but the Pacific and ten million miles

separated the stress of work from Venice Zen. The horizon—you could just see Malibu, hazy along the shoreline—and those thirty degrees kept him going.

45. R. Pappalardo, telephone interview by author, October 27, 2017.

46. R. Pappalardo, "Meeting Goals & Philosophy" (PowerPoint presentation, Europa Project Science Group Meeting #1, JPL, Pasadena, CA, August 2015).

47. "The Mission to Mercury: Mission Design," Johns Hopkins University Applied Physics Laboratory (abbreviated JHUAPL from this point forward), *MESSENGER*, accessed December 26, 2019, http://messenger .jhuapl.edu/the_mission/mission_design.html.

    NASA as an institution seems criminally incapable of preserving its hyperlinks, and thus hell-bent on burying the extraordinary digital record of its missions and activities. Searching the website of the European Space Agency, on the other hand, is like strapping on a time machine. Its gloriously maintained hyperlinks will survive the heat death of the universe.

48. NASA, JHUAPL, and Carnegie Institution of Washington, *Twins Image* (spacecraft image), NASA Image and Video Library, last modified August 2, 2005, https://images.nasa.gov/details-PIA10122.html.

49. NASA, JHUAPL, and Carnegie Institution of Washington, *Galapagos Islands (*spacecraft image), NASA Image and Video Library, last modified August 2, 2005, https://images.nasa.gov/details-PIA10121.html.

50. L. Prockter, email message to C. Niebur regarding SDT membership application, December 21, 2006.

51. P. H. Schultz, "Atmospheric Effects on Ejecta Emplacement and Crater Formation on Venus from Magellan," *Journal of Geophysical Research* 97, no. e10 (1992): 161–83, https://doi.org/10.1029/92JE01508.

52. L. Prockter, interview by author, December 12, 2017.

53. J. W. Head, interview by R. Wright, June 6, 2002, NASA Johnson Space Center Oral History Project, Edited Oral History Transcript, last modified July 16, 2020, https://historycollection.jsc.nasa.gov/JSCHis toryPortal/history/oral_histories/HeadJW/HeadJW_6-6-02.htm.

54. M. Johnson, "The Galileo High Gain Antenna Deployment Anomaly" (presentation, Twenty-Eighth Aerospace Mechanisms Symposium, Cleveland, 1994), https://ntrs.nasa.gov/search.jsp?R=19940028813.

55. W. O'Neil et al., "Performing the Galileo Jupiter Mission with the Low Gain Antenna (LGA)" (presentation, International Astronautical Federation Space Exploration, Graz, Austria, 1993), https://ntrs.nasa.gov/search.jsp?R=20060039129.

56. J. Taylor, K. Cheung, and D. Seo, *DESCANSO Design and Performance Summary Series, Article 5: Galileo Telecommunications* (Pasadena, CA: Deep Space Communications and Navigation Systems Center of Excellence, NASA JPL, CIT, July 2002), https://descanso.jpl.nasa.gov/DPSummary/Descanso5--Galileo_new.pdf.

57. Ibid.

58. D. J. Mudgway, *Uplink-Downlink: A History of the NASA Deep Space Network, 1957–1997*, NASA History Series (Washington, DC: NASA, 2001), https://ntrs.nasa.gov/archive/nasa/casi.ntrs.nasa.gov/20020033033.pdf.

But don't feel like you have to go to the library or bookstore to get this one. The seven-hundred-twenty-two-page book is available for free at the above URL. I learned this only after buying a copy, so you're welcome, Mudgway.

59. NASA, JPL, "Galileo Spacecraft Anomaly Being Investigated," press release, October 12, 1995, https://nssdc.gsfc.nasa.gov/planetary/text/gal_tape.txt.

Galileo is the Apollo 13 of outer planets flagship missions. So many things went wrong, and the Rube Goldberg–like rescue of its data storage and communications capacities make it one of the greatest achievements of the space age. Somebody should write a book about it, but not me.

See also M. R. Johnson and G. C. Levanas, "The Galileo Tape Recorder Rewind Operation Anomaly" (presentation, Thirty-First Aerospace Mechanisms Symposium, Huntsville, AL, 1994), https://ntrs.nasa.gov/search.jsp?R=19970021630.

60. Associated Press, "Data Recorder Malfunctions in Spacecraft," *New York Times*, October 13, 1995, A16.

61. K. Sawyer, "Rewind Almost Unhinges Galileo Mission," *Washington Post*, October 29, 1995, https://www.washingtonpost.com/archive/poli

tics/1995/10/29/rewind-almost-unhinges-galileo-mission/0d550514-4
fa5-49b7-98d8-4011d13c6dc6/?utm_term=.37724af19a33.

62. NASA, JPL, "Galileo on Track After Tape Recorder Recovery," press release, October 26, 1995, https://nssdc.gsfc.nasa.gov/planetary/text/gal_tape_3.txt.

63. Khurana, "Induced Magnetic Fields," 777–80.

64. N. Paulter, *Users' Guide for Hand-Held and Walk-Through Metal Detectors* (Washington, DC: National Institute of Justice, Office of Science and Technology, January 2001), https://www.ncjrs.gov/pdffiles1/nij/184433.pdf.
    See also K. Khurana, interview by author, August 27, 2018.

65. Kivelson et al., "Galileo Magnetometer Measurements," 1340–43.

66. C. Niebur, email message to L. Prockter regarding SDT membership application, December 22, 2006.

67. L. Prockter, email message to C. Niebur regarding SDT membership application, December 22, 2006.

68. J. Morse, *Pathological Aspects of Religion*, vol. 1 (Worcester, MA: Clark University Press, 1906), 259.
    Add this to the list of works I didn't expect to read while researching this book.

69. M. J. Neufeld, "Transforming Solar System Exploration: The Origins of the Discovery Program, 1989–1993," *Space Policy* 30, no. 1 (2014): 5–12, https://doi.org/10.1016/j.spacepol.2013.10.002.

70. N. Panagakos and F. Bristow, NASA Office of Public Affairs, "Voyager 1: Saturn Encounter, Summary of Events," news release no. 80-145, September 1980, https://ntrs.nasa.gov/archive/nasa/casi.ntrs.nasa.gov/19810066669.pdf.
    The timeline of encounter events can be found on pages 5–8.

71. R. D. Launius, "Public Opinion Polls and Perceptions of U.S. Human Spaceflight," *Space Policy* 19, no. 3 (2003): 163–75, https://doi.org/10.1016/S0265-9646(03)00039-0.

72. *Declassifying "Fact of" National Reconnaissance Office's Use of the Space Shuttle as a Launch Vehicle: A Policy Decision Risk Assessment* (Chantilly,

VA: Center for the Study of National Reconnaissance, Office of Policy, National Reconnaissance Office, July 2001), accessed June 14, 2019, https://www.scribd.com/document/348134338/Declassifying-the-Fact-of-the-NRO-s-Use-of-the-Space-Shuttle-as-a-Launch-Vehicle.

Thank you to the inimitable journalist Matt Novak for filing a Freedom of Information Act request for this document and sharing it publicly.

73. B. Hendrickx, "Kidnapping a Soviet Space Station," Space Review, last modified July 14, 2014, http://www.thespacereview.com/article/2554/1.

74. "ESA Takes Its Anger to Washington," *New Scientist* 89, no. 1244 (March 12, 1981): 661.

75. J. M. Logsdon, *Ronald Reagan and the Space Frontier* (Cham, Switz.: Palgrave Macmillan, 2019), 30.

See also J. Schefter, "It's Time to Stop the Space Retreat," *Popular Science* 221, no. 4 (October 1982): 32–45. So many of the Reagan administration's early attempts to derail space science are enumerated in this eye-opening piece.

76. J. M. Logsdon, "The Survival Crisis of the U.S. Solar System Exploration Program in the 1980s," in *Exploring the Solar System: The History and Science of Planetary Exploration,* ed. R. D. Launius (New York: Palgrave Macmillan, 2013), 45–76, https://doi.org/10.1057/9781137273178_3.

This chapter—indeed, anything by John Logsdon—is required reading for anyone interested in how the American space program works.

77. T. O'Toole, "NASA Weighs Abandoning Voyager," *Washington Post,* October 7, 1981, https://www.washingtonpost.com/archive/politics/1981/10/07/nasa-weighs-abandoning-voyager/a3df075c-6f7b-4573-8956-d3852c7fe59a.

78. P. Ulivi and D. M. Harland, *Robotic Exploration of the Solar System,* part 2: *Hiatus and Renewal, 1983–1996* (New York: Springer, 2007).

79. O'Toole, "NASA Weighs Abandoning Voyager."

80. S. Crow, "The Viking Fund," *Icon VI* (June 1981), 17.

This program from the 1981 ICON science-fiction convention in Iowa, manually typed and mimeographed, is especially fun in that it includes a brief article by an unknown, early-career author named George R. R. Martin.

81. "Mariner 6 and 7," Arizona State University, Space Exploration Resources, accessed October 10, 2019, http://ser.sese.asu.edu/M67/mar67 .html.

82. W. von Braun and C. Ryan, "Can We Get to Mars?," *Collier's Weekly*, April 30, 1954, 22–29.

83. R. D. Launius, "Project Apollo: A Retrospective Analysis," NASA History Division, last modified April 21, 2014, https://history.nasa.gov /Apollomon/Apollo.html.

84. M. Wright, "The Disney–Von Braun Collaboration and Its Influence on Space Exploration," NASA's Marshall Space Flight Center History, last modified August 3, 2017, http://www.nasa.gov/centers/marshall /history/vonbraun/disney_article.html.

85. W. von Braun, "Manned Mars Landing" (presentation, Space Task Group, Washington, DC, August 4, 1969), https://www.nasa.gov/sites /default/files/atoms/files/19690804_manned_mars_landing_presenta tion_to_the_space_task_group_by_dr._wernher_von_braun.pdf.

86. Roper Center for Public Opinion Research (blog), last modified March 10, 2015, https://ropercenter.cornell.edu/blog/fly-me-moon-public-and -nasa-blog.
    See also T. W. Smith, "General Social Survey Final Report: Trends in National Spending Priorities, 1973–2014" (presentation, University of Chicago, March 2015), http://www.norc.org/PDFs/GSS%20Re ports/GSS_Trends%20in%20Spending_1973-2014.pdf.

87. NASA, "Viking," press kit, February 1975, https://mars.nasa.gov/mro /odyssey/newsroom/presskits/viking.pdf.

88. Achenbach, "NASA's 1976 Viking Mission to Mars."

89. V. K. McElheny, "Viking Team, Pressed to Supply News, Sees Itself Practicing 'Instant Science,'" *New York Times*, August 18, 1976, https:// www.nytimes.com/1976/08/18/archives/viking-team-pressed-to-sup ply-news-sees-itself-practicing-instant.html.

90. J. N. Wilford, "Mars Chemistry Still Puzzles Scientists," *New York Times*, August 3, 1976, https://www.nytimes.com/1976/08/03/ar chives/mars-chemistry-still-puzzles-scientists.html.

91. W. Sullivan, "Nitrogen, Key to Life, Is Found," *New York Times*, July 21, 1976, https://www.nytimes.com/1976/07/21/archives/nitrogen-key-to -life-is-found.html.

92. V. K. McElheny, "Hunt for Evidence of Life on Mars Is Still a Puzzle," *New York Times*, August 11, 1976, https://www.nytimes .com/1976/08/11/archives/hunt-for-evidence-of-life-on-mars-is-still-a -puzzle.html.

93. "Viking Experiments in Biology on Mars Called Inconclusive," *New York Times*, September 1, 1976, https://www.nytimes.com/1976/09/01 /archives/viking-experiments-in-biology-on-mars-called-inconclusive .html.

94. J. N. Wilford, "New Viking 2 Data Called 'Marginally Positive,' but Scientists Are Still Unsure About Mars Life Tests," *New York Times*, September 24, 1976, https://www.nytimes.com/1976/09/24/archives /new-viking-2-data-called-marginally-positive-but-scientists-are.html.

95. V. K. McElheny, "Technology Hopes for a Mobile *Viking* on Mars," *New York Times*, November 24, 1976, https://www.nytimes.com/1976/11/24 /archives/technology-hopes-for-a-mobile-viking-on-mars-technology -a-chance-to.html.

96. J. N. Wilford, "Tests by Viking 2 Keep Alive Belief in Biological Activity," *New York Times*, September 17, 1976, https://www.nytimes .com/1976/09/17/archives/tests-by-viking-2-keep-alive-belief-in-bio logical-activity.html.

97. "Life on Mars? . . ." *New York Times*, September 20, 1976, https://www .nytimes.com/1976/09/20/archives/life-on-mars.html.

98. J. N. Wilford, "Viking 2 Test Finds No Organic Matter," *New York Times*, October 1, 1976, https://www.nytimes.com/1976/10/01/ar chives/viking-2-test-finds-no-organic-matter-results-termed-prelimi nary.html.

99. G. Williams, "Viking and Friends," *Popular Mechanics*, December 1980, 154.

100. W. H. Patterson and R. A. Heinlein: *In Dialogue with His Century*, vol. 2: *The Man Who Learned Better, 1948–1988*, 1st ed. (New York: Tor Books, 2014), 436.

101. "Passing the Hat for Space," *Boston Globe*, June 28, 1981, 16.

102. M. Toner, "Private Citizens Kick in Money to Keep Space Programs Going," *Wisconsin State Journal* (Madison, WI), May 28, 1981, 3.

103. H. Mark, "New Enterprises in Space," *Bulletin of the American Academy of Arts and Sciences* 28 no. 4 (1975): 14–26, doi:10.2307/3823240.

    Hoooo boy, what a piece! When the incoming deputy administrator of a NASA center says, Unfortunately, the results of space science to date have not been of major significance, you can bet that a lot of scientists need to start polishing their résumés, because this space business is on borrowed time.

104. "Airplane Types," *Travel Insider* (blog), accessed October 10, 2019, https://blog.thetravelinsider.info/airplane-types. An early configuration of the DC-10 could hold everyone.

105. H. Finley, *Space Exploration: Cost, Schedule, and Performance of NASA's Mars Observer Mission—Fact Sheet for the Chairman, Subcommittee on Science, Technology, and Space, Committee on Commerce, Science, and Transportation, U.S. Senate* (Washington, DC: U.S. General Accounting Office, National Security and International Affairs, May 27, 1988), https://www.gao.gov/products/nsiad-88-137fs.

106. E. M. Conway, *Exploration and Engineering: The Jet Propulsion Laboratory and the Quest for Mars* (Baltimore: Johns Hopkins University Press, 2015). This is perhaps the most comprehensive and engaging book out there on how Jet Propulsion Laboratory built a Mars program.

107. A. F. AbuTaha, *The Problem with the Space Shuttle and the Space Program* (Washington, DC: NASA, February 2003), https://www.nasa.gov/pdf /382045main_19%20-%2020090730.11.STS%20Problem%202003.pdf.

108. M. C. Malin et al., "Design and Development of the Mars Observer Camera," *International Journal of Imaging Systems and Technology* 3, no. 2 (1991): 76–91, https://doi.org/10.1002/ima.1850030205.

    And what a camera it was! Engineers had to reinvent inventing to do it, but they came up with a camera without a shutter. Moving parts wear out. Moving parts bring surprises, and nobody likes surprises a hundred million miles from home. The camera worked like a fax machine, would scan Mars one line at a time and algorithmically assemble the image in memory and return it back to Earth.

109. D. S. F. Portree, online interview by author, July 10, 2018.

My kind thanks to David S. F. Portree, the extraordinary historian who, to the best of my knowledge, has done the best and most comprehensive work connecting SDI and Faster-Better-Cheaper. His work can be found at DSFP's Spaceflight History, https://spaceflighthistory .blogspot.com.

110. "Strategic Defense Initiative (SDI)," Atomic Heritage Foundation, last modified July 18, 2018, https://www.atomicheritage.org/history/strate gic-defense-initiative-sdi.

111. A. Wellerstein, "Why Build So Many Nukes? Factors Behind the Size of the Cold War Stockpile" (presentation, Putting the Genie Back in the Bottle: MIT Faculty and Nuclear Arms Reduction, Cambridge, MA, May 2011), http://web.mit.edu/fnl/volume/235/wallslides.pdf.

Mutually Assured Destruction, of course, required both countries to build ever-growing arsenals of increasingly sophisticated doomsday weaponry, keep them on constant alert, and have them pointed perpetually at each other, because if you couldn't *absolutely annihilate your opponent,* destruction was not mutually assured. And since you could never know for certain how many nuclear weapons the enemy actually had, you had to keep making more . . . just in case!

112. D. G. Brennan, "Strategic Alternatives: I," *New York Times,* May 24, 1971, https://www.nytimes.com/1971/05/24/archives/strategic-alter natives-i.html.

113. H. M. Kristensen and R. S. Norris, "Global Nuclear Weapons Inventories, 1945–2013," *Bulletin of the Atomic Scientists* 69, no. 5 (2013): 75–81, https://doi.org/10.1177/0096340213501363.

114. *Soviet Military Power* (Washington, DC: Defense Intelligence Agency, 1983), available at Federation of American Scientists, https://fas.org /irp/dia/product/smp_83_ch2.htm.

115. S. D. Drell, P. J. Farley, and D. Holloway, "Preserving the ABM Treaty: A Critique of the Reagan Strategic Defense Initiative," *International Security* 9, no. 2 (1984): 51, https://doi.org/10.2307/2538668.

116. S. Nozette, *Defense Applications of Near-Earth Resources* (CSI-83-3) (La Jolla, CA: California Space Institute, September 1, 1983), https://apps .dtic.mil/dtic/tr/fulltext/u2/a340021.pdf. This document is one of the most stunning and exciting artifacts in all of space history.

117. H. Smith, "Would a Space-Age Defense Ease Tensions or Create Them?," *New York Times*, March 27, 1983, https://timesmachine.ny times.com/timesmachine/1983/03/27/060428.html?pageNumber=140.

118. J. A. Abrahamson and H. F. Cooper, "What Did We Get for Our $30 Billion Investment in SDI/BMD?" (paper, National Institute for Public Policy, Fairfax, VA, September 1993), http://highfrontier.org/wp-con tent/uploads/2016/08/What-for-30B_.pdf.

119. R. Hale, *Statement of Robert F. Hale, Assistant Director, National Security Division, Congressional Budget Office, Before the Defense Policy Panel and the Subcommittee on Research and Development, Committee on Armed Services, U.S. House of Representatives* (Washington, DC: Congressional Budget Office, March 26, 1987), https://apps.dtic.mil/dtic/tr/fulltext /u2/a529879.pdf.
    The testimony of Robert F. Hale is an excellent place to start if you want to understand funding trends for SDI.

120. National Research Council, *Lessons Learned from the Clementine Mission* (Washington, DC: National Academies Press, 1997), https://doi .org/10.17226/5815.

121. Hale, *Statement of Robert F. Hale*. This is the source of SDI funding figures for the years 1984, 1985, 1986, and 1987.
    See also B. S. Lambeth and K. N. Lewis, *The Strategic Defense Initiative in Soviet Planning and Policy* (R-3550-AF) (Santa Monica, CA: Rand Corporation, January 1988), 101, https://www.rand.org/pubs/re ports/R3550.html. This is the source of the 1988 figure.
    See also *CQ Almanac 1989: 101st Congress, 1st Session* (Thousand Oaks, CA: Sage Publications, 2013). The 1989 *CQ Almanac* is the source of the four-point-one-billion-dollar figure for 1989.
    See also L. Aspin, *National Defense Authorization Act for Fiscal Years 1990 and 1991* (Pub. L. 101–189), 1989. Source of the 1990 figure.
    See also M. Duric, *The Strategic Defence Initiative: U.S. Policy and the Soviet Union*, 1st ed. (London: Routledge, 2017), https://doi .org/10.4324/9781315236926. Source of the 1991 figure.

122. T. Hogan, *Mars Wars: The Rise and Fall of the Space Exploration Initiative*, NASA History Series (Washington, DC: Government Printing Office, 2009), 134, https://history.nasa.gov/sp4410.pdf. This book is quite worth the time of anyone interested in the political

realities of attempting to mount an Apollo-esque human space-flight program.

123. M. Cabbage and W. Harwood, *Comm Check: The Final Flight of Shuttle Columbia* (New York: Free Press, 2004), 221.

124. S. Begley, "The Stars of Mars," *Newsweek,* last modified July 20, 1997, https://www.newsweek.com/stars-mars-174306.

125. Neufeld, "Transforming Solar System Exploration," 5–12.
    See also Conway, *Exploration and Engineering,* 145.

126. Ibid., 49.

127. This is $170 million, adjusted for inflation.

128. E. Weiler, interview by author, September 4, 2017.

129. Ibid.

130. L. Spitzer Jr., "Astronomical Advantages of an Extra-Terrestrial Observatory," *Astronomy Quarterly* 7, no. 3 (1990): 131–42, https://doi.org/10.1016/0364-9229(90)90018-V.

131. J. N. Wilford, "Scientists Assess the Hubble Loss," *New York Times,* June 29, 1990, https://timesmachine.nytimes.com/timesmachine/1990/06/29/issue.html.

132. NASA, "Space Shuttle Mission STS-61," press kit, December 1993, https://www.nasa.gov/pdf/139889main_PressKit12_05.pdf.

133. J. Hester, "Repairing the Hubble Space Telescope," Jeff Hester, last modified December 14, 2015, http://www.jeff-hester.com/reality-straight-up/repairing-the-hubble-space-telescope.

134. E. Weiler, interview by author, September 4, 2017.

135. E. J. Weiler, interview by R. Wright, October 31, 2007, oral history transcript, NASA at 50 Oral History Project, NASA, Johnson Space Center, https://historycollection.jsc.nasa.gov/JSCHistoryPortal/history/oral_histories/NASA_HQ/NAF/WeilerEJ/WeilerEJ_10-31-07.pdf.

136. NASA, "Mars Reconnaissance Orbiter Arrival," press kit, March 2006, https://mars.nasa.gov/mro/files/mro/mro-arrival.pdf.

137. NASA, "Phoenix Landing: Mission to the Martian Polar North," press kit, May 2008, https://www.jpl.nasa.gov/news/press_kits/phoenix-landing.pdf.

138. With apologies to Jerry Pournelle.

139. National Research Council, *New Frontiers in the Solar System: An Integrated Exploration Strategy* (Washington, DC: National Academies Press, 2003), 318, https://doi.org/10.17226/10432.

140. *The Next Great Observatory: Assessing the James Webb Space Telescope— Full Committee,* U.S. House of Representatives, Subcommittee on Space, December 6, 2011, video, 1:43.57, https://science.house.gov /news/videos/watch/the-next-great-observatory-assessing-the-james -webb-space-telescope-full-committee.

141. J. K. Alexander, *Science Advice to NASA: Conflict, Consensus, Partnership, Leadership* (Washington, DC: NASA, Office of Communications, NASA History Division, 2017), 89, https://www.nasa.gov/sites/default /files/atoms/files/275710-science_advice_book_tagged.pdf.

142. National Research Council, *New Frontiers in the Solar System: An Integrated Exploration Strategy.*

143. Ibid., 196.

144. *Major Savings and Reforms in the President's 2006 Budget* (Washington, DC: Office of Management and Budget, 2006), https://www .govinfo.gov/content/pkg/BUDGET-2006-SAVINGS/pdf/BUD GET-2006-SAVINGS.pdf.

145. "Huygens Descent Timeline," European Space Agency, last modified January 14, 2005, https://www.esa.int/Science_Exploration/Space _Science/Cassini-Huygens/Huygens_descent_timeline.

146. In 1680 in the gardens of Versailles, Gaspard Marsy began work on a fountain depicting the agony of the giant Enceladus. He had quite a bit of material to work with.

    Enceladus was born of Gaia and Uranus.

    Gaia first begot Uranus through parthenogenesis, and son sired with mother the Titans, the Hecatonchires, and the Cyclopes. But Uranus hated each brood, burying all but the first back inside of their mother, and she hated that. So she created a jagged adamantium sickle and gave it to her son, the Titan Cronus, and asked him to lie in wait inside of her and, at the right time, to strike vengeance. And did he ever. When descended the god of the sky to bed the goddess of the world, Titans gripped Uranus at the Earth's four corners, and Cronus emerged, raised his blade, and castrated his terrible father. But it gets so much worse, really. The blood shed from the gelded god impregnated Gaia, the splatterhouse union yielding the Giants, the Furies, and the Meliae. (The goolies of Uranus, cast into the sea, made Aphrodite, the goddess of love.)

Thus Cronus assumed the throne of the universe, proving meanwhile as loving a father as his own. He reburied the Hecatonchires and Cyclopes, married his sister, the Titaness Rhea, and together they had six kids of their own. Cronos ate the first five, fearing they would overthrow him as he had Uranus. Rhea, expectant with no. 6, did not like her husband eating their children and asked Gaia how she should handle the problem. The two settled on the old wrap-a-rock switcheroo. Cronos mistakenly swallowed a swaddled stone, and the spared child Zeus was stolen away and raised in a cave by the goat Amalthea. He grew and eventually freed from his father's belly his swallowed siblings, and from the Earth liberated the Hecatonchires and Cyclopes, the latter of whom gave him the gift of lightning. So began the Titanomachy, during which Zeus and siblings—as Uranus feared—waged war against Titans for universal primacy, and they would win, Zeus taking his throne on the mountain Olympus. Later, a cattle dispute (really) ignited the Gigantomachy, an epic war between the gods and the Giants (again, for control of the cosmos). One by one, Zeus and his children vanquished the Giants, Zeus hurling thunderbolts and progeny locking swords. Accounts of the war vary, names and numbers, but famous of the fighters is the goddess Athena, daughter of Zeus, who grappled the giant Enceladus on the Phlegraean plain, interring him at last inside of a mountain. (During that war, Athena was very clearly all out of bubble gum. She dispatched the giant Pallas, flayed him with his own claws, and fashioned a shield from his sloughed skin.) By Gigantomachy's end, the Giants were routed and buried, and the gods secure as rulers of heaven and the Earth, those threats staved off until other gods from other cultures would make mythology of Western polytheism, arcana of orthodoxy, and Renaissance art of religious canon.

But alas, poor Enceladus. When artists conjure the Gigantomachy, it is his defeat above others that meets canvas and chisel. Weary Enceladus, struck down by Athena, patron of Athens, pushed once into rock but interred ever anew in oil, stone, ingot, and ink. Enceladus, who yearned for more, for the freedom of godliness, to summit that mountain! And then in vanquishment, his shameful burial beneath another, his yearning now for freedom from rock, the continents ever pressing down on him, the giant ever writhing: "not yet dead indeed but always dying," imagined the sophist Philostratus the Elder. Enceladus the cypher! What motivated the Athena-slain son of Uranus? In words written through Pierre's eyes, Melville saw a creature in torment, vexed by an incestuous birth and saddled with "that reckless, sky-assaulting mood of his." With metal and masonry for Louis XIV, the sculptor Gaspard Marsy carved for the proto-palace Versailles a fountain in gilded bronze, the giant captured in cruel consequent convulsions. It resides now in a small copse in the lower gardens: Bassin d'Encelade, a bowered wooden trellis surrounding a wide round pool. At water's center, an enormous golden man of muscle and anguish writhes and thrashes, buried in severe black stones, a rock clutched and raised in a liberated hand as if removing it—a futile effort to free himself, the rock just one of infinite raining upon him. His struggle is doomed, his burial imminent, and his eyes and arms and mouth are wide and he is screaming. From the ghastly green water around him, meager fountains splash upward where the rocks rain down, and from open jaws of the giant himself, a focused jet of water blasts eighty feet into the air, dissolving distantly into plumes and raining back down upon him. Such is the fate of Enceladus, inhumed, recumbent and baying an eternal cry.

In all the groves grown and groomed for visitors of Versailles, only Marsy's masterpiece would prove prophetic of a discovery millions of miles and three hundred thirty years away. In the orbit of Saturn, nature would imitate art imitating mythology. Marsy of Cambrais, born in 1624 and adjunct rector of the Academy of Sculpture and Painting in Paris, was right about Enceladus, about the perpetual pressure of uncompromising geology, and the resultant plume gushing from a god aspirant.

147. "Enceladus: Ocean Moon," NASA Solar System Exploration, last modified September 25, 2018, https://solarsystem.nasa.gov/missions /cassini/science/enceladus.

148. R. Cowen, "Saturn's Moon Has Never-Ending Winter," *Nature*, last modified October 3, 2011, https://www.nature.com/news/2011/111003 /full/news.2011.569.html.

149. L. Dajose, M. Wong, and C. Dreier, "NASA's Planetary Science Division Funding and Number of Missions, 2004–2020," Planetary Society,

last modified February 9, 2015, https://www.planetary.org/multimedia /space-images/charts/historical-levels-of-planetary-exploration-fund ing-fy2003-fy2019.html.

150. C. Niebur, *Investments in Europa Mission Concept Studies,* 2010.

151. *The Farthest: Voyager in Space,* dir. E. Reynolds (PBS, 2017), DVD.

152. P. E. Mack, *From Engineering Science to Big Science* (Washington, DC: NASA, 2007), https://history.nasa.gov/SP-4219/Chapter11.html. For a description of the "grand tour" generally.

153. C. Niebur, email message to author regarding 2007 competed flagship, January 11, 2018. See also C. Niebur, telephone interview by author, February 20, 2018. See also F. Bagenal, interview by author, August 23, 2016.

154. F. Bagenal, interview by author, August 23, 2016.

155. S. Niebur, "Children's Museums," *Toddler Planet* (blog), March 10, 2010, https://toddlerplanet.wordpress.com/2010/03/10/childrens-mu seums.

156. NASA, "Susan Niebur (1973–2012)," NASA Solar System Exploration, accessed October 24, 2019, https://solarsystem.nasa.gov/people/1700 /susan-niebur-1973-2012.

157. C. Niebur, interview by author, August 4, 2016.

158. J. N. Wilford, "Pioneer Nearing Asteroid Region," *New York Times,* July 14, 1972, 64.

159. *President's Fiscal Year 2006 Budget Request* (Washington, DC: NASA, 2005), available at https://www.nasa.gov/pdf/107486main_FY06 _high.pdf.

160. Ibid., 130.

161. C. Young, *Consolidated Appropriations Resolution FY2003 Report 108-10* (Pub. L. 108–7), February 20, 2003.

162. M. Griffin, *Statement of Hon. Michael D. Griffin, Administrator, National Aeronautics and Space Administration, Hearing Before the Committee on Science, House of Representatives, 109th Congress, First Session, November 3, 2005* (Washington, DC: U.S. Government Printing Office, 2006),

28–31, https://www.govinfo.gov/content/pkg/CHRG-109hhrg24151/html/CHRG-109hhrg24151.htm.

163. J. Culberson, "Realizing the Promise of Prometheus," *Aerospace America* 43, no. 7 (July 2005).

   Many thanks to Duane Hyland of *Aerospace America,* who kindly provided a copy of this article.

   See also J. Culberson, telephone interview by author, May 30, 2019.

164. "State Runoff Elections," *Del Rio (TX) News Herald,* May 7, 1986, 3C.

165. G. Jones, "Runoff Candidates Back at Work for June 7," *Monitor* (McAllen, TX), May 5, 1986, 3A.

166. Xenophon, *The Works of Xenophon,* vol. 1, trans. H. G. Dakyns (New York: Macmillan, 1890), 149.

167. T. Fleck, "Wareing Thin," *Houston Press,* last modified April 20, 2000, https://www.houstonpress.com/news/wareing-thin-6564910.

168. J. Transahl, *Final Edition: List of Standing Committees and Select Committees and Their Subcommittees of the House of Representatives of the United States Together with Joint Committees of the Congress with an Alphabetical List of the Members and Their Committee Assignments* (Washington, DC: U.S. Government Publishing Office, December 26, 2002).

169. Ibid.

170. J. Tollestrup, *Appropriations Subcommittee Structure: History of Changes from 1920–2011* (Washington, DC: Congressional Research Service, 2011).

   You would *not* believe how boring this stuff is to research.

171. "Katy, a Great Place to Be!," Katy Heritage Society, accessed October 24, 2019, https://www.katyheritagesociety.com/content.asp?secnum=9. Also, the author extends no thanks to the Katy Heritage Society for failing to return my calls and email messages requesting help here. Are you really that busy?

172. J. Thornock, "Katy: West Houston Wonder," *Houston History* 13, no. 2 (March 2016): 25–29.

173. "Historical Population: 1900 to 2013, City of Houston," City of Houston Planning & Development Department, accessed October 24, 2019,

http://www.houstontx.gov/planning/Demographics/docs_pdfs/Cy
/coh_hist_pop.pdf.

174. P. Brinckerhoff, *Building a Legacy: The IH 10 West Katy Freeway Story*,
Katy Freeway: Building the Future, last modified December 2008,
http://www.katyfreeway.org/GrandOpening/Katy_Video_Booklet
.pdf.

175. "Interstate 10 West Construction: Historical Information," West Houston Association, last modified April 12, 2004, https://www.westhous
ton.org/interstate-10-west-construction-historical-information.

176. Southwest Freeway / Transitway Project, *US-59 (Southwest Freeway)
Improvement and Widening, Transitway Project, Harris County: Environmental Impact Statement* (Austin: Texas State Department of Highways
and Public Transportation, 1985), FHWA-TX-EIS-85-01-F.

177. G. Goodin, telephone interview by author, October 10, 2017.
    Ginger Goodin is director of the Transportation Policy Research
Center and a senior research engineer at the Texas A&M Transportation Institute.

178. Brinckerhoff, *Building a Legacy*.

179. J. Culberson, interview by author, October 4, 2016.

180. B. Shuster, *Transportation Equity Act for the 21st Century* (Pub. L. 105-
178), June 9, 1998.

181. Katy Freeway Public Information Office, "Governor Rick Perry, U.S.
Congressman John Culberson, and Federal Highway Administrator
Thomas J. Madison Join TxDOT in Celebrating Completion of the
Katy Freeway Reconstruction Program," press release, October 27,
2008, http://www.newsrouter.com/NewsRouter_Uploads/56/news_re
lease.asp?intRelease_ID=4315&intAcc_ID=56.

182. D. Adams, "Mars Exploration Rover Airbag Landing Loads Testing
and Analysis" (presentation, Forty-Fifth AIAA/ASME/ASCE/AHS/
ASC Structures, Structural Dynamics, and Materials Conference,
Palm Springs, CA, April 22, 2004), https://doi.org/10.2514/6.2004
-1795.
    See also B. Nelson, telephone interview by author, February 8, 2019.
    Bill Nelson was the engineering manager for the Mars Exploration
Rover project, and he helped describe the landings generally.

See also NASA, "Mars Exploration Rover Launches," press kit, June 2003, https://www.jpl.nasa.gov/news/press_kits/merlaunch.pdf.

183. NASA, "Mars Exploration Rover Launches."

184. Adams, "Mars Exploration Rover Airbag Landing."

185. J. Culberson, telephone interview by author, June 26, 2019.

186. G. Webster, "Jupiter Radiation Belts Harsher Than Expected," NASA Solar System Exploration, last modified March 28, 2001, https://solarsystem.nasa.gov/news/12227/jupiter-radiation-belts-harsher-than-expected.

187. A. Dressler, *HST and Beyond: Exploration and the Search for Origins: A Vision for Ultraviolet-Optical-Infrared Space Astronomy* (Washington, DC: Association of Universities for Research in Astronomy, 1996).

188. National Research Council, *A Scientific Assessment of a New Technology Orbital Telescope* (Washington, DC: National Academies Press, 1995), vii, https://doi.org/10.17226/9295.

189. U.S. Government Accountability Office, *James Webb Space Telescope: Actions Needed to Improve Cost Estimate and Oversight of Test and Integration, Report to Congressional Committees* (Washington, DC: U.S. Government Accountability Office, December 3, 2012), 3–4, https://www.gao.gov/products/gao-13-4.

190. Space Studies Board, *New Frontiers in the Solar System: An Integrated Exploration Strategy* (Washington, DC: National Academies Press, 2003), 5.

191. NASA Office of Inspector General, *Audit Report: NASA's Management of the Mars Science Laboratory Project* (Report IG-11-19) (Washington, DC: NASA Inspector General, June 8, 2011), https://oig.nasa.gov/docs/IG-11-019.pdf.

192. NASA, "Mars Science Laboratory Launch," press kit, November 2001, https://www.jpl.nasa.gov/news/press_kits/MSLLaunch.pdf.
See also NASA, "Mars Exploration Rover Launches."

193. A. Stern, "NASA's Black Hole Budgets," *New York Times*, November 23, 2008, A23.

194. A. Cohen, *Report of the 90-Day Study on Human Exploration of the Moon and Mars* (Report TM-102999) (Washington, DC: NASA, November 1989), https://ntrs.nasa.gov/archive/nasa/casi.ntrs.nasa.gov/19910017741.pdf.

195. Associated Press, "A Station in Space Predicted," *New York Times,* July 19, 1983, C3.

196. W. J. Broad, "House Vote Sets Stage for Conflict Between Two Allies in Space Program," *New York Times,* June 8, 1991, A7.

197. M. Smith, "NASA's Space Station Program: Evolution and Current Status," (presentation, House Science Committee, Washington, DC, April 4, 2001).

198. Broad, "House Vote Sets Stage for Conflict," A7.

199. J. M. Logsdon and J. R. Millar, "US-Russian Cooperation in Human Spaceflight: Assessing the Impacts," *Space Policy* 17, no. 3 (2001): 171–78, https://doi.org/10.1016/S0265-9646(01)00021-2.

200. J. Stromberg, "Why NASA Is Utterly Dependent on Russia," Vox, last modified May 5, 2014, https://www.vox.com/2014/5/5/5674744/how-nasa-became-utterly-dependent-on-russia-for-space-travel.

201. W. E. Leary, "Outpost in Space; Space Station, Long a Dream, to Soar at Last," *New York Times,* November 16, 1998, A1.

202. Ibid.

203. Broad, "House Vote Sets Stage for Conflict," A7.

204. Ibid.

205. T. Reichhardt, "Unstoppable Force," *Nature* 426, no. 6965 (2003): 380–81, https://doi.org/10.1038/426380a.

206. R. R. Sims and W. I. Sauser, eds., *Experiences in Teaching Business Ethics: Contemporary Human Resource Management* (Charlotte, NC: Information Age, 2011), 262.

207. United States Postal Service, "NASA's Breathtaking Planet Images Get Stamps of Approval," news release, May 31, 2016, https://webcache.googleusercontent.com/search?q=cache:J5auoQVrXtoJ:https://blueearth.usps.gov/news/national-releases/2016/pr16_042.pdf+&cd=1&hl=en&ct=clnk&gl=us.

208. CAW-CAW! RAAAR!

209. K. Chang, "The Long, Strange Trip to Pluto, and How NASA Nearly Missed It," *New York Times*, July 19, 2015, A1.

210. T. May, interview by author, August 11, 2016.

211. Ibid., April 3, 2018.
    See also A. Stern and D. H. Grinspoon, *Chasing New Horizons: Inside the Epic First Mission to Pluto*, 1st ed. (New York: Picador, 2018), 124.

212. T. May, interview by author, August 11, 2016.

213. Ibid.

214. A. Stern, interview by author, August 24, 2016.

215. Ibid., August 26, 2016.

216. J. Langmaier and J. Elliott, *Assessment of Alternative Europa Mission Architectures* (JPL-08-1) (Pasadena, CA: NASA, JPL, January 2008), http://citeseerx.ist.psu.edu/viewdoc/download?doi=10.1.1.541.8006&rep=rep1&type=pdf.
    Of the studies listed in the text, Karla was involved with the 1998 internal all-solar Europa Orbiter study; led the 2001 Europa Orbiter study; led the Europa Orbiter Alternate Missions study in 2001; led the Europa Orbiter Competitive study in 2002; was technical lead for the nonfission option of the Jupiter Icy Moons Tour study in 2002; was spacecraft manager for JIMO in 2005; and led the Europa Explorer study in 2007.

217. C. Niebur, interview by author, August 4, 2016.

218. B. Balmer, J. Gregory, and M. Godwin, "'Brain Drain' Debate in the United Kingdom, c. 1950–1970" (data collection), UK Data Service, 2009, http://doi.org/10.5255/UKDA-SN-6099-1.

219. Committee Appointed by the Council of the British Royal Society, "Emigration of Scientists from the United Kingdom," *Contemporary Physics* 4, no. 4 (1963): 304–5, https://doi.org/10.1080/00107516308247984.

220. U.S. Congress, *Congressional Record*, 89th Cong., 1st sess., 1965, pt. 18, p. 24238, accessed from https://www.govinfo.gov/content/pkg/GPO-CRECB-1965-pt18/pdf/GPO-CRECB-1965-pt18-4.pdf.

221. B. Lovell, "A British 'Brain' Explains the 'Brain Drain'; An Eminent Scientist Examines the Reasons Why So Many of His Colleagues Are Emigrating to the United States—to the Consternation of Britain's Scientific Community," *New York Times,* March 22, 1964, 13.

222. K. Clark, email message to author regarding Cassini project development, July 4, 2019.

223. R. Staehle, "Ice & Fire: Missions to the Most Difficult Solar System Destinations . . . on a Budget," *Acta Astronautica* 45, nos. 4–9 (1999): 423–39, https://doi.org/10.1016/S0094-5765(99)00162-9.

224. Ibid.

225. "Galileo Venus Flyby," NASA, JPL, CIT, last modified February 7, 1990, http://www.jpl.nasa.gov/news/news.php?feature=5703.

See also "Galileo Heads Towards Second Gravity Assist," NASA, JPL, CIT, last modified December 4, 1990, http://www.jpl.nasa.gov /news/news.php?feature=5741.

226. R. W. Farquhar, *Fifty Years on the Space Frontier: Halo Orbits, Comets, Asteroids, and More* (Denver: Outskirts Press, 2011).

Books on orbital dynamics are not generally fun to read, but the late Bob Farquhar's out-of-print memoirs are well worth tracking down. Farquhar, who worked at the Applied Physics Laboratory, was a pioneer in the field and had a mischievous streak that kept him crosswise with NASA headquarters for much of his storied career. He was also a fellow paratrooper. Airborne, brother!

227. "About Plutonium-238," NASA Radioisotope Power Systems, accessed October 24, 2019, https://rps.nasa.gov/about-rps/about-plutonium-238.

228. *60 Minutes,* "5-4-3-2-1 Liftoff" (transcript), produced by Don Hewitt, featuring Steve Kroft, aired October 5, 1997, on CBS-TV, available at Proposition One online, http://prop1.org/2000/cassini/971005fl.htm.

No matter how bad I might make this story seem, it was way, *way* worse than that.

229. D. Netburn, "'OK. Let's Do It!' An Oral History of How NASA's Cassini Mission to Saturn Came to Be," *Los Angeles Times,* September 12, 2017, https://www.latimes.com/science/la-sci-cassini-oral-histo ry-20170912-htmlstory.html.

230. T. Reichhardt, "Designs on Europa Unfurl," *Nature* 437, no. 8 (2005): https://doi.org/10.1038/437008a.

231. Not everyone agreed with this philosophy, as we shall see later in the story. NASA and its governing documents are unambiguous on who is in charge of a flagship-class mission: the project manager.

232. C. Phillips, interview by author, September 14, 2017.

233. L. Prockter, interview by author, January 15, 2018.

234. L. M. Prockter, "Folds on Europa: Implications for Crustal Cycling and Accommodation of Extension," *Science* 289, no. 5481 (2000): 941–43, https://doi.org/10.1126/science.289.5481.941.

235. G. C. Collins et al., "Evaluation of Models for the Formation of Chaotic Terrain on Europa," *Journal of Geophysical Research: Planets* 105, no. e1 (2000): 1709–16, https://doi.org/10.1029/1999JE001143.

236. G. V. Hoppa, "Formation of Cycloidal Features on Europa," *Science* 285, no. 5435 (1999): 1899–1902, https://doi.org/10.1126/science.285.5435.1899.

237. T. Kane, "The Decline of American Engagement: Patterns in US Troop Deployments" (Economics working paper 16101, Hoover Institution, Stanford University, Stanford, CA, 2016).

238. C. Greeley, interview by author, February 7, 2017.

239. Ibid. Calculated for inflation and rising real estate prices, its value as of February 7, 2017, was $1,870,629. Incidentally, the door is still red and the exterior still gray.

240. R. Greeley, curriculum vitae, February 2, 2009.

241. K. Cole, "Galileo Takes Close-Ups of Icy Europa," *Los Angeles Times*, December 17, 1997, https://www.latimes.com/archives/la-xpm-1997 -dec-17-me-64998-story.html.

242. R. T. Pappalardo, *Strawman Europa SDT List*, January 8, 2007.

243. S. Niebur, "Not Good," *Toddler Planet* (blog), June 16, 2007, https:// toddlerplanet.wordpress.com/2007/06/16/not-good.

244. S. Niebur, "Goodbye Hair," *Toddler Planet* (blog), July 27, 2007, https:// toddlerplanet.wordpress.com/2007/07/27/goodbye-hair.

245. S. Niebur, "A Difficult Day," *Toddler Planet* (blog), August 1, 2007, https://toddlerplanet.wordpress.com/2007/08/01/a-difficult-day.

246. K. Clark et al., *2007 Europa Explorer Mission Study: Final Report* (JPL D-41283) (Pasadena, CA: NASA, JPL, CIT, November 1, 2007), available at Lunar and Planetary Institute, https://www.lpi.usra.edu /opag/europa_explorer_public_report_1.pdf.

247. K. Clark, telephone interview by author, March 23, 2018.

248. T. May, interview by author, February 2, 2017.

249. B. W. Stauffer et al., "A Friend Acting Strangely: An Exhibition on Climate Change in the Arctic" (abstract ED21B-1212), *American Geophysical Union, Fall Meeting, Abstracts,* December 2003, available at Smithsonian Astrophysical Observatory/NASA Astrophysics Data System, https://ui.adsabs.harvard.edu/abs/2003AGUFMED21B1212S /abstract.

250. CBS / Associated Press, "Did Smithsonian Alter Climate Change Show?," CBS News, last modified May 21, 2007, https://www.cbsnews .com/news/did-smithsonian-alter-climate-change-show.

251. J. Grimaldi and J. Trescott, "Scientists Fault Climate Exhibit Changes," *Washington Post,* November 16, 2007, http://www.washingtonpost.com /wp-dyn/content/article/2007/11/15/AR2007111502550.html.

252. S. S. Board and National Research Council, *Grading NASA's Solar System Exploration Program: A Midterm Report* (Washington, DC: National Academies Press, 2008), http://nap.edu/12070.

253. Stern and Grinspoon, *Chasing New Horizons,* 155.

254. D. F. Everett, "Engineering a Successful Mission: Lessons from the Lunar Reconnaissance Orbiter," *Proceedings of the 2011 Aerospace Conference,* Big Sky, MT, 2011, 1–2, https://doi.org/10.1109/AERO.2011.5747273.

255. Stern, "NASA's Black Hole Budgets."

256. B. Berger, "Kepler Team Cuts Costs, Avoids Cancellation," *SpaceNews,* last modified July 25, 2007, https://spacenews.com/kepler-team-cuts -costs-avoids-cancellation.

257. M. Mewhinney, NASA Ames Research Center, "NASA Ames' Kepler Mission Selected for Discovery Program," news release 01-107AR,

December 21, 2001, https://www.nasa.gov/centers/ames/news/releases /2001/01_107AR.html.

258. J. R. Minkel, "Stern Steps Down as NASA Science Chief After Mars Budget Dustup," *Scientific American*, March 27, 2008, https://www.sci entificamerican.com/article/stern-steps-down-as-head.

See also K. Tobin, "Mixed Signals from NASA About Fate of Mars Rover," CNN, last modified March 24, 2008, http://edition.cnn .com/2008/TECH/space/03/24/nasa.mars.rover.

See also K. Cowing, "Give Us What We Want or We'll Shoot the Cute Little Rover," NASAWatch.com, last modified March 25, 2008, http://www.nasawatch.com/archives/2008/03/give-us-what-we-want -or-well-shoot-the-cute-little-rover.html.

See also J. Foust, "Mars Rover Funding Cuts: Will There Be a Backlash?," Space Politics, last modified March 24, 2008, http://www .spacepolitics.com/2008/03/24/mars-rover-funding-cuts-will-there -be-a-backlash.

259. Stern and Grinspoon, *Chasing New Horizons*, 142.
See also T. May, interview by author, May 11, 2016.
See also A. Stern, interview by author, August 26, 2016.

260. Stern and Grinspoon, *Chasing New Horizons*, 144–47.
See also A. Stern, interview by author, August 26, 2016.
See also T. May, interview by author, August 11, 2016.

261. A. Stern, telephone interview by author, August 26, 2016.

262. T. Gavin, telephone interview by author, December 29, 2016.
See also T. May, interview by author, May 3, 2018.

263. Reichhardt, "Unstoppable Force," 380–81.

264. T. May, interview by author, February 2, 2017.
See also E. Weiler, interview by author, September 4, 2017.

265. "Icy, Patterned Ground on Mars," NASA, JPL, CIT, last modified May 25, 2008, https://www.jpl.nasa.gov/spaceimages/details.php?id= PIA10682.

See also W. Harwood, "Satellite Orbiting Mars Imaged Descending Phoenix," Spaceflight Now, last modified May 26, 2008, https://space flightnow.com/mars/phoenix/080526mrochute.html.

266. T. May, interview by author, April 3, 2018. As of July 7, 2019, the NASA job search portal still lists "a sense of daring" as a sought trait of agency employees.

See also "Careers at NASA," NASA, accessed October 24, 2019, https://nasajobs.nasa.gov/default.htm.

267. M. Petrovich, email message to author regarding the building count, July 10, 2019.

I manually counted the buildings as well on an untitled but detailed map of Jet Propulsion Laboratory made circa 2003.

268. P. Cargill et al., *Cosmic Vision: Space Science for Europe 2015–2025* (BR-247) (Noordwijk, Neth.: European Space Agency, October 2005), http://www.esa.int/esapub/br/br247/br247.pdf.

269. Planning and Coordination Office: Directorate of the Scientific Programme, "First Call for Missions for the Cosmic Vision Plan 2015–2025 (CV1): Programmatic Context" (PowerPoint presentation, European Space Agency headquarters, Paris, November 2007), https://www.lpi.usra.edu/opag/nov_2007_meeting/presentations/cosmic_visions.pdf.

270. "What is ESA?," European Space Agency, accessed October 24, 2019, http://www.esa.int/About_Us/Corporate_news/What_is_ESA.

271. NASA Office of Inspector General, *NASA's International Partnerships: Capabilities, Benefits, and Challenges* (IG-16-020) (Washington, DC: May 5, 2016), 24, https://www.oversight.gov/sites/default/files/oig-reports/IG-16-020.pdf.

272. Ibid.

273. W. Faulconer, email message to author regarding JPL-APL partnership for Europa mission, July 10, 2019.

See also C. Elachi and R. Roca, *Memorandum of Agreement (MOA) Between the California Institute of Technology, Jet Propulsion Laboratory (JPL), and the Johns Hopkins University, Applied Physics Laboratory (APL),* September 15, 2006.

274. T. Magner, telephone interview by author, December 21, 2018.

275. A. Lawler, "Space Science: A Space Race to the Bottom Line," *Science* 311, no. 5767 (2006): 1540–43, https://doi.org/10.1126/science.311.5767.1540.

276. NASA, *President's FY2007 Budget Request* (Washington, DC: NASA, 2006), https://www.nasa.gov/pdf/142458main_FY07_budget_full.pdf.

277. W. Huntress Jr. and L. Friedman, "Commentary: NASA's 2007 Budget Proposal: No Real Vision," Space.com, last modified February 14, 2006, https://www.space.com/2048-commentary-nasas-2007-budget-pro posal-real-vision.html.

278. F. Wolf, House Committee Report 109-118: *Science, State, Justice, Commerce, and Related Agencies Appropriations Bill, Fiscal Year 2006* (Washington, DC: U.S. House of Representatives, 2005).

   In this subcommittee appropriations report, the "committee urges NASA to consider incorporating a nonnuclear Europa mission as part of its fiscal year 2007 budget request."

   See also F. Wolf, House Report 109-272: *Making Appropriations for Science, the Departments of State, Justice, Commerce, and Related Agencies for the Fiscal Year Ending September 30, 2006, and for Other Purposes* (Washington, DC: U.S. House of Representatives, 2005).

   See also F. Wolf, House Report 109-520: *Science, State, Justice, Commerce, and Related Agencies Appropriations Bill, Fiscal Year 2007* (Washington, DC: U.S. House of Representatives, 2006).

   This report sets aside $15 million to initiate planning for an orbiter/lander mission to Europa.

279. J. Culberson, "An Urgent Personal Letter from Congressman John Culberson to American Scientists and Engineers," press release, March 9, 2006, available at SpaceRef, http://images.spaceref.com /news/2006/2006.03.09.Culberson.letter.pdf.

280. Outer Planets Assessment Group, "Europa's Priority in NASA, NRC, and Community Document," Lunar and Planetary Institute, last modified 2006, https://www.lpi.usra.edu/opag/europa_priority.pdf.

281. C. Shupla, email message to author regarding LPSC 2006 numbers, September 24, 2018.

282. "In Memoriam: Harold Masursky, 1923–1990," *Lunar and Planetary Information Bulletin*, no. 100 (November 2004): 5, https://www.lpi.usra .edu/publications/newsletters/lpib/lpib100.pdf.

283. J. Lunine, "Masursky Lecture: Beyond the Asteroid Belt: Where to Go Next in the Solar System and Why" (PowerPoint presentation, Thirty-

Seventh Lunar and Planetary Science Conference, the Woodlands, TX, March 2006).

See also "37th Lunar and Planetary Science Conference: The Conference in Review," *LPI Bulletin* 106 (May 2006): 4–6.

284. L. David, "Scientist Urges 'Clean Sheet' Approach to Outer Planets Exploration," SpaceNews.com, April 24, 2006, https://spacenews.com /scientist-urges-clean-sheet-approach-outer-planets-exploration.

285. L. David, "Researcher Touts Saturn's Titan as New Exploration Goal," Space.com, March 14, 2006, https://www.space.com/2152-researcher -touts-saturns-titan-exploration-goal.html.

286. E. Lakdawalla, "LPSC: Thursday—The Moons of Jupiter and the Future of Outer Planet Exploration," Planetary Society, last modified March 18, 2006, https://www.planetary.org/blogs/emily-lakda walla/2007/0500.html.

287. "Christmas 2004: Cassini Delivers a Very Special Gift," NASA Solar System Exploration, last modified December 20, 2016, https://solarsys tem.nasa.gov/news/12980/christmas-2004-cassini-delivers-a-very-spe cial-gift.

"Christmas Day" might also be "Christmas Eve," depending on the time zone.

288. NASA, "Cassini Launch," press kit, October 1997, https://solarsystem .nasa.gov/resources/17320/cassini-launch-press-kit.

See the press kit for a general description of the Huygens landing.

See also R. Lorenz and J. Mitton, *Titan Unveiled: Saturn's Mysterious Moon Explored* (Princeton, NJ: Princeton University Press, 2010), 135–48.

289. "Huygens," European Space Agency Science & Technology, last modified September 1, 2019, https://sci.esa.int/web/cassini-huygens /-/47052-huygens.

290. NASA, "Cassini Launch."

291. J.-P. Lebreton et al., "Results from the Huygens Probe on Titan," *Astronomy and Astrophysics Review* 17, no. 2 (2009): 149–79, https://doi .org/10.1007/s00159-009-0021-5. See page 152.

292. E. Lakdawalla, "They Were the First, and the Last, to Hear from Huygens," Planetary Society, last modified February 7, 2005, https://www

.planetary.org/blogs/emily-lakdawalla/2005/20050207-they-were-the
-first-and-the-last-to-hear-from-huygens.html.

293. D. Bressan, "Dante's Inferno: The Geology of Hell," History of Geology, last modified June 17, 2016, http://historyofgeology.fieldofscience
.com/2016/06/dantes-inferno-geology-of-hell.html.

294. Dante, *Inferno*, trans. R. Hollander (New York: Anchor Books, 2002). The salient text can be found in Canto XII.

295. J. Leary et al., *Titan Explorer Flagship Mission Study* (Laurel, MD: JHUAPL, January 2008), available at Lunar and Planetary Institute, https://www.lpi.usra.edu/opag/Titan_Explorer_Public_Report.pdf.

296. Lorenz and Mitton, *Titan Unveiled*, 243–44.
This also presents a first-person account of the transition to a Lunine-led Titan Saturn System Mission. Note also that Ralph's books on Titan and the exploration thereof are among the most compelling works of science writing in the genre, and are highly recommended.
See also R. Lorenz, telephone interview by author, September 14, 2017.
See also H. Waite, telephone interview by author, October 4, 2017.

297. D. Linick and C. Briggs, "Aerocapture Technology" (presentation, Venus Exploration Analysis Group, Greenbelt, MD, May 8, 2008), https://www.lpi.usra.edu/vexag/may2008/presentations/24munk.pdf.

298. F. Trevens, "After Rockettes—Rocky Bottom of High Tor?" *Art Times,* April 2005, https://www.arttimesjournal.com/dance/Apr_05_Francine
_Trevens_Rockettes.htm. For a general account of being a Rockette.
See also A. Bodine, "Here's How Much the Iconic Rockettes Get Paid to Do 4 Shows a Day, Change Costumes in 78 Seconds, and Kick up to 1,200," *Business Insider,* last modified December 4, 2017, https://
www.businessinsider.com/heres-how-much-rockettes-get-paid-to
-kick-1200-times-a-day-2017-12.

299. C. Sagan, letter to J. Lunine, April 18, 1974.

300. A. Simon, telephone interview by author, May 24, 2018.
See also J. Lunine, telephone interview by author, August 15, 2017.
See also L. Prockter, telephone interview by author, September 6, 2018.

See also H. Waite, telephone interview by author, October 4, 2017.

See also R. Pappalardo, telephone interview by author, June 2, 2018.

Literally everyone I interviewed who was involved in either study just could not overstate how bad things got between the two teams.

301. NASA, JPL, "Cassini Finds Lakes on Titan's Arctic Region," news release 06-274, NASA Newsroom, July 26, 2006, https://www.nasa.gov /home/hqnews/2006/jul/HQ_06274_cassini_lakes_titan.html.

302. Science@NASA, "New Lakes Discovered on Titan," NASA Science Mission Directorate, last modified October 12, 2007, https://science .nasa.gov/science-news/science-at-nasa/2007/12oct_titan.

303. K. Reh et al., *Titan Saturn System Mission Study: Final Report* (Washington, DC: NASA and the European Space Agency, October 4, 2017).

304. NASA Science Mission Directorate, "NASA Offers Pre-screening of Principal Investigator Revised Requirements for New Frontiers Opportunity," February 26, 2008, available at SpaceRef, http://www.spaceref .com/news/viewsr.html?pid=27162.

305. S. Niebur, "Double Mastectomy," *Toddler Planet* (blog), January 11, 2011, https://toddlerplanet.wordpress.com/2008/01/11/double-mastectomy.

306. S. Niebur, "Or What's a Heaven For," *Toddler Planet* (blog), August 13, 2007, https://toddlerplanet.wordpress.com/2007/08/13/or-whats-a -heaven-for.

307. S. Niebur, "I Had a Bad Scan," *Toddler Planet* (blog), March 13, 2010, https://toddlerplanet.wordpress.com/2010/03/13/i-had-a-bad-scan.

308. S. Niebur, "Regret Turns to Action," *Toddler Planet* (blog), March 14, 2008, https://toddlerplanet.wordpress.com/2008/03/14/regret-turns-to -action.

309. The site remains active: https://womeninplanetaryscience.wordpress .com.

310. S. Niebur, "Flight Missions," Women in Planetary Science, last modified March 15, 2008, https://womeninplanetaryscience.wordpress .com/2008/03/15/flight-missions.

311. NASA, "Cassini Launch."

312. NASA, "NASA and ESA Prioritize Outer Planet Missions," press release, February 18, 2009, https://www.nasa.gov/topics/solarsystem/features/20090218.html.

313. G. McCartney, email message to K. Patel regarding interview, July 17, 2019.

314. K. Clark, telephone interview by author, August 21, 2018.
    See also K. Patel, telephone interview by author, November 14, 2018.
    Though this chapter is told largely from Karla's perspective, the narrative of her ousting is drawn from hours of interviews with Karla, Keyur, and multiple members of the science and engineering teams, some of whom spoke on background. I asked Keyur point-blank about the circumstances of Karla's exit from the project, the staffing issues, and the general negativity throughout, and he repeatedly pleaded ignorance of what happened: "I don't think I have any of the inside knowledge of what transpired there within the team and such," he said.
    When I pointed out that he was her direct supervisor, he responded, "I am trying to think if Karla was actually within this organization. If she was, remember: *direct supervision* is an interesting term. It is, the director for—between the two of us, Rick [Grammier, former director for solar system exploration at JPL] and myself, we would have been the supervision for this thing. But any conversation I've had with Karla, the things that you're bringing up about internal turmoil on the team, I don't remember her saying anything like that. I'll remind you that you're talking about a memory that is ten years old."
    Multiple members of the science team expressed skepticism at this, but the reader is invited to judge based on his above comments in full.

315. *Destination Moon*, dir. I. Pichel (Eagle-Lion Films, 1950).

316. Ibid.

317. Ibid.

318. Ibid.

319. J. F. Kennedy, "John F. Kennedy Moon Speech, Rice Stadium, September 12, 1961," text, audio, and movie clips, NASA, accessed October 24, 2019, https://er.jsc.nasa.gov/seh/ricetalk.htm.

320. "Attention . . . Engineers and Scientists," Career Center advertisement, *Philadelphia Inquirer*, September 18, 1962, 4.

See also "Engineers, Scientists: Which of 67 Leading Firms Will Offer You the Greatest Degree of Professional Advancement?" Career Center advertisement, *Philadelphia Inquirer*, September 19, 1962, 50, https://www.newspapers.com/clip/24743018/the_philadelphia_inquirer.

321. T. Gavin, telephone interview by author, December 29, 2016.

322. B. Momsen, "Mariner IV—First Flyby of Mars—Some Personal Experiences," personal blog of Bill Momsen, last modified December 14, 2002, https://web.archive.org/web/20021214075710/http://home.earthlink.net:80/~nbrass1/mariner/miv-2.htm.

323. S. M. Krimigis and D. Venkatesan, "The Radial Gradient of Interplanetary Radiation Measured by Mariners 4 and 5," *Journal of Geophysical Research* 74, no. 16 (1969): 4129–45, https://doi.org/10.1029/JA074i016p04129.

324. R. Pyle, "Alone in the Darkness: Mariner 4 to Mars, 50 Years Later," Caltech, last modified July 14, 2015, https://www.caltech.edu/about/news/alone-darkness-mariner-4-mars-50-years-later-47324.

Mars is hard, and it took a long time for Jet Propulsion Laboratory to develop the institutional knowledge necessary for mission success. One example of the sheer challenge of early exploration came from Mariner 4. The spacecraft navigated outer space by locking first on the sun and then on the shining star Canopus in the constellation Carina. Like lumber-hewn liners of old, the stars would steer her by. Once launched, however, Canopus proved . . . slippery for a star. It kept . . . moving around? *There it is!* said the spacecraft, and went the wrong way, and *there it is!* and another course correction—*there it*—this continued, and the engineers at Jet Propulsion Laboratory were vexed and bemused. They were puzzled and anxious. Astrophysics really did not provide for this sort of behavior in a star. It was probably a spacecraft issue. The engineers sharpened their pencils and carried their ones, and found finally that what Mariner 4 marked as a star was in fact a chip of paint that followed the spacecraft after separation from the rocket. (Physics did allow for that.) The fleck floated and flipped this way or that, and every so often it would grab a glint of sunlight, and an ersatz Canopus captured and confused computer guidance. But who would have guessed, even at Jet Propulsion Laboratory! Whose hallway wanderers knew no A's-sans-plus. No one certainly would now ever forget, enriching the lab's institutional knowledge base with marginalia that must (and could only) be hard learned.

325. JPL, "Voyager—Planetary Voyage," JPL, accessed October 24, 2019, https://voyager.jpl.nasa.gov/mission/science/planetary-voyage.

326. R. Dawe and J. Arnett, *Thermoelectric Outer Planets Spacecraft (TOPS) Electronic Packaging and Cabling Development Summary Report* (JPL-TM-33-716) (Pasadena, CA: JPL, 1974).

327. A. Butrica, "Voyager: The Grand Tour of Big Science," NASA History Series, 1998, https://history.nasa.gov/SP-4219/Chapter11.html.

328. Homer, *The Iliad*, 27th ed., trans. W. H. D. Rouse (Nashville, TN: Thomas Nelson, 1938).

329. JPL, *Design, Verification/Validation and Operations Principles for Flight Systems* (JPL-D-17868) (Pasadena, CA: JPL, 2001).
        See also D. Linick and C. Briggs, "Developing the JPL Engineering Processes" (presentation, Space 2004 Conference and Exhibit, San Diego, CA, 2004), https://doi.org/10.2514/6.2004-6129.

330. T. Gavin, *Flight Project Practices* (D-58032) (Pasadena, CA: JPL, 2002).

331. T. Gavin, telephone interview by author, December 29, 2016.

332. K. Clark, email message to the Europa team regarding her situation, June 21, 2010.

333. R. Pappalardo, email message to Europa Jupiter System Mission team regarding gift for Karla, August 9, 2010.

334. Woodlands Waterway Marriott representative, telephone interview by author, September 20, 2018.

335. C. Shupla, email message to author regarding LPSC 2006 numbers, September 24, 2018.

336. S. Squyres, "Vision and Voyages for Planetary Science in the Decade 2013–2022" (PowerPoint presentation, Forty-Second Lunar and Planetary Science Conference, the Woodlands, TX, March 2011).

337. E. Weiler, email message to C. Kennel regarding a planetary science Decadal Survey, December 5, 2008.

338. Ibid.

339. E. Hand, "Steve Squyres on Planetary Priorities," *Nature,* last updated March 26, 2009, https://doi.org/10.1038/news.2009.195.

340. J. Spencer, *Mission Concept Study: Planetary Science Decadal Survey, Jupiter Europa Orbiter Component of EJSM* (Pasadena, CA: NASA, JPL, March 2010).

341. G. Schilling, "ESA on Countdown to Flagship Mission Selection," *Nature,* last updated February 4, 2011, https://doi.org/10.1038/news.2011.74.

342. European Space Agency, "Cosmic Vision Call for Proposals: Missions Selected," European Space Agency Science & Technology, last modified September 1, 2019, https://sci.esa.int/web/cosmic-vision/-/46510-cosmic-vision?fbodylongid=2152.

343. M. Dougherty et al., "Europa Jupiter System Mission–Laplace" (PowerPoint presentation, Cosmic Vision 2015–2025 Plan L-Class Missions Presentation, Institut Océanographique de Paris, 2011).
   See also M. Dougherty et al., "Europa Jupiter System Mission–Laplace" (MP3 presentation, Cosmic Vision 2015–2025 Plan L-Class Missions Presentation, Institut Océanographique de Paris, 2011).

344. Lunar and Planetary Institute, "Planetary Decadal Survey Briefing" (Livestream presentation, Forty-Second Lunar and Planetary Science Conference, the Woodlands, TX, March 2011).
   See also Squyres, "Vision and Voyages for Planetary Science."

345. F. Bagenal, interview by author, August 23, 2016.
   See also J. Foust, "Europa Mission Planning for Possible Budget Cuts in 2017," *SpaceNews,* last modified August 17, 2016, https://spacenews.com/europa-mission-planning-for-possible-budget-cuts-in-2017.
   See also F. Bagenal, telephone interview by author, August 24, 2018.

346. F. Bagenal, interview by author, August 23, 2016.
   See also Foust, "Europa Mission Planning."

347. F. Bagenal, telephone interview by author, August 24, 2018.
   See also R. Pappalardo, interview by author, September 20, 2016.

348. National Research Council, *Vision and Voyages for Planetary Science in the Decade 2013–2022* (Washington, DC: National Academies Press, 2011), 16, https://doi.org/10.17226/13117.

349. Ibid., 5.

350. Mars Mid-Range Rover Science Analysis Group, "Final Report of the Mars Mid-Range Rover Science Analysis Group (MRR-SAG)," *Astrobiology* 10, no. 2 (October 14, 2009): 127–63, http://doi.org/10.1089 /ast.2010.0462.

   See also F. Bagenal, email message to B. Hood regarding Europa mission planning for possible budget cuts in 2017, August 23, 2016.

351. National Academies, "Spring 2011 Meetings of the Aeronautics and Space Engineering Board and Space Studies Board Agenda, April 5th– 7th, 2011" (April 4, 2011).

352. NASA, *President's FY2012 Budget Request* (Washington, DC: NASA, 2011), retrieved from https://www.nasa.gov/pdf/516675main_NASA FY12_Budget_Estimates-508.pdf.

353. R. Pappalardo, email message to T. Gavin and K. Patel regarding notes from space studies board meeting, April 11, 2011.

   See also R. Pappalardo, email message to author regarding SSB and Squyres on Europa (picking up the pieces following the Decadal), October 26, 2018.

354. R. Pappalardo, telephone interview by author, June 2, 2018.

   See also R. Pappalardo, email message to T. Gavin and K. Patel regarding notes from space studies board meeting, April 11, 2011.

   See also R. Pappalardo, email message to author regarding SSB and Squyres on Europa (picking up the pieces following the Decadal), October 26, 2018.

355. C. Niebur, telephone interview by author, October 16, 2018.

356. Stern and Grinspoon, *Chasing New Horizons*, 67.

357. E. Weiler, interview by author, September 4, 2017.

358. R. Pappalardo, email message to D. Senske, L. Prockter, T. Magner, and G. Garner regarding first cut at a Europa SDT charter to send to Curt, April 5, 2011.

359. D. E. Smith, "A Budget Phasing Approach to Europa Jupiter System Mission Science" (white paper, NASA Goddard Space Flight Center, 2010).

360. R. Pappalardo, email message to A. Stern regarding Europa SDT, April 25, 2011.

361. D. D. Blankenship et al., *Feasibility Study and Design Concept for an Orbiting Ice-Penetrating Radar Sounder to Characterize in Three-Dimensions the Europan Ice Mantle Down to (and Including) Any Ice/Ocean Interface* (technical report no. 184, University of Texas Institute for Geophysics, Austin, 1999), 26.

See also D. Blankenship, telephone interview by author, November 16, 2018.

362. D. Blankenship, interview by author, March 5, 2019.

363. R. Pappalardo, email message to R. Greeley regarding Europa split payload, May 5, 2011.

364. Google Maps Street View, "2650 Haslett Road, East Lansing, Michigan," Google, accessed October 23, 2019. This sign can be found at approximately 2650 Haslett Road driving eastbound from Haslett into East Lansing.

Alternately, driving into town from the west on East Michigan Avenue, you get the same message. You would *absolutely not believe* how much research went into this book. See Google Maps Street View, "330 E Michigan Avenue, East Lansing, Michigan," Google, accessed October 23, 2019.

365. "Legislator Details—Lester J. Allen [Data]," Library of Michigan, accessed October 23, 2019, https://mdoe.state.mi.us/legislators/Legislator/LegislatorDetail/3015.

366. "Legislator Details—Richard John Allen [Data]," Library of Michigan, accessed October 23, 2019, https://mdoe.state.mi.us/legislators/Legislator/LegislatorDetail/4439.

367. L. Kestenbaum, "The Political Graveyard: Index to Politicians: Allen, O to R," Political Graveyard, accessed October 23, 2019, http://politicalgraveyard.com/bio/allen7.html#464.63.54.

368. NASA, "Biographical Data: Sally K. Ride," last modified July 2012, https://www.nasa.gov/sites/default/files/atoms/files/ride_sally.pdf.

369. Ride was not aboard.

370. "Valentina Tereshkova," Smithsonian National Air and Space Museum, accessed October 23, 2019, https://airandspace.si.edu/people/histori cal-figure/valentina-tereshkova.

    Valentina Tereshkova remains the only woman to fly solo in space and also the youngest, going up at age twenty-six. She just has one hell of a story: daughter of a farmworker, earned her degree through a correspondence program, worked in a tire factory and a textile mill, trained secretly to be a competitive parachutist, as one does. When the Soviet cosmonaut corps selected her for training, she had to enlist in the air force, and no favoritism here: she started as a private, clawed her way up. This wasn't theater, either; she orbited Earth *forty-eight* times (John Glenn did three) and by the time she returned to terra firma had logged more space time than every man in the NASA astronaut corps—combined.

371. NASA, "Biographical Data: John Herschel Glenn, Jr.," last modified December 2016, https://www.nasa.gov/sites/default/files/atoms/files /glenn-j.pdf.

    See also T. Wolfe, *The Right Stuff*, 2nd ed. (New York: Picador, 2008). Look, when we're talking about Glenn, it always comes back to Tom Wolfe's masterpiece.

372. D. Shribman, "With Few Words, Glenn Withdraws," *New York Times*, March 17, 1984, A8.

373. D. Brandt-Erichsen, "Brief History of the L5 Society," *Ad Astra*, December 1994.

374. "Biography of Wernher Von Braun," NASA's Marshall Space Flight Center History, last modified August 3, 2017, nasa.gov/centers/mar shall/history/vonbraun/bio.html.

375. E. Thompson and J. Davis, "Daniel S. Goldin," NASA History Division, last modified November 4, 2009, https://history.nasa.gov/dan _goldin.html.

376. M. Burke, "Medical Research Investment Takes Off for Fisk Johnson," *Journal Times* (Racine, WI), January 23, 2002, https://journal times.com/medical-research-investment-takes-off-for-fisk-johnson/arti cle_09bdfb27-1093-5b2b-8f22-60090a6899f5.html.

    See also D. Lockney, "Bioreactors Drive Advances in Tissue Engineering," NASA Spinoff, accessed October 22, 2019, https://spinoff .nasa.gov/Spinoff2011/hm_1.html.

See also "Administrator Daniel S. Goldin's Accomplishments," NASA History Division, accessed March 26, 2020, https://history .nasa.gov/goldin-accomplishments.pdf.

377. "What is ALH 84001?" Lunar and Planetary Institute, accessed October 22, 2019, https://www.lpi.usra.edu/lpi/meteorites/The_Meteorite .shtml.

378. D. S. McKay et al., "Search for Past Life on Mars: Possible Relic Biogenic Activity in Martian Meteorite ALH84001," *Science* 273, no. 5277 (1996): 924–30, https://doi.org/10.1126/science.273.5277.924.

379. Ibid.

380. L. Garver, interview by author, August 2, 2016. This covers her other conversations with Dan Goldin as well.

381. W. Clinton, "President Clinton Statement Regarding Mars Meteorite Discovery" (statement, the White House, Washington, DC, August 7, 1996), https://www2.jpl.nasa.gov/snc/clinton.html.

382. J.-M. Garcia-Ruiz, "Morphological Behavior of Inorganic Precipitation Systems" (paper, Instruments, Methods, and Missions for Astrobiology II, Denver, CO, July 1999), https://doi.org/10.1117/12.375088.

383. R. Baalke, "Mars Meteorites," JPL, accessed October 22, 2019, https:// www2.jpl.nasa.gov/snc/

384. NASA, "Space Shuttle Mission STS-90," press kit, April 1998, https:// www.nasa.gov/home/hqnews/presskit/1998/STS-90_presskit.txt.

385. L. Garver, interview by author, August 2, 2016.

386. M. Wines, "Russia Scraps Space Plans of Pop Star," *New York Times,* September 4, 2002, E1.

See also Associated Press, "Lance Bass Officially Kicked Off Space Flight," *Billboard,* last modified September 9, 2002, https://www.bill board.com/articles/news/74272/lance-bass-officially-kicked-off-space -flight.

387. L. Garver, interview by author, August 2, 2016.

388. White House, "President Bush Announces New Vision for Space Exploration Program," news release, January 14, 2004, https://history .nasa.gov/Bush%20SEP.htm.

389. B. Weinraub, "Bush May Back Manned Flights to Moon and Mars,"
      New York Times, July 18, 1989, A14.

      See also G. H. W. Bush, "Remarks on the 20th Anniversary of the
      Apollo 11 Moon Landing," American Presidency Project, accessed
      March 12, 2020, https://www.presidency.ucsb.edu/documents/re
      marks-the-20th-anniversary-the-apollo-11-moon-landing.

      In 1989, on the twentieth anniversary of Apollo 11, newly elected
      president George H. W. Bush wanted to get the space program back on
      track, and he wanted to do it in a major address. He didn't run for office
      as a "space president," but the problem had festered long enough—the
      dearth of science missions (not a single science spacecraft launched
      during the Reagan administration), the shuttle hemorrhaging billions,
      and NASA still space-station-less and thus devoid of purpose—it was
      as good a time as any to get things back in order.

      "In 1961," he said, "it took a crisis—the space race—to speed things
      up. Today we don't have a crisis; we have an opportunity. To seize this
      opportunity, I'm not proposing a 10-year plan like Apollo; I'm propos-
      ing a long-range, continuing commitment. First, for the coming de-
      cade, for the 1990s: Space Station Freedom, our critical next step in
      all our space endeavors. And next, for the new century: Back to the
      Moon; back to the future. And this time, back to stay. And then a jour-
      ney into tomorrow, a journey to another planet: a manned mission to
      Mars."

      This was the first serious presidential commitment to deep space ex-
      ploration since Kennedy's "We Choose to Go to the Moon" speech. The
      details would follow. There would be a long-term increase in NASA's
      budget—it would be double by Y2K. In ten years' time, space station
      Freedom would fly. In twenty years, astronauts would set up a perma-
      nent base on the moon. In thirty years—by the fiftieth anniversary of
      the One Giant Leap—boot prints on Mars.

      In a sense, it was the same program laid out by Wernher von Braun
      decades earlier.

      Regardless, it did not succeed in placing humans on the moon or
      Mars. It found virtually no support among the public, but not to worry,
      said Congress: We were never going to fund it anyway. A recession soon
      set in, Bush lost reelection (space was not a factor), the Soviet Union
      collapsed, space station *Freedom* was replaced, &c. The program was
      more flare than wildfire.

      And yet! It was a triumph if only symbolically of how far the Mars program had come. A decade earlier, science fiction fans were rattling cups to fund the Viking mission when NASA could no longer pay its bills. Now the president of the United States wanted to double NASA's budget and put *humans* on Mars. That was progress. Moreover, it placed the agency on Mars footing. For the duration of the Bush presidency, studies proliferated at national laboratories, aerospace companies, and NASA centers. The moment it became clear that We Were Going to Mars, Jet Propulsion Laboratory slapped on the agency's desk a little something it had been working on called the Mars Rover Sample Return. What size rover do you fine ladies and gents want? We have one that weighs about the same as a kangaroo, and gee willikers, is this thing a beaut. Ever see a kangaroo? Put a pair of boxing gloves on them and get out of the way, let—me—tell—you. Can go about a hundred meters. The rover [*laughter*], not the kangaroo. I don't know how far kangaroos can go. This thing won't be able to grab much sample, though. This is Mars on the cheap. But you know [*wink*], I really see you in something a little bigger. You want something bigger. You do, don't you? I can always tell. How about this. One ton. Thing's a doggone bison! A bison! Right there on the prairie. Nothing more American than that. We'd send it to Mangala Vallis, that old water channel on Mars. The rover! Not the bison! Got you again [*nudge*]. The channel was created by overflows! Water just gushing in! You believe that? Mars, what can I say! You don't like that, we've got another model here—one ton, just like that one—we are not skimping here, let me tell you—and you could land this baby just about anywhere. We'd send an orbiter first. Nineteen ninety-six! You guys find a spot you like, and we'll launch this baby in 'ninety-eight. Wait a minute. Hold on. You look like the kind of official who wants a little more monster for the money. [*leans in, speaks softly*] I've got something here. I—no, you don't want this one. It's too much to handle. It's—well, look, OK—I'll tell you. Can you keep a secret? We call it the Godzilla rover. Really! Could climb straight up, one-point-five meters. You know how tall that is? That's the height of a refrigerator. Quite a discovery if we found a fridge on Mars, am I right? So we'd send this thing to Valles Marineris—they call it the Grand Canyon of Mars. Makes the real Grand Canyon look like a kiddie pool. Stretches over an entire face of the planet! Makes Mars look like a cracked egg. You think you'll find some interesting samples there? [*slaps back*] The real question is what *won't* you find there. Now let's talk numbers . . .

390. Review of U.S. Human Spaceflight Plans Committee, *Seeking a Human Spaceflight Program Worthy of a Great Nation* (Washington, DC: NASA, October 2009), 50, https://www.nasa.gov/pdf/396093main_HSF _Cmte_FinalReport.pdf.

391. Ibid., 86.

392. B. Iannotta, "Key U.S. Senator Cautions Obama on NASA Pick," Space .com, last modified January 14, 2009, https://www.space.com/6313 -key-senator-cautions-obama-nasa-pick.html.
    This being his first . . .

393. R. Block, "Bill Nelson and Co. Take Down Obama's NASA Front-runner," *Orlando Sentinel,* last modified March 29, 2009, https://web .archive.org/web/20090329162433/http://blogs.orlandosentinel.com /news_space_thewritestuff/2009/03/bill-nelson-and-co-take-down -obamas-nasa-frontrunner.html.
    . . . his second . . .

394. S. M. Powell, "Lampson Seen as Contender to Lead NASA," *Houston Chronicle,* April 1, 2009, https://www.chron.com/neigh borhood/baytown-news/article/Lampson-seen-as-contender-to-lead -NASA-1591164.php.
    . . . and his third choice.

395. Review of U.S. Human Spaceflight Plans Committee, *Seeking a Human Spaceflight Program Worthy of a Great Nation,* 7, 135.

396. J. Matson, "Phased Out: Obama's NASA Budget Would Cancel Constellation Moon Program, Privatize Manned Launches," *Scientific American,* last modified February 1, 2010, https://www.scientificameri can.com/article/nasa-budget-constellation-cancel.

397. B. H. Obama, "Barack Obama Speech, Kennedy Space Center, April 15, 2010," text, NASA, accessed October 22, 2019, https://www.nasa .gov/about/obama_ksc_pod.html.
    See also L. Garver, interview by author, August 2, 2016.

398. J. Rockefeller, *National Aeronautics and Space Administration Authorization Act of 2010* (Pub. L. 111–267), 2010.

399. J. Davis, "To Mars, with a Monster Rocket: How Politicians and Engineers Created NASA's Space Launch System," *Planetary Society*

(blog), October 3, 2016, https://www.planetary.org/blogs/jason-davis/2016/20161003-horizon-goal-part-4.html.

400. Review of U.S. Human Spaceflight Plans Committee, *Seeking a Human Spaceflight Program Worthy of a Great Nation,* 7, 135.

    NASA, *Preliminary Report Regarding NASA's Space Launch System and Multi-Purpose Crew Vehicle* (Washington, DC: NASA, January 2011), https://www.nasa.gov/pdf/510449main_SLS_MPCV_90-day_Report.pdf.

401. Ibid.

402. K. Chang, "For NASA, Longest Countdown Awaits," *New York Times,* January 25, 2011, D1.

403. F. Newport, "Americans Want Space Shuttle Program to Go On" (poll, Gallup, Washington, DC, February 3, 2003), https://news.gallup.com/poll/7708/Americans-Want-Space-Shuttle-Program.aspx.

    See also T. Redburn, "The Times Poll: 3 out of 4 Back Shuttle Program," *Los Angeles Times,* February 28, 1986, https://www.latimes.com/archives/la-xpm-1986-02-28-mn-12716-story.html.

    See also Zogby International, "IBOPE Zogby Poll: 6 in 10 Disagree with Ending Space Shuttle & Fear Others Will Surpass U.S. in Exploration," press release, July 29, 2011, http://www.spaceref.com/news/viewpr.html?pid=34231.

404. J. McBride, *What Ever Happened to Orson Welles? A Portrait of an Independent Career* (Lexington: University Press of Kentucky, 2006).

405. T. May, interview by author, August 11, 2016.
    See also ibid., February 2, 2017.
    See also ibid., April 3, 2018.

406. Rockefeller, *National Aeronautics and Space Administration Authorization Act of 2010.*

407. Primarily K. Robinson, interview by author, April 3, 2018.
    Thank you to Kimberly Robinson of Marshall Space Flight Center, who absolutely did not have time to speak with me, for teaching me how to build a rocket from scratch. It was like learning the ways of the Force from Obi-Wan Kenobi.
    See also T. May, interview by author, August 11, 2016.

See also ibid., February 2, 2017.

See also ibid., April 3, 2018.

See also D. Hitt, telephone interview by author, July 11, 2016.

408. M. Garcia, "NASA Applies Insights for Manufacturing of Orion Heat Shield," press release, September 24, 2015, http://www.nasa .gov/feature/nasa-applies-insights-for-manufacturing-of-orion-space craft-heat-shield.

409. S. Creech, P. Sumrall, and C. Cockrell, *Ares V Overview and Status* (Washington, DC: NASA, 2009), https://ntrs.nasa.gov/archive/nasa /casi.ntrs.nasa.gov/20090042947.pdf.

See also C. Bergin, "NASA Report Favors SD HLV for SLS, Complains Agency Can't Afford 2016 Target," *NASASpaceFlight.com* (blog), January 12, 2011, https://www.nasaspaceflight.com/2011/01/nasa-re port-favors-sd-hlv-sls-complains-cant-afford-2016.

See also NASA, *Preliminary Report Regarding NASA's Space Launch System and Multi-Purpose Crew Vehicle* (Washington, DC: NASA, January 2011), https://www.nasa.gov/pdf/510449main_SLS_MPCV_90 -day_Report.pdf.

See also C. Bergin, "SLS: Studies Focusing on SD HLV Versus RP-1 with F-1 Engines," *NASASpaceFlight.com* (blog), March 24, 2011, https://www.nasaspaceflight.com/2011/03/sls-studies-focusing-sd -hlv-versus-rp-1-f-1-engines.

See also K. Robinson, interview by author, April 3, 2018.

410. W. D. Woods, *How Apollo Flew to the Moon*, 2nd ed. (New York: Springer, 2011), 79.

411. L. Hutchinson, "How NASA Brought the Monstrous F-1 'Moon Rocket' Engine Back to Life," Ars Technica, last modified April 15, 2013, https://arstechnica.com/science/2013/04/how-nasa-brought-the -monstrous-f-1-moon-rocket-back-to-life.

412. T. May, interview by author, February 2, 2017.

413. U.S. Congress, House of Representatives, Committee on Science, Space, and Technology, *A Review of NASA's Space Launch System Hearing*, 112th Cong., 1st sess., 2011, 6.

See also NASA, *Preliminary Report Regarding NASA's Space Launch System and Multi-Purpose Crew Vehicle*.

414. NASA, *FY2013 President's Budget Request* (Washington, DC: NASA, 2012), retrieved from https://www.nasa.gov/sites/default/files/659660main _NASA_FY13_Budget_Estimates-508-rev.pdf.

415. NASA, *FY2014 President's Budget Request* (Washington, DC: NASA, 2013), retrieved from https://www.nasa.gov/pdf/750614main_NASA _FY_2014_Budget_Estimates-508.pdf.

416. NASA, *FY2015 President's Budget Request* (Washington, DC: NASA, 2014), retrieved from https://www.nasa.gov/sites/default/files/files/508 _2015_Budget_Estimates.pdf.

417. J. Ryba, "Space Shuttle Launch and Landing," NASA, last modified September 11, 2012, https://www.nasa.gov/mission_pages/shuttle /launch/index.html.

418. T. May, interview by author, April 3, 2018.
     See also T. Gavin, telephone interview by author, December 29, 2016.

419. B. Goldstein, telephone interview by author, November 16, 2019.

420. R. Pappalardo, interview by author, March 7, 2017.

421. S. Holaday, email message to R. Pappalardo regarding Ron Greeley, October 27, 2011.

422. C. Greeley, interview by author, February 7, 2017.

423. C. Greeley, email message to author, December 13, 2013.

424. Ibid., December 8, 2013.

425. G. Webster, "'Greeley Haven' Is Winter Workplace for Mars Rover," NASA: Spirit and Opportunity: Mars Exploration Rovers, last modified January 5, 2012, https://www.nasa.gov/mission_pages/mer/news /mer20120105.html.

426. R. Greeley, *Introduction to Planetary Geomorphology* (Cambridge: Cambridge University Press, 2013), i–vi.

427. Europa Study Team, *Europa Study 2012 Report* (JPL D-71990) (Washington, DC: NASA, May 1, 2012), https://europa.nasa.gov/re sources/63/europa-study-2012-report.

428. Ibid., B-25.

429. L. Prockter, "Europa Mission Studies" (presentation, Outer Planets Assessment Group, St. Louis, MO, March 2012).

430. R. Pappalardo, "SRS Research Leave: Proposal" (paper, JPL, Pasadena, CA, April 10, 2012).

431. R. Pappalardo, telephone interview by author, December 29, 2017.

432. J. Kane, "I Had an Epiphany," *New Yorker,* October 3, 2005, 81.

433. K. Clark et al., *Jupiter Europa Orbiter Mission Study 2008: Final Report* (Pasadena, CA: NASA, JPL, January 2009).

434. K. Clark et al., *2007 Europa Explorer Mission Study: Final Report* (JPL D-41283) (Pasadena, CA: NASA, JPL, November 2007).

435. J. R. R. Tolkien, *The Lord of the Rings,* 50th anniversary 1 vol. ed. (Boston: Houghton Mifflin, 2005).

436. R. Pappalardo, "SRS Research Leave: Accomplishments" (paper, JPL, Pasadena, CA, February 19, 2013).

437. C. Niebur, telephone interview by author, October 16, 2018.

438. D. Blankenship, personal journal entry, March 20, 2012.

439. *Meeting Findings* (Atlanta, GA: Outer Planets Assessment Group, January 11, 2013), https://www.lpi.usra.edu/opag/meetings/jan2013/MeetingReportl.pdf.

440. R. Pappalardo and L. Prockter, joint interview by author, October 9, 2018.

441. H. Rogers, *Consolidated Appropriations Act, 2013* (Pub. L. 113-6), March 26, 2013.

442. S. Streeter, "Appropriations Bills: What Is Report Language?" Congressional Research Reports, accessed October 16, 2019, http://congressionalresearch.com/98-558/document.php?study=Appropriations+Bills+What+is+Report+Language.

443. Mikulski, B. *Consolidated and Further Continuing Appropriations Act H.R. 933 Explanatory Statement,* March 11, 2013.

444. H. Rogers, *Consolidated Appropriations Act, 2013* (Pub. L. 113-6), March 26, 2013.

445. J. Culberson, telephone interview by author, August 1, 2019.
See also J. Casani and H. Eisen, *Europa Lander Mission Study: Space-craft & Mission Description Document* (Pasadena, CA: JPL, June 2011).

446. J. Culberson, telephone interview by author, May 29, 2019.

447. J. Green, interview by author, May 6, 2017.

448. L. Garver, interview by author, August 2, 2016.

449. NASA, *FY2013 President's Budget Request* (Washington, DC: NASA, 2012), retrieved from https://www.nasa.gov/sites/default/files/659660 main_NASA_FY13_Budget_Estimates-508-rev.pdf.

450. C. Dreier, C. "[Updated] Senate Bill Restores $223 million to NASA's Planetary Science Division," *Planetary Society* (blog), March 27, 2013, http://www.planetary.org/blogs/casey-dreier/2013/20130312-proposed -senate-bill-restores-223-million-to-planetary-science.html.

451. J. Green memorandum to Jet Propulsion Laboratory Director for Solar System Exploration, April 22, 2013.

452. Europa Enhancement Science Definition Team, *Europa Summer Study Final Report, Part 1* (Pasadena, CA: JPL, December 2012).

453. Ibid.

454. D. Senske, email message to R. Pappalardo and L. Prockter regarding draft email to Europa Science Advisory Group, January 8, 2013.

455. NASA, "Europa Formulation Key Accomplishments in FY12–FY13" (PowerPoint slide, NASA Headquarters, Washington, DC, 2013).

456. D. Leone, "NASA's Europa Mission Concept Progresses on the Back Burner," SpaceNews.com, July 22, 2013, https://spacenews .com/36388nasas-europa-mission-concept-progresses-on-the-back -burner.
See also V. Thomas, email to ICEE teams regarding the rules of the competition, October 1, 2013.

457. S. Niebur, "Waiting for Test Results," *Toddler Planet* (blog), January 3, 2011, https://toddlerplanet.wordpress.com/2011/01/03/waiting-for -test-results.

458. S. Niebur, "Perspective Shift," *Toddler Planet* (blog), January 20, 2011, https://toddlerplanet.wordpress.com/2011/01/20/perspective-shift.

459. S. Niebur, "Us Lucky," *Toddler Planet* (blog), January 22, 2011, https://toddlerplanet.wordpress.com/2011/01/22/us-lucky.

460. S. Niebur, "My Baby Is Four Today," *Toddler Planet* (blog), January 14, 2011, https://toddlerplanet.wordpress.com/2011/01/14/my-baby-is-four-today.

461. S. Niebur, "How Did We Get Here?," *Toddler Planet* (blog), January 22, 2012, https://toddlerplanet.wordpress.com/2012/01/22/how-did-we-get-here.

462. Division for Planetary Sciences, "2012 Annual Meeting Schedule," 2012, https://dps.aas.org/dps_meetings_archive/2012/block.pdf.

463. Division for Planetary Sciences, "2012 Prize Recipients," 2012, https://dps.aas.org/prizes/2012.

464. K. Retherford, email message to C. Niebur regarding discovery of Europa's water vapor plumes, December 5, 2013.

465. C. Niebur, email message to K. Retherford regarding discovery of Europa's water vapor plumes, December 6, 2013.

466. K. Retherford, email message to C. Niebur regarding discovery of Europa's water vapor plumes, December 6, 2013.

467. L. Roth et al., "Transient Water Vapor at Europa's South Pole," *Science* 343, no. 6167 (2014): 171–74, https://doi.org/10.1126/science.1247051.

468. "Hubble Space Telescope Optics System," NASA, last modified November 12, 2019, http://www.nasa.gov/content/goddard/hubble-space-telescope-optics-system.

469. L. Roth, telephone interview by author, January 18, 2019.

470. K. Retherford, telephone interview by author, December 6, 2018.

471. Ibid.
     See also L. Roth, telephone interview by author, January 18, 2019.

472. C. Niebur, interview by author, May 6, 2017.

473. J. Green, interview by author, May 6, 2017.

474. "Hubble Space Telescope Sees Evidence of Water Vapor Venting off Jupiter Moon," NASA, last modified August 7, 2017, http://www.nasa.gov/content/goddard/hubble-europa-water-vapor.

475. Wright Brothers National Memorial North Carolina, "Announcing Details of the 50th Anniversary of the Apollo 11 Moon Landing Event," press release, July 16, 2019, https://www.nps.gov/wrbr/planyourvisit/apollo-11-50th.htm.

476. A. Hamer, "Neil Armstrong Took a Piece of the North Carolina Outer Banks to the Moon," Curiosity, last modified October 13, 2018, https://curiosity.com/topics/neil-armstrong-took-a-piece-of-the-north-carolina-outer-banks-to-the-moon-curiosity.

477. W. Guthrie, "This Land Is Your Land," on *Folksay: American Ballads and Dances* (New York: Asch Record Company, 1944).

478. J. Salute, telephone interview by author, October 25, 2018.

479. T. Gavin, telephone interview by author, January 18, 2017.

480. R. Pappalardo, telephone interview by author, December 21, 2018.

481. Ibid.

482. B. Goldstein, telephone interview by author, November 16, 2018.

483. Nor, for the record, was there any doubt at NASA headquarters about who was in charge: the project manager.

484. R. Pappalardo and B. Cooke, "Europa Clipper OPAG Update" (PowerPoint presentation, Outer Planets Assessment Group, Arlington, VA, July 2013), https://www.lpi.usra.edu/opag/jul2013/presentations/Clipper_Summary.pdf.

485. K. Clark et al., *Jupiter Europa Orbiter Mission Study 2008: Final Report* (Pasadena, CA: NASA, JPL, January 2009).

486. A. Frick et al., "Overview of Current Capabilities and Research and Technology Developments for Planetary Protection," *Advances in Space Research* 54, no. 2 (2014): 221–40, https://doi.org/10.1016/j.asr.2014.02.016.

487. National Academies of Sciences, Engineering, and Medicine, *Review and Assessment of Planetary Protection Policy Development Processes* (Washington, DC: National Academies Press, 2018), 48, https://doi.org/10.17226/25172.

488. NASA, "Phoenix Launch: Mission to the Martian Polar North," press kit, August 2007, https://www.jpl.nasa.gov/news/press_kits/phoenix-launch-presskit.pdf.
    See also Frick, "Overview of Current Capabilities," 221–40.

489. NASA, "Galaxy Evolution Explorer Launch," press kit, April 2003, https://www.jpl.nasa.gov/news/press_kits/galex.pdf.

490. B. Cooke, interview by author, March 8, 2017.

491. B. Cooke, telephone interview by author, November 15, 2018.

492. NASA, "Juno Launch," press kit, August 2011, https://www.jpl.nasa .gov/news/press_kits/JunoLaunch.pdf.

493. B. Goldstein, telephone interview by author, November 16, 2018.

494. B. Cooke, telephone interview by author, November 15, 2018.

495. T. Stern and S. DeLapp, "Techniques for Magnetic Cleanliness on Spacecraft Solar Arrays" (presentation, Second International Energy Conversion Engineering Conference, Providence, RI, 2004), https:// doi.org/10.2514/6.2004-5581.

496. R. Pappalardo, telephone interview by author, December 21, 2018.

497. R. T. Pappalardo, "Europa Exploration" (presentation, Europa Clipper & SLS Partnerships Briefing, NASA Headquarters, Washington, DC, January 24, 2014).

498. M. Smith, "Are the Days of NASA's Science Flagship Missions Over?," December 4, 2013, available at SpacePolicyOnline.com, https://space policyonline.com/news/are-the-days-of-nasas-science-flagship-mis sions-over.

499. B. Pershing, "Frank Wolf to Retire After 17 Terms in Congress; N. Va. Seat Will Be a Prime Battleground in 2014," *Washington Post*, December 17, 2013, https://www.washingtonpost.com/local/virginia-politics /frank-wolf-to-retire-after-17-terms-in-congress-northern-va-seat-to -be-a-battleground-in-2014/2013/12/17/712bb608-6749-11e3-a0b9 -249bbb34602c_story.html.

500. E. Berger, "Love Planetary Science? Dying to Explore Europa's Oceans? Meet the Man Who Can Make It Happen," *Houston Chronicle*, December 20, 2013, https://blog.chron.com/sciguy/2013/12/love-plan etary-science-dying-to-explore-europas-oceans-meet-the-man-who -can-make-it-happen.

501. L. Smith, *Consolidated Appropriations Act, 2014* (Pub. L. 113–76), January 14, 2014.

502. C. Niebur, *Investments in Europa Mission Concept Studies*, 2010.

503. C. Bolden, email message to Europa stakeholders regarding Europa Clipper & SLS partnerships briefing, December 31, 2013.

504. C. Niebur, email message to author regarding directing MFB, December 6, 2018.

505. C. Niebur, personal journal entry, 2016.

506. Ibid.

507. Many thanks to Jim Green, the chief scientist of NASA, for patiently walking me through each slide of his presentation to the SIP. Ancillary details of the meeting come from the following interviews and documents:
    J. Salute, telephone interview by author, October 25, 2018.
    C. Niebur, telephone interview by author, October 19, 2018.
    C. Niebur, personal journal entry, 2016.
    J. Green, interview by author, May 16, 2017.
    C. Niebur, interview by author, May 16, 2017.
    J. Salute and C. Niebur, ongoing email messages to and from author, 2016–2019.

508. P. K. Dick and L. Sutin, *The Shifting Realities of Philip K. Dick: Selected Literary and Philosophical Writings*, 1st Vintage ed. (New York: Vintage Books, 1995).

509. I. Kant and I. Johnston, *Universal Natural History and Theory of the Heavens: Or, an Essay on the Constitution and the Mechanical Origin of the Entire Structure of the Universe Based on Newtonian Principles* (Arlington, VA: Richer Resources Publications, 2009).

510. I. Kant and M. J. Gregor, *Critique of Practical Reason*, rev. ed. (Cambridge: Cambridge University Press, 2015).

511. A. Losch, "Kant's Wager. Kant's Strong Belief in Extra-Terrestrial Life, the History of This Question and Its Challenge for Theology Today," *International Journal of Astrobiology* 15, no. 4 (2016): 261–70, https://doi .org/10.1017/S1473550416000112.

512. Kant and Gregor, *Critique of Practical Reason*.

513. 2 Kings 19:35 (King James Version).

514. Daniel 10:5–6 (King James Version).

515. Luke 1:11–13 (King James Version).

516. Revelation 22:8–9 (King James Version).

517. Genesis 1:28 (King James Version).

518. Consolmagno and Mueller, *Would You Baptize an Extraterrestrial?*

519. C. Niebur, email message to author, August 5, 2016.

520. R. T. Pappalardo et al. "Does Europa Have a Subsurface Ocean? Evaluation of the Geological Evidence," *Journal of Geophysical Research: Planets* 104, no. e10 (1999): 24015–55, https://doi.org/10.1029/1998JE000628.

521. R. T. Pappalardo et al., "Geological Evidence for Solid-State Convection in Europa's Ice Shell," *Nature* 391, no. 6665 (1998): 365–68, https://doi.org/10.1038/34862.

522. J. Salute, telephone interview by author, October 25, 2018.

523. L. Prockter, email message to the Europa science team regarding thanks and goodbye, June 25, 2014. This was sent to the entire science definition team.

524. C. Niebur, email message to the Europa science team regarding thanks and goodbye, June 25, 2014. This was sent to the entire science definition team as part of an email chain with Louise.

525. C. Dreier, "[Updated] To Europa! . . . Slowly. First Impressions of NASA's New Budget Request," *Planetary Society* (blog), March 7, 2014, http://www.planetary.org/blogs/casey-dreier/2014/0304-first-impressions-of-the-2015-nasa-budget-request.html.

526. R. Pappalardo, email message to the Europa science team regarding Culberson's news, November 20, 2014. This correspondence was sent to the internal pre-project science team.

527. A general narrative was constructed from interviews with:
    C. Niebur, telephone interview by author, October 19, 2018.
    J. Salute, telephone interview by author, October 28, 2018.
    L. Prockter and R. Pappalardo, interview by author, October 9, 2018.
    R. Pappalardo, telephone interview by author, December 21, 2018.
    L. Prockter, interview by author, February 6, 2018.

528. R. Pappalardo, email message to author, December 20, 2018. Robert Pappalardo provided photographs of the document signing, celebration, and key players described in the text.

529. R. Pappalardo, telephone interview by author, December 21, 2018. See also R. Pappalardo, email message to author, December 20, 2018. See also R. Pappalardo, email message to author regarding the monolith, December 20, 2018. Robert Pappalardo provided photographs of the Monolith's delivery and relocation.

530. ThinkGeek, "Monolith Action Figure," accessed October 11, 2019, http://www.thinkgeek.com/stuff/41/monolith-action-figure.shtml.

531. R. Pappalardo, email message to E. Rios regarding monolith status report, June 13, 2015.

532. E. Rios, email message to R. Pappalardo regarding monolith status report, June 20, 2015. See also E. Rios, email message to R. Pappalardo regarding monolith status report, June 24, 2015.

533. A. C. Clarke, *2001: A Space Odyssey* (New York: Penguin Books, 2016). I have one of those key chains sitting on my desk, courtesy of Robert Pappalardo. The description is accurate.

534. R. Pappalardo, *PSG1 Opening* (QuickTime movie, JPL, 2015).

535. Skeletons and More, email message to R. Pappalardo regarding Skeletons and More aged left femur bone order, June 19, 2015. Original receipt provided by Robert Pappalardo.

536. R. T. Pappalardo, W. B. McKinnon, and K. Khurana, eds., *Europa* (Tucson: University of Arizona Press, 2009), 220.

537. S. Katterhorn, interview by author, March 20, 2018.

538. S. A. Kattenhorn and L. M. Prockter, "Evidence for Subduction in the Ice Shell of Europa," *Nature Geoscience* 7 (2014): 762–7.

539. L. Prockter, interview by author, February 6, 2018.

Index